NATURAL SELECTION AND TROPICAL NATURE ESSAYS ON DESCRIPTIVE AND THEORETICAL BIOLOGY

BY

ALFRED RUSSEL WALLACE

NEW EDITION WITH CORRECTIONS AND ADDITIONS

1895

British Library Cataloguing-in-Publication Data
A catalogue record for this book is available from the
British Library

Contents

TROPICAL NATURE AND OTHER ESSAYS

Alfred Russel Wallace

Alfred Russel Wallace was born on 8th January 1823 in the village of Llanbadoc, in Monmouthshire, Wales.

At the age of five, Wallace's family moved to Hertford where he later enrolled at Hertford Grammar School. He was educated there until financial difficulties forced his family to withdraw him in 1836. He then boarded with his older brother John before becoming an apprentice to his eldest brother, William, a surveyor. He worked for William for six years until the business declined due to difficult economic conditions.

After a brief period of unemployment, he was hired as a master at the Collegiate School in Leicester to teach drawing, map-making, and surveying. During this time he met the entomologist Henry Bates who inspired Wallace to begin collecting insects. He and bates continued exchanging letters after Wallace left teaching to pursue his surveying career. They corresponded on prominent works of the time such as Charles Darwin's *The Voyage of the Beagle* (1839) and Robert Chamber's *Vestiges of the Natural History of Creation* (1844).

Wallace was inspired by the travelling naturalists of the day and decided to begin his exploration career collecting specimens in the Amazon rainforest. He explored the Rio Negra for four years, making notes on the peoples and languages he encountered as well as the geography, flora, and fauna. On his return voyage his ship, Helen, caught fire and he and the crew were stranded for ten days before being picked up by the Jordeson, a brig travelling from Cuba to London. All of his specimens aboard Helen had been lost.

After a brief stay in England he embarked on a journey to the Malay Archipelago (now Singapore, Malaysia, and Indonesia). During this eight year period he collected more than 126,000 specimens, several thousand of which represented new species to science. While travelling, Wallace refined his thoughts about evolution and in 1858 he outlined his theory of natural selection in an article he sent to Charles Darwin. This was published in the same year along with Darwin's own theory. Wallace eventually published an account of his travels *The Malay Archipelago* in 1869, and it became one of the most popular books of scientific exploration in the 19th century.

Upon his return to England, in 1862, Wallace became a staunch defender of Darwin's landmark work *On the Origin of Species* (1859). He wrote responses to those critical of the theory of natural selection, including 'Remarks on the Rev. S. Haughton's Paper on the Bee's Cell, And on the Origin of Species' (1863) and 'Creation by Law' (1867). The former of these was particularly pleasing to Darwin. Wallace also published important papers such as 'The Origin of Human Races and the Antiquity of Man Deduced from the Theory of 'Natural Selection'' (1864) and books, including the much cited *Darwinism* (1889).

Wallace made a huge contribution to the natural sciences and he will continue to be remembered as one of the key figures in the development of evolutionary theory.

Wallace died on 7th November 1913 at the age of 90. He is buried in a small cemetery at Broadstone, Dorset, England.

PREFACE

THE present volume consists mainly of a reprint of two volumes of essays—*Contributions to the Theory of Natural Selection,* which appeared in 1870, which a second edition in 1871, and has now been many years out of print; and, *Tropical Nature and Other Essays,* which appeared in 1878.

In preparing a new edition of these works to appear as a single volume I have thought it advisable to omit two essays—that on "The Malayan Papilionidæ" as being too technical for general readers, and that on "The Distribution of Animals as indicating Geographical Changes," which contains nothing that is not more fully treated in my other works. Another essay—"By - Paths in the Domain of Biology"—has also been partly omitted, one portion of it forming a short chapter on "The Antiquity and Origin of Man," while another portion has been incorporated in the chapter on "The Colours of Animals and Sexual Selection." More than compensating for these omissions are two new chapters—"The Antiquity of Man in North America" and "The Debt of Science to Darwin."

Many corrections and some important additions have been made to the text, the chief of which are indicated in the table given below; and to facilitate reference the two original works have separate headings, and form Parts I. and II. of the present volume.

ALTERATIONS IN THE SECOND EDITION OF CONTRIBUTIONS, ETC.

1ST ED.	2D ED.		PRESENT VOLUME.
221	221	Additional facts as to birds acquiring the song of other species	105
223	223A 223B	Mr. Spruce's remarks on young birds pairing with old	107
228	228A 228B	Pouchet's observations on a change in the nests of swallows	omitted
229	—	Passage omitted about nest of Golden Crested Warbler, which had been inserted on Rennie's authority, but has not been confirmed by any later observers.	
261	261	Daines Barrington, on importance of protection to the female bird	138
	372	Note A	205
	372B	Note B	209

ADDITIONAL MATTER IN THE PRESENT VOLUME.

NATURAL SELECTION.

Additional facts by Leroy, Spalding, Lowne, and Dixon on the Nest-Building and other Instincts of Birds
Dr. Abbott on Nesting of Baltimore Oriole
Professor Jeitteles and Mr. Henry Reeks on Alterations in Mode of Nest-Building

TROPICAL NATURE.

Note on Dr. Shufeldt's Investigations into the Affinities of Swifts and Humming-Birds

THE ANTIQUITY OF MAN IN NORTH AMERICA.

(Additional Chapter)

THE DEBT OF SCIENCE TO DARWIN.

(Additional Chapter)

PARKSTONE, DORSET,

March 1891.

ESSAYS ON NATURAL SELECTION

I

ON THE LAW WHICH HAS REGULATED THE INTRODUCTION OF NEW SPECIES[1]

Geographical Distribution dependent on Geologic Changes

EVERY naturalist who has directed his attention to the subject of the geographical distribution of animals and plants must have been interested in the singular facts which it presents. Many of these facts are quite different from what would have been anticipated, and have hitherto been considered as highly curious, but quite inexplicable. None of the explanations attempted from the time of Linnæus are now considered at all satisfactory; none of them have given a cause sufficient to account for the facts known at the time, or comprehensive enough to include all the new facts which have since been, and are daily being, added. Of late years, however, a great light has been thrown upon the subject by geological investigations, which have shown that the present state of the earth and of the organisms now inhabiting it is but the last stage of a long and uninterrupted series of changes which it has undergone, and consequently, that to endeavour to explain and account for its present condition without any reference to those changes (as has frequently been done) must lead to very imperfect and erroneous conclusions.

[1] This article, written at Sarawak in February 1855 and published in the *Annals and Magazine of Natural History*, September 1855, was intended to show that some form of evolution of one species from another was needed in order to explain the various classes of facts here indicated; but at that time no means has been suggested by which the actual change of species could have been brought about.

The facts proved by geology are briefly these: That during an immense but unknown period the surface of the earth has undergone successive changes; land has sunk beneath the ocean, while fresh land has risen up from it; mountain chains have been elevated; islands have been formed into continents, and continents submerged till they have become islands; and these changes have taken place, not once merely, but perhaps hundreds, perhaps thousands of times.—That all these operations have been more or less continuous but unequal in their progress, and during the whole series the organic life of the earth has undergone a corresponding alteration. This alteration also has been gradual, but complete; after a certain interval not a single species existing which had lived at the commencement of the period. This complete renewal of the forms of life also appears to have occurred several times.—That from the last of the geological epochs to the present or historical epoch, the change of organic life has been gradual: the first appearance of animals now existing can in many cases be traced, their numbers gradually increasing in the more recent formations, while other species continually die out and disappear, so that the present condition of the organic world is clearly derived by a natural process of gradual extinction and creation of species from that of the latest geological periods. We may therefore safely infer a like gradation and natural sequence from one geological epoch to another.

Now, taking this as a fair statement of the results of geological inquiry, we see that the present geographical distribution of life upon the earth must be the result of all the previous changes, both of the surface of the earth itself and of its inhabitants. Many causes, no doubt, have operated of which we must ever remain in ignorance, and we may, therefore, expect to find many details very difficult of explanation, and in attempting to give one, must allow ourselves to call into our service geological changes which it is highly probable may have occurred, though we have no direct evidence of their individual operation.

The great increase of our knowledge within the last twenty years, both of the present and past history of the organic world, has accumulated a body of facts which should afford a sufficient foundation for a comprehensive law embracing and explaining them all, and giving a direction to new researches. It is about ten years since the idea of such a law suggested itself to the writer of this essay, and he has since taken every opportunity of testing it by all the newly-ascertained facts with which he has become acquainted, or has been able to observe himself. These have all served to convince him of the correctness of his hypothesis. Fully to enter into such a subject would occupy much space, and it is only in consequence of some views having been lately promulgated, he believes, in a wrong direction, that he now ventures to present his ideas to the public, with only such obvious illustrations of the arguments and results as occur to him in a place far removed from all means of reference and exact information.

A Law deduced from well-known Geographical and Geological Facts

The following propositions in Organic Geography and Geology give the main facts on which the hypothesis is founded.

GEOGRAPHY

1. Large groups, such as classes and orders, are generally spread over the whole earth, while smaller ones, such as families and genera, are frequently confined to one portion, often to a very limited district.

2. In widely distributed families the genera are often limited in range; in widely distributed genera well-marked groups of species are peculiar to each geographical district.

3. When a group is confined to one district, and is rich in species, it is almost invariably the case that the most closely allied species are found in the same locality or in closely adjoining localities, and that therefore the natural sequence of the species by affinity is also geographical.

4. In countries of a similar climate, but separated by a wide sea or lofty mountains, the families, genera, and species of the one are often represented by closely allied families, genera, and species peculiar to the other.

GEOLOGY

5. The distribution of the organic world in time is very similar to its present distribution in space.

6. Most of the larger and some small groups extend through several geological periods.

7. In each period, however, there are peculiar groups, found nowhere else, and extending through one or several formations.

8. Species of one genus, or genera of one family occurring in the same geological time, are more closely allied than those separated in time.

9. As, generally, in geography no species or genus occurs in two very distant localities without being also found in intermediate places, so in geology the life of a species or genus has not been interrupted. In other words, no group or species has come into existence twice.

10. The following law may be deduced from these facts: *Every species has come into existence coincident both in space and time with a pre-existing closely allied species.*

This law agrees with, explains, and illustrates all the facts connected with the following branches of the subject: 1st, The system of natural affinities. 2d, The distribution of animals and plants in space. 3d, The same in time, including all the phenomena of representative groups, and those which Professor Forbes supposed to manifest polarity. 4th, The phenomena of rudimentary organs. We will briefly endeavour to show its bearing upon each of these.

The Form of a true system of Classification determined by this Law

If the law above enunciated be true, it follows that the natural series of affinities will also represent the order in which the several species came into existence, each one having had for its immediate antitype a closely allied species existing at the time of its origin. It is evidently possible that two or three distinct species may have had a common antitype, and that each of these may again have become the antitypes from which other closely allied species were created. The effect of this would be, that so long as each species has had but one new species formed on its model, the line of affinities will be simple, and may be represented by placing the several species in direct succession in a straight line. But if two or more species have been independently formed on the plan of a common antitype, then the series of affinities will be compound, and can only be represented by a forked or many-branched line. Now, all attempts at a Natural classification and arrangement of organic beings show that both these plans have obtained in creation. Sometimes the series of affinities can be well represented for a space by a direct progression from species to species or from group to group, but it is generally found impossible so to continue. There constantly occur two or more modifications of an organ or modifications of two distinct organs, leading us on to two distinct series of species, which at length differ so much from each other as to form distinct genera or families. These are the parallel series or representative groups of naturalists, and they often occur in different countries, or are found fossil in different formations. They are said to have an analogy to each other when they are so far removed from their common antitype as to differ in many important points of structure, while they still preserve a family resemblance. We thus see how difficult it is to determine in every case whether a given relation is an analogy or an affinity, for it is evident that as we go back along the parallel or divergent series, towards the common antitype, the analogy which existed between the two groups becomes an affinity. We are also made aware of the difficulty of arriving at a true classification, even in a small and perfect group; in the actual state of nature it is almost impossible, the species being so numerous and the modifications of form and structure so varied, arising probably from the immense number of species which have served as antitypes for the existing species, and thus produced a complicated branching of the lines of affinity, as intricate as the twigs of a gnarled oak or the vascular system of the human body. Again, if we consider that we have only fragments of this vast system, the stem and main branches being represented by extinct species of which we have no knowledge, while a vast mass of limbs and boughs and minute twigs and scattered leaves is what we have to place in order, so as to determine the true position which each originally occupied with regard to the others, the whole difficulty of the true Natural System of classification becomes apparent to us.

We shall thus find ourselves obliged to reject all those systems of classification which arrange species or groups in circles, as well as those which fix a definite number for the divisions of each group. The latter class have been very generally rejected by naturalists, as contrary to nature, notwithstanding the ability with which they have been advocated; but the circular system of affinities seems to have obtained a deeper hold, many eminent naturalists having to some extent adopted it. We have, however, never been able to find a case in which the circle has been closed by a direct and close affinity. In most cases a palpable analogy has been substituted, in others the affinity is very obscure or altogether doubtful. The complicated branching of the lines of affinities in extensive groups must also afford great facilities for giving a show of probability to any such purely artificial arrangements. Their death-blow was given by the admirable paper of the lamented Mr. Strickland, published in the *Annals of Natural History*, in which he so clearly showed the true synthetical method of discovering the Natural System.

Geographical Distribution of Organisms

If we now consider the geographical distribution of animals and plants upon the earth, we shall find all the facts beautifully in accordance with, and readily explained by, the present hypothesis. A country having species, genera, and whole families peculiar to it, will be the necessary result of its having been isolated for a long period, sufficient for many series of species to have been created on the type of pre-existing ones, which, as well as many of the earlier-formed species, have become extinct, and thus made the groups appear isolated. If in any case the antitype had an extensive range, two or more groups of species might have been formed, each varying from it in a different manner, and thus producing several representative or analogous groups. The Sylviadæ of Europe and the Sylvicolidæ of North America, the Heliconidæ of South America and the Euplœas of the East, the group of Trogons inhabiting Asia and that peculiar to South America, are examples that may be accounted for in this manner.

Such phenomena as are exhibited by the Galapagos Islands, which contain little groups of plants and animals peculiar to themselves, but most nearly allied to those of South America, have not hitherto received any, even a conjectural explanation. The Galapagos are a volcanic group of high antiquity, and have probably never been more closely connected with the continent than they are at present. They must have been first peopled, like other newly-formed islands, by the action of winds and currents, and at a period sufficiently remote to have had the original species die out, and the modified prototypes only remain. In the same way we can account for the separate islands having each their peculiar species, either on the supposition that the same original emigration peopled the whole of the islands with the same species from which differently modified prototypes were created, or that the islands were successively peopled from each other, but that new species have been created in each on the plan of the pre-existing ones. St. Helena is a similar case of a very ancient island having obtained an entirely peculiar, though limited, flora. On the other hand, no example is known of an island which can be proved geologically to be of very recent origin (late in the Tertiary, for instance), and yet possesses generic or family groups, or even many species peculiar to itself.

When a range of mountains has attained a great elevation, and has so remained during a long geological period, the species of the two sides at and near their bases will be often very different, representative species of some genera occurring, and even whole genera being peculiar to one side only, as is remarkably seen in the case of the Andes and Rocky Mountains. A similar phenomenon occurs when an island has been separated from a continent at a very early period. The shallow sea between the Peninsula of Malacea, Java, Sumatra, and Borneo was probably a continent or large island at an early epoch, and may have become submerged as the volcanic ranges of Java and Sumatra were elevated; the organic results we see in the very considerable number of species of animals common to some or all of these countries, while at the same time a number of closely allied representative species exist peculiar to each, showing that a considerable period has elapsed since their separation. The facts of geographical distribution and of geology may thus mutually explain each other in doubtful cases, should the principles here advocated be clearly established.

In all those cases in which an island has been separated from a continent, or raised by volcanic or coralline action from the sea, or in which a mountain-chain has been elevated in a recent geological epoch, the phenomena of peculiar groups or even of single representative species will not exist. Our own island is an example of this, its separation from the continent being geologically very recent, and we have consequently scarcely a species which is peculiar to it; while the Alpine range, one of the most recent mountain elevations, separates faunas and floras which scarcely differ more than may be due to climate and latitude alone.

The series of facts alluded to in Proposition (3), of closely allied species in rich groups being found geographically near each other, is most striking and important. Mr. Lovell Reeve has well exemplified it in his able and interesting paper on the Distribution of the Bulimi. It is also seen in the Humming-birds and Toucans, little groups of two or three closely allied species being often found in the same or closely adjoining districts, as we have had the good fortune of personally verifying. Fishes give evidence of a similar kind: each great river has its peculiar genera, and in more extensive genera its groups of closely allied species. But it is the same throughout Nature; every class and order of animals will contribute similar facts. Hitherto no attempt has been made to explain these singular phenomena, or to show how they have arisen. Why are the genera of Palms and of Orchids in almost every case confined to one hemisphere? Why are the closely allied species of brown-backed Trogons all found in the East, and the green-backed in the West? Why are the Macaws and the Cockatoos similarly restricted? Insects furnish a countless number of analogous examples—the Goliathi of Africa, the Ornithopteræ of the Indian Islands, and Heliconidæ of South America, the Danaidæ of the East, and in all the most closely allied species found in geographical proximity. The question forces itself upon every thinking mind, Why are these things so? They could not be as they are had no law regulated their creation and dispersion. The law here enunciated not merely explains but necessitates the facts we see to exist, while the vast and long-continued geological changes of the earth readily account for the exceptions and apparent discrepancies that here and there occur. The writer's object in putting forward his views in the present imperfect manner is to submit them to the test of other minds, and to be made aware of all the facts supposed to be inconsistent with them. As his hypothesis is one which claims acceptance solely as explaining and connecting facts which exist in nature, he expects facts alone to be brought to disprove it, not *à priori* arguments against its probability.

Geological Distribution of the Forms of Life

The phenomena of geological distribution are exactly analogous to those of geography. Closely allied species are found associated in the same beds, and the change from species to species appears to have been as gradual in time as in space. Geology, however, furnishes us with positive proof of the extinction and production of species, though it does not inform us how either has taken place. The extinction of species, however, offers but little difficulty, and the *modus operandi* has been well illustrated by Sir C. Lyell in his admirable *Principles*. Geological changes, however gradual, must occasionally have modified external conditions to such an extent as to have rendered the existence of certain species impossible. The extinction would in most cases be effected by a gradual dying-out, but in some instances there might have been a sudden destruction of a species of limited range. To discover how the extinct species have from time to time been replaced by new ones down to the very latest geological period, is the most difficult, and at the same time the most interesting problem in the natural history of the earth. The present inquiry, which seeks to eliminate from known facts a law which has determined, to a certain degree, what species could and did appear at a given epoch, may, it is hoped, be considered as one step in the right direction towards a complete solution of it.

High Organisation of very ancient Animals consistent with this Law

Much discussion has of late years taken place on the question whether the succession of life upon the globe has been from a lower to a higher degree of organisation. The admitted facts seem to show that there has been a general, but not a detailed progression. Mollusca and Radiata existed before Vertebrata, and the progression from Fishes to Reptiles and Mammalia, and also from the lower mammals to the higher, is indisputable. On the other hand, it is said that the Mollusca and Radiata of the very earliest periods were more highly organized than the great mass of those now existing, and that the very first fishes that have been discovered are by no means the lowest organised of the class. Now it is believed the present hypothesis will harmonise with all these facts, and in a great measure serve to explain them; for though it may appear to some readers essentially a theory of progression, it is in reality only one of gradual change. It is, however, by no means difficult to show that a real progression in the sale of organisation is perfectly consistent with all the appearances, and even with apparent retrogression, should such occur.

Returning to the analogy of a branching tree, as the best mode of representing the natural arrangement of species and their successive creation, let us suppose that at an early geological epoch any group (say a class of the Mollusca) has attained to a great richness of species and a high organisation. Now let this great branch of allied species, by geological mutations, be completely or partially destroyed. Subsequently a new branch springs from the same trunk—that is to say, new species are successively created, having for their antitypes the same lower organised species which had served as the antitypes for the former group, but which have survived the modified conditions which destroyed it. This new group being subject to these altered conditions, has modifications of structure and organisation given to it, and becomes the representative group of the former one in another geological formation. It may, however, happen, that though later in time, the new series of species may never attain to so high a degree of organisation as those preceding it, but in its turn become extinct, and give place to yet another modification from the same root, which may be of higher or lower organisation, more or less numerous in species, and more or less varied in form and structure, than either of those which preceded it.

Again, each of these groups may not have become totally extinct, but may have left a few species, the modified prototypes of which have existed in each succeeding period, a faint memorial of their former grandeur and luxuriance. Thus every case of apparent retrogression may be in reality a progress, though an interrupted one: when some monarch of the forest loses a limb, it may be replaced by a feeble and sickly substitute. The foregoing remarks appear to apply to the case of the Mollusca, which, at a very early period, had reached a high organisation and a great development of forms and species in the testaceous Cephalopoda. In each succeeding age modified species and genera replaced the former ones which had become extinct, and as we approach the present era, but few and small representatives of the group remain, while the Gasteropods and Bivalves have acquired an immense preponderance. In the long series of changes the earth has undergone, the process of peopling it with organic beings has been continually going on, and whenever any of the higher groups have become nearly or quite extinct, the lower forms which have better resisted the modified physical conditions have served as the antitypes on which to found the new races. In this manner alone, it is believed, can the representative groups at successive periods, and the risings and fallings in the scale of organisation, be in every case explained.

Objections to Forbes' Theory of Polarity

The hypothesis of polarity, recently put forward by Professor Edward Forbes to account for the abundance of generic forms at a very early period and at present, while in the intermediate epochs there is a gradual diminution and impoverishment, till the minimum occurred at the confines of the Palæozoic and Secondary epochs, appears to us quite unnecessary, as the facts may be readily accounted for on the principles already laid down. Between the Palæozoic and Neozoic periods of Professor Forbes there is scarcely a species in common, and the greater parts of the genera and families also disappear, to the replaced by new ones. It is almost universally admitted that such a change in the organic world must have occupied a vast period of time. Of this interval we have no record; probably because the whole area of the early formations now exposed to our researches was elevated at the end of the Palæozoic period, and remained so through the interval required for the organic changes which resulted in the fauna and flora of the Secondary period. The records of this interval are buried beneath the ocean which covers three-fourths of the globe. Now it appears highly probable that a long period of quiescence or stability in the physical conditions of a district would be most favourable to the existence of organic life in the greatest abundance, both as regards individuals and also as to variety of species and generic group, just as we now find that the places best adapted to the rapid growth and increase of individuals also contain the greatest profusion of species and the greatest variety of forms,—the tropics in comparison with the temperate and arctic regions. On the other hand, it seems no less probable that a change in the physical conditions of a district, even small in amount if rapid, or even gradual if to a great amount, would be highly unfavourable to the existence of individuals, might cause the extinction of many species, and would probably be equally unfavourable to the creation of new ones. In this too we may find an analogy with the present state of our earth, for it has been shown to be the violent extremes and rapid changes of physical conditions, rather than the actual mean state in the temperate and frigid zones, which renders them less prolific than the tropical regions, as exemplified by the great distance beyond the tropics to which tropical forms penetrate when the climate is equable, and also by the richness in species and forms of tropical mountain regions which principally differ from the temperate zone in the uniformity of their climate. However this may be, it seems a fair assumption that during a period of geological repose the new species which we know to have been created would have appeared, that the creations would then exceed in number the extinctions, and therefore the number of species would increase. In a period of geological activity, on the other hand, it seems probable that the extinctions might exceed the creations, and the number of species consequently diminish. That such effects did take place in connection with the causes to which we have imputed them, is shown in the case of the Coal formation, the faults and contortions of which show a period of great activity and violent convulsions, and it is in the formation immediately succeeding this that the poverty of forms of life is most apparent. We have then only to suppose a long period of somewhat similar action during the vast unknown interval at the termination of the Palæozoic period, and then a decreasing violence or rapidity through the Secondary period, to allow for the gradual repopulation of the earth with varied forms, and the whole of the facts are explained.[1] We thus have a clue to the increase of the forms of life during certain periods, and their decrease during others, without recourse to any causes but those we know to have existed, and to effects fairly deducible from them. The precise manner in which the geological changes of the early formations were effected is so extremely obscure, that when we can explain important facts by a retardation at one time and an acceleration at another of a process which we know from its nature and from observation to have been unequal,—a cause so simple may surely be preferred to one so obscure and hypothetical as polarity.

I would also venture to suggest some reasons against the very nature of the theory of Professor Forbes. Our knowledge of the organic world during any geological epoch is necessarily very imperfect. Looking at the vast numbers of species and groups that have been discovered by geologists, this may be doubted; but we should compare their numbers not merely with those that now exist upon the earth, but with a far larger amount. We have no reson for believing that the number of species on the earth at any former period was much less than at present; at all events the aquatic portion, with which geologists have most acquaintance, was probably often as great or greater. Now we know that there have been many complete changes of species; new sets of organisms have many times been introduced in place of old ones which have become extinct, so that the total amount which have existed on the earth from the earliest geological period must have borne about the same proportion to those now living, as the whole human race who have lived and died upon the earth to the population at the present time. Again, at each epoch, the whole earth was, no doubt, as now, more or less the theatre of life, and as the successive generations of each species died, their exuviæ and preservable parts would be deposited over every portion of the then existing seas and oceans, which we have reason for supposing to have been more, rather than less, extensive than at present. In order then to understand our possible knowledge of the early world and its inhabitants, we must compare, not the area of the whole field of our geological researches with the earth's surface, but the area of the examined portion of each formation separately with the whole earth. For example, during the Silurian period all the earth was Silurian, and animals were living and dying and depositing their remains more or less over the whole area of the globe, and they were probably (the species at least) nearly as varied in different latitudes and longitudes as at present. What proportion do the Silurian districts bear to the whole surface of the globe, land and sea (for far more extensive Silurian districts probably exist beneath the ocean than above it), and what portion of the known Silurian districts has been actually examined for fossils? Would the area of rock actually laid open to the eye be the thousandth or the tenthousandth part of the earth's surface? Ask the same question with regard to the Oolite or the Chalk, or even to particular beds of these when they differ considerably in their fossils, and you may then get some notion of how small a portion of the whole we know.

[1] Professor Ramsay has since shown that a glacial epoch probably occurred at the time of the Permian formation, which will more satisfactorily account for the comparative poverty of species.

But yet more important is the probability, nay, almost the certainty, that whole formations containing the records of vast geological periods are entirely buried beneath the ocean, and for ever beyond our reach. Most of the gaps in the geological series may thus be filled up, and vast numbers of unknown and unimaginable animals, which might help to elucidate the affinities of the numerous isolated groups which are a perpetual puzzle to the zoologist, may there be buried, till future revolutions may raise them in their turn above the waters, to afford materials for the study of whatever race of intelligent beings may then have succeeded us. These considerations must lead us to the conclusion that our knowledge of the whole series of the former inhabitants of the earth is necessarily most imperfect and fragmentary,—as much so as our knowledge of the present organic world would be, wee we forced to make our collections and observations only in spots equally limited in area and in number with those actually laid open for the collection of fossils. Now, the hypothesis of Professor Forbes is essentially one that assumes to a great extent the completeness of our knowledge of the whole series of organic beings which have existed on the earth. This appears to be a fatal objection to it, independently of all other considerations. It may be said that the same objections exist against every theory on such a subject, but this is not necessarily the case. The hypothesis put forward in this paper depends in no degree upon the completeness of our knowledge of the former condition of the organic world, but takes what facts we have as fragments of a vast whole, and deduces from them something of the nature and proportions of that whole which we can never know in detail. It is founded upon isolated groups of facts, recognizes their isolation, and endeavours to deduce from them the nature of the intervening protions.

Rudimentary Organs

Another important series of facts, quite in accordance with, and even necessary deductions from, the law now developed, are those of rudimentary organs. That these really do exist, and in most cases have no special function in the animal economy, is admitted by the first authorities in comparative anatomy. The minute limbs hidden beneath the skin in many of the snake-like lizards, the anal hooks of the boa constrictor, the complete series of jointed finger-bones in the paddle of the Manatus and whale, are a few of the most familiarinstances. In botany a similar class of facts has been long recognised. Abortive stamens, rudimentary floral envelopes and undeveloped carpels, are of the most frequent occurrence. To every thoughtful naturalist the question must arise, What are these for? What have they to do with the great laws of creation? Do they not teach us something of the system of Nature? If each species has been created independently, and without any necessary relations with pre-existing species, what do these rudiments, these apparent imperfections mean?

There must be a cause for them; they must be the necessary results of some great natural law. Now, if, as it has been endeavoured to be shown, the great law which has regulated the peopling of the earth with animal and vegetable life is, that every change shall be gradual; that no new creature shall be formed widely differing from anything before existing; that in this, as in everything else in nature, there shall be gradation and harmony,—then these rudimentary organs are necessary, and are an essential part of the system of nature. Ere the higher Vertebrata were formed, for instance, many steps were required, and many organs had to undergo modifications from the rudimental condition in which only they had as yet existed. We still see remaining an antitypal sketch of a wing adapted for flight in the scaly flapper of the penguin, and limbs first concealed beneath the skin, and then weakly protruding from it, wee the necessary gradations before others should be formed fully adapted for locomotion.[1] Many more of these modifications should we behold, and more complete series of them, had we a view of all the forms which have ceased to live. The great gaps that exist between fishes, reptiles, birds, and mammals would then, no doubt, be softened down be intermediate groups, and the whole organic world would be seen to be an unbroken and harmonious system.

Conclusion

It has now been shown though most briefly and imperfectly, how the law that "*Every species has come into existence coincident both in time and space with a pre-existing closely allied species,*" connects together and renders intelligible a vast number of independent and hitherto unexplained facts. The natural system of arrangement of organic beings, their geographical distribution, their geological sequence, the phenomena of representative and substituted groups in all their modifications, and the most singular peculiarities of anatomical structure, are all explained and illustrated by it, in perfect accordance with the vast mass of facts which the researches of modern naturalists have brought together, and, it is believed,

[1] The theory of Natural Selection has now taught us that these are not the steps by which limbs have been formed; and that most rudimentary organs have been produced by abortion, owing to disuse, as explained by Mr. Darwin.

not materially opposed to any of them. It also claims a superiority over previous hypotheses, on the ground that it not merely explains, but necessitates what exists. Granted the law, and many of the most important facts in Nature could not have been otherwise, but are almost as necessary deductions from it as are the elliptic orbits of the planets from the law of gravitation.

INTRODUCTORY NOTE TO CHAPTER II IN PRESENT EDITION

As this chapter sets forth the main features of a theory identical with that discovered by Mr. Darwin many years before but not then published, and as it has thus an historical interest, a few words of personal statement may be permissible. After writing the preceding paper the question of *how* changes of species could have been brought about was rarely out of my mind, but no satisfactory conclusion was reached till February 1858. At that time I was suffering from a rather severe attack of intermittent fever at Ternate in the Moluccas, and one day while lying on my bed during the cold fit, wrapped in blankets, though the thermometer was at 88° F., the problem again presented itself to me, and something led me to think of the "positive checks" described by Malthus in his "Essay on Population," a work I had read several years before, and which had made a deep and permanent impression on my mind. These checks—war, disease, famine and the like—must, it occurred to me, act on animals as well as on man. Then I thought of the enormously rapid multiplication of animals, causing these checks to be much more effective in them than in the case of man; and while pondering vaguely on this fact there suddenly flashed upon me the *idea* of the survival of the fittest—that the individuals removed by these checks must be on the whole inferior to those that survived. In the two hours that elapsed before my ague fit was over I had thought out almost the whole of the theory, and the same evening I sketched the draft of my paper, and in the two succeeding evenings wrote it out in full, and sent it by the next post to Mr. Darwin. Up to this time the only letters I had received from him were those printed in the second volume of his *Life and Letters,* (vol. ii. pp. 95 and 108), in which he speaks of its being the twentieth year since he "opened his first note-book on the question how and in what way do species and varieties differ from each other," and after referring to oceanic islands, the means of distribution of landshells, etc., added: "My work, on which I have now been at work more or less for twenty years, *will not fix or settle anything;* but I hope it will aid by giving a large collection of facts, with one definite end." The words I have italicised, and the whole tone of his letters, led me to conclude that he had arrived at no definite view as to the origin of species, and I fully anticipated that my theory would be new to him, because it seemed to me to settle a great deal. The immediate result of my paper was that Darwin was induced at once to prepare for publication his book on the *Origin of Species* in the condensed form in which it appeared, instead of waiting an indefinite number of years to complete a work on a much larger scale which he had partly written, but which in all probability would not have carried conviction to so many persons in so short a time. I feel much satisfaction in having thus aided in bringing about the publication of this celebrated book, and with the ample recognition by Darwin himself of my independent discovery of "natural selection." (See *Origin of Species,* 6th ed., introduction, p. 1, and *Life and Letters,* vol. ii. chap. iv., pp. 115–129 and 145.)

II

ON THE TENDENCY OF VARIETIES TO DEPART INDEFINITELY FROM THE ORIGINAL TYPE

Instability of Varieties supposed to prove the permanent distinctness of Species

ONE of the strongest arguments which have been adduced to prove the original and permanent distinctness of species is, that *varieties* produced in a state of domesticity are more or less unstable, and often have a tendency, if left to themselves, to return to the normal form of the parent species; and this instability is considered to be a distinctive peculiarity of all varieties, even of those occurring among wild animals in a state of nature, and it constitute a provision for preserving unchanged the originally created distinct species.

In the absence or scarcity of facts and observations as to *varieties* occurring among wild animals, this argument has had great weight with naturalists, and has led to a very general and somewhat prejudiced belief in the stability of species. Equally general, however, is the belief in what are called "permanent or true varieties,"—races of animals which continually propagate their like, but which differ so slightly (although constantly) from some other race, that the one is considered to be a *variety* of the other. Which is the *variety* and which the original *species,* there is generally no means of determining, except in those rare cases in which the one race has been known to produce an offspring unlike itself and resembling the other. This, however, would seem quite incompatible with the "permanent invariability of species," but the difficulty is overcome by assuming that such varieties have strict limits, and can never again vary further from the original type, although they may return to it, which, from the analogy of the domesticated animals, is considered to be highly probable, if not certainly proved.

It will be observed that this argument rests entirely on the assumption that *varieties* occurring in a state of nature are in all respects analogous to or even identical with those of domestic animals, and are governed by the same laws as regards their permanence or further variation. But it is the object of the present paper to show that this assumption is altogether false, that there is a general principle in nature which will cause many *varieties* to survive the parent species, and to give rise to successive variations departing further and further from the original type, and which also produces, in domesticated animals, the tendency of varieties to return to the parent form.

The Struggle for Existence

The life of wild animals is a struggle for existence. The full exertion of all their faculties and all their energies is required to preserve their own existence and provide for that of their infant offspring. The possibility of procuring food during the least favourable seasons, and of escaping the attacks of their most dangerous enemies, are the primary conditions which determine the existence both of individuals and of entire species. These conditions will also determine the population of a species; and by a careful consideration of all the circumstances we may be enabled to comprehend, and in some degree to explain, what at first sight appears so inexplicable—the excessive abundance of some species, while others closely allied to them are very rare.

The Law of Population of Species

The general proportion that must obtain between certain groups of animals is readily seen. Large animals cannot be so abundant as small ones; the carnivora must be less numerous than the herbivora; eagles and lions can never be so plentiful as pigeons and antelopes; and the wild asses of the Tartarian deserts cannot equal in numbers the horses of the more luxuriant prairies and pampas of America. The greater or less fecundity of an animal is often considered to be one of the chief causes of its abundance or scarcity; but a consideration of the facts will show us that it really has little or nothing to do with the matter. Even the least prolific of animals would increase rapidly if unchecked, whereas it is evident that the animal population of the globe must be stationary, or perhaps, through the influence of man, decreasing. Fluctuations there may be; but permanent increase, except in restricted localities, is almost impossible. For example, our own observation must convince us that birds do not go on increasing every year in a geometrical ratio, as they would do were there not some powerful check to their natural increase. Very few birds produce less than two young ones each year, while many have six, eight, or ten; four will certainly be below the average; and if we suppose that each pair produce young only four times in their life, that will also be below the average, supposing them not to die either by violence or want of food. Yet at this rate how tremendous would be the increase in a few years from a single pair! A simple calculation will show that in fifteen years each pair of birds would have increased to nearly ten millions![1] whereas we have no reason to believe that the number of the birds of any country increases at all in fifteen or in one hundred and fifty years. With such powers of increase the population must have reached its limits, and have become stationary, in a very few years after the origin of each species.

It is evident, therefore, that each year an immense number of birds must perish—as many in fact as are born; and as on the lowest calculation the progeny are each year twice as numerous as their parents, it follows that, whatever be the average number of individuals existing in any given country, *twice that number must perish annually,*—a striking result, but one which seems at least highly probable, and is perhaps under rather than over the truth. It would therefore appear that, so far as the continuance of the species and the keeping up the average number of individuals are concerned, large broods are superfluous. On the average all above *one* become food for hawks and kites, wild cats or weasels, or perish of cold and hunger as winter comes on. This is strikingly proved by the case of particular species; for we find that their abundance in individuals bears no relation whatever to their fertility in producing offspring.

[1] This is under estimated. The number would really amount to more than two thousand millions!

Perhaps the most remarkable instance of an immense bird population is that of the passenger pigeon of the United States, which lays only one, or at most two eggs, and is said to rear generally but one young one. Why is this bird so extraordinarily abundant, while others producing two or three times as many young are much less plentiful? The explanation is not difficult. The food most congenial to this species, and on which it thrives best, is abundantly distributed over a very extensive region, offering such differences of soil and climate, that in one part or another of the area the supply never fails. The bird is capable of a very rapid and long-continued flight, so that it can pass without fatigue over the whole of the district it inhabits, and as soon as the supply of food begins to fail in one place is able to discover a fresh feeding-ground. This example strikingly shows us that the procuring a constant supply of wholesome food is almost the sole condition requisite for ensuring the rapid increase of a given species, since neither the limited fecundity nor the unrestrained attacks of birds of prey and of man are here sufficient to check it. In no other birds are these peculiar circumstances so strikingly combined. Either their food is more liable to failure, or they have not sufficient power of wing to search for it over an extensive area, or during some season of the year it becomes very scarce, and less wholesome substitutes have to be found; and thus, though more fertile in offspring, they can never increase beyond the supply of food in the least favourable seasons.

Many birds can only exist by migrating, when their food becomes scarce, to regions possessing a milder, or at least a different climate, though, as these migrating birds are seldom excessively abundant, it is evident that the countries they visit are still deficient in a constant and abundant supply of wholesome food. Those whose organisation does not permit them to migrate when their food becomes periodically scarce, can never attain a large population. This is probably the reason why woodpeckers are scarce with us, while in the tropics they are among the most abundant of solitary birds. Thus the house sparrow is more abundant than the redbreast, because its food is more constant and plentiful,—seeds of grasses being preserved during the winter, and our farm-yards and stubble-fields furnishing an almost inexhaustible supply. Why, as a general rule, are aquatic, and especially sea-birds, very numerous in individuals? Not because they are more prolific than others, generally the contrary; but because their food never fails, the sea-shores and river-banks daily swarming with a fresh supply of small mollusca an crustacea. Exactly the same laws will apply to mammals. Wild cats are prolific and have few enemies; why then are they never as abundant as rabbits? The only intelligible answer is, that their supply of food is more precarious. It appears evident, therefore, that so long as a country remains physically unchanged, the numbers of its animal population cannot materially increase. If one species does so, some others requiring the same king of food must diminish in proportion. The numbers that die annually must be immense; and as the individual existence of each animal depends upon itself, those that die must be the weakest—the very young, the aged, and the diseased—while those that prolong their existence can only be the most perfect in health and vigour—those who are best able to obtain food regularly, and avoid their numerous enemies. It is, as we commenced by remarking, "a struggle for existence," in which the weakest and least perfectly organized must always succumb.

The Abundance or Rarity of a Species dependent upon its more or less perfect Adaptation to the Conditions of Existence

It seems evident that what takes place among the individuals of a species must also occur among the several allied species of a group,—viz., that those which are best adapted to obtain a regular supply of food, and to defend themselves against the attacks of their enemies and the vicissitudes of the seasons, must necessarily obtain and preserve a superiority in population; while those species which, from some defect of power or organisation, are the least capable of counteracting the vicissitudes of food-supply, etc., must diminish in numbers, and, in extreme cases, become altogether extinct. Between these extremes the species will present various degrees of capacity for ensuring the means of preserving life; and it is thus we account for the abundance or rarity of species. Our ignorance will generally prevent us from accurately tracing the effects to their causes; but could we become perfectly acquainted with the organisation and habits of the various species of animals, and could we measure the capacity of each for performing the different acts necessary to its safety and existence under all the varying circumstances by which it is surrounded, we might be able even to calculate the proportionate abundance of individuals which is the necessary result.

If now we have succeeded in establishing these two points—1st, *that the animal population of a country is generally stationary, being kept down by a periodical deficiency of food, and other checks;* and, 2d, *that the comparative abundance or scarcity of the individuals of the several species is entirely due to their organisation and resulting habits, which, rendering it more difficult to procure a regular supply of food and to provide for their personal safety*[1] *in some cases than in others, can only be balanced by a difference in the population which have to exist in a given area*—we shall be in a condition to proceed to the consideration of *varieties,* to which the preceding remarks have a direct and very important application.

1 "And that of their offspring" should have been added. But it must be remembered that the writer had no opportunity of correcting the proofs of this paper.

Useful Variations will tend to Increase; useless or hurtful Variations to Diminish

Most or perhaps all the variations from the typical form of a species must have some definite effect, however slight, on the habits or capacities of the individuals. Even a change of colour might, by rendering them more or less distinguishable, affect their safety; a greater or less development of hair might modify their habits. More important changes, such as an increase in the power or dimensions of the limbs or any of the external organs, would more or less affect their mode of procuring food or the range of country which they could inhabit. It is also evident that most changes would affect, either favourably or adversely, the powers of prolonging existence. An antelope with shorter or weaker legs must necessarily suffer more from the attacks of the feline carnivora; the passenger pigeon with less powerful wings would sooner or later be affected in its powers of procuring a regular supply of food; and in both cases the result must necessarily be a diminution of the population of the modified species. If on the other hand, any species should produce a variety having slightly increased powers of preserving existence, that variety must inevitably in time acquire a superiority in numbers. These results must follow as surely as old age, intemperance, or scarcity of food produce an increased mortality. In both cases there may be many individual exceptions: but on the average the rule will invariably be found to hold good. All varieties will therefore fall into two classes—those which under the same conditions would never reach the population of the parent species, and those which would in time obtain and keep a numerical superiority. Now, let some alteration of physical conditions occur in the district—a long period of drought, a destruction of vegetation by locusts, the irruption of some fresh carnivorous animal seeking "pastures new"— any change in fact tending to render existence more difficult to the species in question, and tasking its utmost powers to avoid complete extermination,—it is evident that, of all the individuals composing the species, those forming the least numerous and most feebly organised variety would suffer first, and, were the pressure severe, must soon become extinct. The same causes continuing in action, the parent species would next suffer, would gradually diminish in numbers, and with a recurrence of similar unfavourable conditions might also become extinct. The superior variety would then alone remain, and on a return to favourable circumstances would rapidly increase in numbers and occupy the place of the extinct species and variety.

Superior Varieties will ultimately Extirpate the original Species

The *variety* would now have replaced the *Species,* of it would be a more perfectly developed and more organised form. It would be in all respects better a to secure its safely, and to prolong its individual existence and that of the race. Such a variety *could not* return to the original form; for that form is an inferior one, and could never compete with it for existence. Granted, therefore, a "tendency" to reproduce the original type of the species, still the variety must ever remain preponderant in numbers, and under adverse physical conditions *again alone survive.* But this new, improved, and populous race might itself, in course of time, give rise to new varieties, exhibiting several diverging modifications of forms, any of which, tending to increase the facilities for preserving existence, must, by the same general law, in their turn become predominant. Here, then, we have *progression and continued divergence* deduced from the general laws which regulate the existence of animals in a state of nature, and from the undisputed fact that varieties do frequently occur. It is not, however, contended that this result would be invariable; a change of physical conditions in the district might at times materially modify it, rendering the race which had been the most capable of supporting existence under the former conditions now the least so, and even causing the extinction of the newer and, for a time, superior race, while the old or parent species and its first inferior varieties continued to flourish. Variations in unimportant parts might also occur, having no perceptible effect on the life-preserving powers; and the varieties so furnished might run a course parallel with the parent species, either giving rise to further variations or returning to the former type. All we argue for is, that certain varieties have a tendency to maintain their existence longer than the original species, and this tendency, must itself felt; for though the doctrine of chances or averages can never be trusted on a limited scale, yet, if applied to high numbers, the results come nearer to what theory demands, and, as we approach to an infinity of examples, become strictly accurate. Now the scale on which nature works is so vast—the numbers of individuals and the periods of time with which she deals approach so near to infinity—that any cause, however slight, and however liable to be veiled and counteracted by accidental circumstances, must in the end produce its full legitimate results.

The Partial Reversion of Domesticated Varieties explained

Let us now turn to domesticated animals, and inquire how varieties produced among them are affected by the principles here enunciated. The essential difference in the condition of wild and domestic animals is this,—that among the former, their well-being and very existence depend upon the full exercise and healthy condition of all their senses and physical powers, whereas, among the latter, these are only partially exercised, and in some cases are absolutely unused. A wild animal has to search, and often to labour, for every mouthful of food—to exercise sight, hearing, and smell in seeking it, and in avoiding dangers, in procuring shelter from the inclemency of the seasons, and in providing for the subsistence and safety of its offspring. There is no muscle of its body that is not called into daily and hourly activity; there is no sense or faculty that is not strengthened by continual exercise. The domestic animal, on the other hand, has food provided for it, is sheltered, and often confined, to guard it against the vicissitudes of the seasons, is carefully secured from the attacks of its natural enemies, and seldom even rears its young without human assistance. Half of its senses and faculties become quite useless, and the other half are but occasionally called into feeble exercise, while even its muscular system is only irregularly brought into action.

Now when a variety of such an animal occurs having increased power or capacity in any organ or sense, such increase is totally useless, is never called into action, and may even exist without the animal ever becoming aware of it. In the wild animal, on the contrary, all its faculties and powers being brought into full action for the necessities of existence, any increase becomes immediately available, is strengthened by exercise, and must even slightly modify the food, the habits, and the whole economy of the race. It creates as it were a new animal, one of superior powers, and which will necessarily increase in numbers and outlive those which are inferior to it.

Again, in the domesticated animal all variations have an equal chance of continuance; and those which would decidedly render a wild animal unable to compete with its fellows and continue its existence are no disadvantage whatever in a state of domesticity. Our quickly fattening pigs, short-legged sheep, pouter pigeons, and poodle dogs could never have come into existence in a state of nature, because the very first steps towards such inferior forms would have led to the rapid extinction of the race; still less could they now exist in competition with their wild allies. The great speed but slight endurance of the racehorse, the unwieldly strength of the ploughman's team, would both be useless in a state of nature. If turned wild on the pampas, such animals would probably soon become extinct, or under favourable circumstances might each gradually lose those extreme qualities which would never be called into action, and in a few generations revert to a common type, which must be that in which the various powers and faculties are so proportioned to each other as to be best adapted to procure food and secure safety,—that in which, by the full exercise of every part of its organisation, the animal can alone continue to live. Domestic varieties, when turned wild, *must* return to something near the type of the original wild stock, *or become altogether extinct.*[1]

We see, then, that no inferences as to the permanence of varieties in a state of nature can be deduced from the observations of those occurring among domestic animals. The two are so much opposed to each other in every circumstance of their existence, that what applies to the one is almost sure not to apply to the other. Domestic animals are abnormal, irregular, artificial; they are subject to variations which never occur, in a state of nature: their very existence depends altogether on human care—so far are many of them removed from that just proportion of faculties, that true balance of organisation, by means of which alone an animal left to its own resources can preserve its existence and continue its race.

[1] That is, they will vary, and the variations which tend to adapt them to the wild state, and therefore approximate them to wild animals, will be preserved. Those individuals which do not vary sufficiently will perish.

Lamarck's Hypothesis very different from that now advanced

The hypothesis of Lamarck—that progressive changes in species have been produced by the attempts of animals to increase the development of their own organs, and thus modify their structure and habits—has been repeatedly and easily refuted by all writers on the subject of varieties and species, and it seems to have been considered that when this was done the whole question has been finally settled; but the view here developed renders such an hypothesis quite unnecessary, by showing that similar results must be produced by the action of principles constantly at work in nature. The powerful retractile talons of the falcon and the cat tribes have not been produced or increased by the volition of those animals; but among the different varieties which occurred in the earlier and less highly organised forms of these groups, *those always survived longest which had the greatest facilities for seizing their prey.* Neither did the giraffe acquire its long neck by desiring to reach the foliage of the more lofty shrubs, and constantly stretching its neck for the purpose, but because any varieties which occurred among its antitypes with a longer neck than usual *at once secured a fresh range of pasture over the same ground as their shorter-necked companions, and on the first scarcity of food were thereby enabled to outlive them.* Even the peculiar colours of many animals, more especially of insects, so closely resembling the soil or leaves or bark on which they habitually reside, are explained on the same principle; for though in the course of ages varieties of many tints may have occurred, *yet those races having colours best adapted to concealment from their enemies would inevitably survive the longest.* We have also here an acting cause to account for that balance so often observed in nature,—a deficiency in one set of organs always being compensated by an increased development of some others—powerful wings accompanying weak feet, or great velocity making up for the absence of defensive weapons; for it has been shown that all varieties in which an unbalanced deficiency occurred could not long continue their existence. The action of this principle is exactly like that of the centrifugal governor of the steam-engine, which checks and corrects any irregularities almost before they become evident; and in like manner no unbalanced deficiency in the animal kingdom can ever reach any conspicuous magnitude, because it would make itself felt at the very first step, by rendering existence difficult and extinction almost sure soon to follow. An origin such as is here advocated will also agree with the peculiar character of the modifications of form and structure which obtain in organised beings—the many lines of divergence from a central type, the increasing efficiency and power of a particular organ through a succession of allied species, and the remarkable persistence of unimportant parts, such as colour, texture of plumage and hair, form of horns or crests, through a series of species differing considerably in more essential characters. It also furnishes us with a reason for that "more specialised structure" which Professor Owen states to be a characteristic of recent compared with extinct forms, and which would evidently be the result of the progressive modification of any organ applied to a special purpose in the animal economy.

Conclusion

We believe we have now shown that there is an tendency in nature to the continued progression of certain classes of *varieties* further and further from the original type—a progression to which there appears no reason to assign any definite limits—and that the same principle which produces this result in a state of nature will also explain why domestic varieties have a tendency, when they become wild, to revert to the original type. This progression, by minute steps, in various directions, but always checked and balanced by the necessary conditions, subject to which alone existence can be preserved, may, it is believed, be followed out so as to agree with all the phenomena presented by organised beings, their extinction and succession in past ages, and all the extraordinary modifications of form, instinct, and habits which they exhibit.

III

MIMICRY, AND OTHER PROTECTIVE RESEMBLANCES AMONG ANIMALS[1]

THERE is no more convincing proof of the truth of a comprehensive theory than its power of absorbing and finding a place for new facts, and its capability of interpreting phenomena which had been previously looked upon as unaccountable anomalies. It is thus that the law of universal gravitation and the undulatory theory of light have become established and universally accepted by men of science. Fact after fact has been brought forward as being apparently inconsistent with them, and one after another these very facts have been shown to be the consequences of the laws they were at first supposed to disprove. A false theory will never stand this test. Advancing knowledge brings to light whole groups of facts which it cannot deal with, and its advocates steadily decrease in numbers, notwithstanding the ability and scientific skill with which it may have been supported. The great name of Edward Forbes did not prevent his theory of "Polarity in the distribution of Organic beings in Time" from dying a natural death; but the most striking illustration of the behaviour of a false theory is to be found in the "Circular and Quinarian System" of classification propounded by MacLeay, and developed by Swainson, with an amount of knowledge and ingenuity that has rarely been surpassed. This theory was eminently attractive, both from its symmetry and completeness, and from the interesting nature of the varied analogies and affinities which it brought to light and made use of. The series of Natural History volumes in *Lardner's Cabinet Cyclopœdia,* in which Mr. Swainson developed it in most departments of the animal kingdom, made it widely known; and in fact for a long time these were the best and almost the only popular text-books for the rising generation of naturalists. It was favourably received too by the older school, which was perhaps rather an indication of its unsoundness. A considerable number of well-known naturalists either spoke approvingly of it, or advocated similar principles, and for a good many years it was decidedly in the ascendant. With such a favourable introduction, and with such talented exponents, it must have become established if it had had any germ of truth in it; yet it quite died out in a few short years; its very existence is now a matter of history; and so rapid was its fall that its talented creator, Swainson, perhaps lived to be the last man who believed in it.

[1] First published in the *Westminster Review,* July 1867; reprinted in 1870 with additions and corrections.

Such is the course of a false theory. That of a true one is very different, as may be well seen by the progress of opinion on the subject of Natural Selection. In less than eight years *The Origin of Species* has produced conviction in the minds of a majority of the most eminent living men of science. New facts, new problems, new difficulties as they arise are accepted, solved, or removed by this theory; and its principles are illustrated by the progress and conclusions of every well established branch of human knowledge. It is the object of the present chapter to show how it has recently been applied to connect together and explain a variety of curious facts which had long been considered as inexplicable anomalies.

Importance of the Principle of Utility

Perhaps no principle has ever been announced so fertile in results as that which Mr. Darwin so earnestly impresses upon us, and which is indeed a necessary deduction from the theory of Natural Selection, namely—that none of the definite facts or organic nature, no special organ, no characteristic form of marking, no peculiarities of instinct or of habit, no relations between species or between groups of species—can exist, but which must now be or once have been *useful* to the individuals or the races which possess them.

This great principle gives us a clue which we can follow out in the study of many recondite phenomena, and leads us to seek a meaning and a purpose of some definite character in minutiæ which we should otherwise be almost sure to pass over as insignificant or unimportant.

Popular Theories of Colour in Animals

The adaptation of the external colouring of animals to their conditions of life has long been recognised, and had been imputed either to an originally created specific peculiarity, or to the direct action of climate, soil, or food. Where the former explanation has been accepted it has completely checked inquiry, since we could never get any further than the fact of the adaptation. There was nothing more to be known about the matter. The second explanation was soon found to be quite inadequate to deal with all the varied phases of the phenomena, and to be contradicted by many well known facts. For example, wild rabbits are always of gray of brown tints well suited for concealment among grass and fern. But when these rabbits are domesticated, without any change of climate or food, they vary into white or black, and these varieties may be multiplied to any extent, forming white or black races. Exactly the same thing has occurred with pigeons; and in the case of rats and mice, the white variety has not been shown to be at all dependent on alteration of climate, food, or other external conditions. In many cases the wings of an insect not only assume the exact tint of the bark or leaf it is accustomed to rest on, but the form and veining of the leaf or the exact rugosity of the bark is imitated; and these detailed modifications cannot be reasonably imputed to climate or to food, since in many cases the species does not feed on the substance it resembles, and when it does, no reasonable connection can be shown to exist between the supposed causes and the effect produced. It was reserved for the theory of Natural Selection to solve all these problems, and many others which were not at first supposed to be directly connected with them. To make these latter intelligible, it will be necessary to give a sketch of the whole series of phenomena which may be classed under the head of useful or protective resemblances.

Importance of Concealment as Influencing Colour

Concealment, more or less complete, is useful to many animals, and absolutely essential to some. Those which have numerous enemies from which they cannot escape by rapidity of motion find safety in concealment. Those which prey upon others must also be so constituted as not to alarm them by their presence or their approach, or they would soon die of hunger. Now it is remarkable in how many cases nature gives this boon to the animal, by colouring it with such tints as may best serve to enable it to escape from its enemies or to entrap its prey. Desert animals as a rule are desert- coloured. The lion is a typical example of this, and must be almost invisible when crouched upon the sand or among desert rocks and stones. Antelopes are all more or less sandy-coloured. The camel is pre-eminently so. The Egyptian cat and the Pampas cat are sandy are earth-coloured. The Australian kangaroos are of the same tints, and the original colour of the wild horse is supposed to have been a sandy or clay-colour.

The desert birds are still more remarkably protected by their assimilative hues. The stonechats, the larks, the quails, the goatsuckers and the grouse, which abound in the North African and Asiatic deserts, are all tinted and mottled so as to resemble with wonderful accuracy the average colour and aspect of the soil in the district they inhabit. The Rev. H. Tristram, in his account of the ornithology of North Africa in the first volume of the *Ibis,* says: "In the desert, where neither trees, brushwood, nor even undulation of the surface afford the slightest protection to its foes, a modification of colour which shall be assimilated to that of the surrounding country is absolutely necessary. Hence *without exception* the upper plumage of *every bird,* whether lark, sylvain, or sand-grouse, and also the fur of *all the smaller mammals,* and the skin of *all the snakes and lizards,* is of one uniform isabelline or sand colour." After the testimony of so able an observer it is unnecessary to adduce further examples of the protective colours of desert animals.

Almost equally striking are the cases of arctic animals possessing the white colour that best conceals them upon snowfields and icebergs. The polar bear is the only bear that is white, and it lives constantly among snow and ice. The arctic fox, the ermine, and the alpine hare change to white in winter only, because in summer white would be more conspicuous than any other colour, and therefore a danger rather than a protection; but the American polar hare, inhabiting regions of almost perpetual snow, is white all the year round. Other animals inhabiting the same Northern regions do not, however, change colour. The sable is a good example, for throughout the severity of a Siberian winter it retains its rich brown fur. But its habits are such that it does not need the protection of colour, for it is said to be able to subsist on fruits and berries in winter, and to be so active upon the trees as to catch small birds among the branches. So also the woodchuck of Canada has a dark-brown fur; but then it lives in burrows and frequents river banks, catching fish and small animals that live in or near the water.

Among birds, the ptarmigan is a fine example of protective colouring. Its summer plumage so exactly harmonises with the lichen-coloured stones among which it delights to sit, that a person may walk through a flock of them without seeing a single bird; while in winter its white plumage is an almost equal protection. The snow-bunting, the jer-falcon, and the snowy owl are also white-coloured birds inhabiting the arctic regions, and there can be little doubt but that their colouring is to some extent protective.

Nocturnal animals supply us with equally good illustrations. Mice, rats, bats and moles possess the least conspicuous of hues, and must be quite invisible at times when any light colour would be instantly seen. Owls and goatsuckers are of those dark mottled tints that will assimilate with bark and lichen, and thus protect them during the day, and at the same time be inconspicuous in the dusk.

It is only in the tropics, among forests which never lose their foliage, that we find whole groups of birds whose chief colour is green. The parrots are the most striking example, but we have also a group of green pigeons in the East; and the barbets, leaf-thrushes, bee-eaters, white-eyes, turacos, and several smaller groups, have so much green in their plumage as to tend greatly to conceal them among the foliage.

Special Modifications of Colour

The conformity of tint which has been so far shown to exist between animals and their habitations is of a somewhat general character; we will now consider the cases of more special adaptation. If the lion is enabled by his sandy colour readily to conceal himself by merely crouching down upon the desert, how, it may be asked, do the elegant markings of the tiger, the jaguar, and the other large cats, agree with this theory? We reply that these are generally cases of more or less special adaptation. The tiger is a jungle animal, and hides himself among tufts of grass or of bamboos, and in these positions the vertical stripes with which his body is adorned must so assimilate with the vertical stems of the bamboo as to assist greatly in concealing him from his approaching prey.[1] How remarkable it is that besides the lion and tiger, almost all the other large cats are arboreal in their habits, and almost all have ocellated or spotted skins, which must certainly tend to blend them with the background of foliage; while the one exception, the puma, has a ashy brown uniform fur, and has the habit of clinging so closely to a limb of a tree while waiting for his prey to pass beneath as to be hardly distinguishable from the bark.

Among birds, the ptarmigan, already mentioned, must be considered a remarkable case of special adaptation. Another is a South American goatsucker (Caprimulgus rupestris), which rests in the bright sunshine on little bare rocky islets in the Upper Rio Negro, where its unusually light colours so closely resemble those of the rock and sand that it can scarcely be detected till trodden upon.

The Duke of Argyll, in his *Reign of Law,* has pointed out the admirable adaptation of the colours of the woodcock to its protection. The various browns and yellows and pale ash-colour that occur in fallen leaves are all reproduced in its plumage, so that when, according to its habit, it rests upon the ground under trees, it is almost impossible to detect it. In snipes the colours are modified so as to be equally in harmony with the prevalent forms and colours of marshy vegetation. Mr. J. M. Lester, in a paper read before the Rugby School Natural History Society, observes: "The wood-dove, when perched amongst the branches of its favourite *fir,* is scarcely discernible; whereas, were it among some lighter foliage, the blue and purple tints in its plumage would far sooner betray it. The robin redbreast too, although it might be thought that the red on its breast made it much easier to be seen, is in reality not at all endangered by it, Since it generally contrives to get among some russet or yellow fading leaves, where the red matches very well with the autumn tints, and the brown of the rest of the body with the bare branches."

[1] This suggestion has been since confirmed. See *Darwinism,* p. 199.

Reptiles offer us many similar examples. The most arboreal lizards, the iguanas, are as green as the leaves they feed upon, and the slender whip-snakes are rendered almost invisible as they glide among the foliage by a similar coloration. How difficult it is sometimes to catch sight of the little green tree-frogs sitting on the leaves of a small plant enclosed in a glass case in the Zoological Gardens; yet how much better concealed must they be among the fresh green damp foliage of a marshy forest. There is a North American frog found on lichen-covered rocks and walls, which is so coloured as exactly to resemble them, and as long as it remains quiet would certainly escape detection. Some of the geckos which cling motionless on the trunks of trees in the tropics are of such curiously marbled colours as to match exactly with the bark they rest upon.

In every part of the tropics there are tree-snakes that twist among boughs and shrubs, or lie coiled up on the dense masses of foliage. These are of many distinct groups, and comprise both venomous and harmless genera; but almost all of them are of a beautiful green colour, sometimes more or less adorned with white or dusky bands and spots. There can be little doubt that this colour is doubly useful to them, since it will tend to conceal them from their enemies, and will lead their prey to approach then unconscious of danger. Dr. Gunther informs me that there is only one genus of true arboreal snakes (Dipsas) whose colours are rarely green, but are of various shades of black, brown, and olive, and these are all nocturnal reptiles, and there can be little doubt conceal themselves during the day in holes, so that the green protective tint would be useless to them, and they accordingly retain the more usual reptilian hues.

Fishes present similar instances. Many flat fish, as for example the flounder and the skate, are exactly the colour of the gravel or sand on which they habitually rest. Among the marine flower gardens of an Eastern coral reef the fishes present every variety of gorgeous colour, while the river fish even of the tropics rarely if ever have gay or conspicuous markings. A very curious case of this kind of adaptation occurs in the sea-horses (Hippocampus) of Australia, some of which bear long foliaceous appendages resembling seaweed, and are of a brilliant red colour; and they are known to live among seaweed of the same hue, so that when at rest they must be quite invisible. There are now in the aquarium of the Zoological Society some slender green pipe-fish which fasten themselves to any object at the bottom by their prehensile tails, and float about with the current, looking exactly like some simple cylindrical algæ.

It is, however, in the insect world that this principle of the adaptation of animals to their environment is most fully and strikingly developed. In order to understand how general this is, it is necessary to enter somewhat into details, as we shall thereby be better able to appreciate the significance of the still more remarkable phenomena we shall presently have to discuss. It seems to be in proportion to their sluggish motions or the absence of other means of defence, that insects possess the protective colouring. In the tropics there are thousands of species of insects which rest during the day clinging to the bark of dead or fallen trees; and the greater portion of these are delicately mottled with gray and brown tints, which, though symmetrically disposed and infinitely varied, yet blend so completely with the usual colours of the bark, that at two or three feet distance they are quite undistinguishable. In some cases a species is known to frequent only one species of tree. This is the case with the common South American long-horned beetle (Onychocerus scorpio), which, Mr. Bates informed me, is found only on a rough-barked tree, called Tapiribá, on the Amazon. It is very abundant, but so exactly does it resemble the bark in colour and rugosity, and so closely does it cling to the branches, that until it moves it is absolutely invisible! An allied species (O. concentricus) is found only at Pará, on a distinct species of tree, the bark of which it resembles with equal accuracy. Both these insects are abundant, and we may fairly conclude that the protection they derive from this strange concealment is at least one of the causes that enable the race to flourish.

Many of the species of Cicindela, or tiger beetle, will illustrate this mode of protection. Our common Cicindela campestris frequents grassy banks, and is of a beautiful green colour, while C. maritima, which is found only on sandy seashores, is of a pale bronzy yellow, so as to be almost invisible. A great number of the species found by myself in the Malay islands are similarly protected. The beautiful Cicindela gloriosa, of a very deep velvety green colour, was only taken upon wet mossy stones in the bed of a mountain stream, where it was with the greatest difficulty detected. A large brown species (C. heros) was found chiefly on dead leaves in forest paths; and one which was never seen except on the wet mud of salt marshes was of a glossy olive so exactly the colour of the mud as only to be distinguished, when the sun shone, by its shadow! Where the sandy beach was coralline and nearly white, I found a very pale Cicindela; wherever it was volcanic and black, a dark species of the same genus was sure to be met with.

There are in the East small beetles of the family Buprestidæ which generally rest on the midrib of a leaf, and the naturalist often hesitates before picking them off, so closely do they resemble pieces of bird's dung. Kirby and Spence mention the small beetle Onthophilus sulcatus as being like the seed of an umbelliferous plant; and another, a small weevil, which is much persecuted by predatory beetles of the genus Harpalus, is of the exact colour of loamy soil, and was found to be particularly abundant in loam pits. Mr. Bates mentions a small beetle (Chlamys pilula) which was undistinguishable by the eye from the dung of caterpillars, while some of the Cassidæ, from their hemispherical forms and pearly gold colour, resemble glittering dew-drops upon the leaves.

A number of our small brown and speckled weevils at the approach of any object roll off the leaf they are sitting on, at the same time drawing in their legs and antennæ, which fit so perfectly into cavities for their reception that the insect becomes a mere oval brownish lump, which it is hopeless to look for among the similarly coloured little stones and earth pellets among which it lies motionless.

The distribution of colour in butterflies and moths respectively is very instructive from this point of view. The former have all their brilliant colouring on the upper surface of all four wings, while the under surface is almost always soberly coloured, and often very dark and obscure. The moths on the contrary have generally their chief colour on the hind wings only, the upper wings being of dull, somber, and often imitative tints, and these generally conceal the hind wings when the insects are in repose. This arrangement of the colours is therefore eminently protective, because the butterfly always rests with his wings raised so as to conceal the dangerous brilliancy of his upper surface. It is probable that if we watched their habits sufficiently we should find the under surface of the wings of butterflies very frequently imitative and protective. Mr. T. W. Wood has pointed out that the little orange-tip butterfly often rests in the evening on the green and white flower heads of an umbelliferous plant, the wild chervil,[1] and that when observed in this position the beautiful green and white mottling of the under surface completely assimilates with the flower heads and renders the creature very difficult to be seen. It is probable that the rich dark colouring of the under side of our peacock, tortoiseshell, and red-admiral butterflies answers a similar purpose.

Two curious South American butterflies that always settle on the trunks of trees (Gynecia dirce and Callizona acesta) have the under surface curiously striped and mottled, and when viewed obliquely must closely assimilate with the appearance of the furrowed bark of many kinds of trees. But the most wonderful and undoubted case of protective resemblance in a butterfly which I have ever seen, is that of the common Indian Kallima inachis, and its Malayan ally, Kallima paralekta. The upper surface of these insects is very striking and showy, as they are of a large size, and are adorned with a broad band of rich orange on a deep-bluish ground. The under side is very variable in colour, so that out of fifty specimens no two can be found exactly alike, but every one of them will be of some shade of ash or brown or ochre, such as are found among dead, dry, or decaying leaves. The apex of the upper wings is produced into an acute point, a very common form in the leaves of tropical shrubs and trees, and the lower wings are also produced into a short narrow tail. Between these two points runs a dark curved line exactly representing the midrib of a leaf, and from this radiate on each side a few oblique lines, which serve to indicate the lateral veins of a leaf. These marks are more clearly seen on the outer portion of the base of the wings, and on the inner side towards the middle and apex, and it is very curious to observe how the usual marginal and transverse striæ of the group are here modified and strengthened so as to become adapted for an imitation of the venation of a leaf. We come now to a still more extraordinary part of the imitation, for we find representations of leaves in every stage of decay, variously blotched and mildewed and pierced with holes, and in many cases irregularly covered with powdery blackdots gathered into patches and spots, so closely resembling the various kinds of minute fungi that grow on dead leaves that it is impossible to avoid thinking at first sight that the butterflies themselves have been attacked by real fungi.

[1] Anthriscus sylvestris.

But this resemblance, close as it is, would be of little use if the habits of the insect did not accord with it. If the butterfly sat upon leaves or upon flowers, or opened its wings so as to expose the upper surface, or exposed and moved its head and antennæ as many other butterflies do, its disguise would be of little avail. We might be sure, however, from the analogy of many other cases, that the habits of the insect are such as still further to aid its deceptive garb; but we are not obliged to make any such supposition, since I myself had the good fortune to observe scores of Kallima paralekta, in Sumatra, and to capture many of them, and can vouch for the accuracy of the following details. These butterflies frequent dry forests and fly very swiftly. They were never seen to settle on a flower or a green leaf; but were many times lost sight of in a bush or tree of dead leaves. On such occasions they were generally searched for in vain, for while gazing intently at the very spot where one had disappeared, it would often suddenly dart out, and again vanish twenty of fifty yards farther on. On one or two occasions the insect was detected reposing, and it could then be seen how completely is assimilates itself to the surrounding leaves. It sits on a nearly upright twig, the wings fitting closely back to back, concealing the antennæ and head, which are drawn up between their bases. The little tails of the hind wing touch the branch, and form a perfect stalk to the leaf, which is supported in its place by the claws of the middle pair of feet, which are slender and inconspicuous. The irregular outline of the wings gives exactly the perspective effect of a shrivelled leaf. We thus have size, colour, form, markings, and habits all combining together to produce a disguise which may be said to be absolutely perfect; and the protection which it affords is sufficiently indicated by the abundance of the individuals that possess it.

The Rev. Joseph Greene has called attention to the

striking harmony between the colours of those British moths which are on the wing in autumn and winter, and the prevailing tints of nature at those seasons. In autumn various shades of yellow and brown prevail, and he shows that out of fifty-two species that fly at this season, no less than forty-two are of corresponding colours. Orgyia antiqua, O. gonostigma, the genera Xanthia, Glæa, and Ennomos are examples. In winter, gray and silvery tints prevail, and the genus Chematobia and several species of Hybernia which fly during this season are of corresponding hues. No doubt if the habits of moths in a state of nature were more closely observed, we should find many cases of special protective resemblance. A few such have already been noticed. Agriopis aprilina, Acronycta psi, and many other moths which rest during the day on the north side of the trunks of trees, can with difficulty be distinguished from the gray and green lichens that cover them. The lappet moth (Gastropacha querci) closely resembles both in shape and colour a brown dry leaf; and the well-known buff-tip moth, when at rest, is like the broken end of a lichen-covered branch. There are some of the small moths which exactly resemble the dung of birds dropped on leaves, and on this point Mr. A. Sidgwick, in a paper read before the Rugby School Natural History Society, gives the following original observation: "I myself have more than once mistaken Cilix compressa, a little white and gray moth, for a piece of bird's dung dropped upon a leaf, and *vice versâ* the dung for the moth. Bryophila Glandifera and Perla are the very image of the mortar walls on which they rest; and only this summer, in Switzerland, I amused myself for some time in watching a moth, probably Larentia tripunctaria, fluttering about quite close to me, and then alighting on a wall of the stone of the district which it so exactly matched as to be quite invisible a couple of yards off." There are probably hosts of these resemblances which have not been observed, owing to the difficulty of finding many of the species in their stations of natural repose. Caterpillars are also similarly protected. Many exactly resemble in tint the leaves they feed upon; Others are like little brown twigs, and many are so strangely marked or humped, that when motionless they can hardly be taken to be living creatures at all. Mr. Andrew Murray has remarked how closely the larva of the peacock moth (Saturnia pavoniaminor) harmonises in its ground colour with that of the young buds of heather on which it feeds, and that the pink spots with which it is decorated correspond with the flowers and flower-buds of the same plant.

The whole order of Orthoptera, grasshoppers, locusts, crickets, etc., are protected by their colours harmonising with that of the vegetation or the soil on which they live, and in no other group have we such striking examples of special resemblance. Most of the tropical Mantidæ and Locustidæ are of the exact tint of the leaves on which they habitually repose, and many of them in addition have the veinings of their wings modified so as exactly to imitate that of a leaf. This is carried to the furthest possible extent in the wonderful genus, Phyllium, the "walking lead" in which not only are the wings perfect imitations of leaves in every detail, but the thorax and legs are flat, dilated, and leaf-like; so that when the living insect is resting among the foliage on which it feeds, the closest observation is often unable to distinguish between the animal and the vegetable.

The whole family of the Phasmidæ, or spectres, to which this insect belongs, is more or less imitative, and a great number of the species are called "walking-stick insects," from their singular resemblance to twigs and branches. Some of these are a foot long and as thick as one's finger, and their whole colouring, form, rugosity, and the arrangement of the head, legs, and antennæ are such as to render them absolutely identical in appearance with dead sticks. They hang loosely about shrubs in the forest, and have the extraordinary habit of stretching out their legs unsymmetrically, so as to render the deception more complete. One of these creatures obtained by myself in Borneo (Ceroxylus laceratus) was covered over with foliaceous excrescences of a clear olive green colour, so as exactly to resemble a stick grown over by a creeping moss or jungermannia. The Dyak who brought it me assured me it was grown over with moss although alive, and it was only after a most minute examination that I could convince myself it was not so.

We need not adduce any more examples to show how important are the details of form and of colouring in animals, and that their very existence may often depend upon their being by these means concealed from their enemies. This kind of protection is found apparently in every class and order, for it has been noticed wherever we can obtain sufficient knowledge of the details of an animal's life-history. It varies in degree, from the mere absence of conspicuous colour or a general harmony with the prevailing tints of nature, up to such a minute and detailed resemblance to inorganic or vegetable structures as to realise the talisman of the fairy tale, and to give its possessor the power of rendering itself invisible.

Theory of Protective Colouring

We will now endeavour to show how these wonderful resemblances have most probably been brought about. Returning to the higher animals, let us consider the remarkable fact of the rarity of white colouring in the mammalia or birds of the temperate or tropical zones in a state of nature. There is not a single white land-bird or quadruped in Europe, except the few arctic or alpine species, to which white is a protective colour. Yet in many of these creatures there seems to be no inherent tendency to avoid white, for directly they are domesticated white varieties arise, and appear to thrive as well as others. We have white mice and rats, white cats, horses, dogs and cattle, white poultry, pigeons, turkeys and ducks, and white rabbits. Some of these animals have been domesticated for a long period, others only for a few centuries; but in almost every case in which an animal has been thoroughly domesticated, parti-coloured and white varieties are produced and become permanent.

It is also well known that animals in a state of nature produce white varieties occasionally. Blackbirds, starlings, and crows are occasionally seen white, as well as elephants, deer, tigers, hares, moles, and many other animals; but in no case is a permanent white race produced. Now there are no statistics to show that the normal-coloured parents produce white offspring oftener under domestication than in a state of nature, and we have no right to make such an assumption if the facts can be accounted for without it. But if the colours of animals do really, in the various instances already adduced, serve for their concealment and preservation, then white or any other conspicuous colour must be hurtful, and must in most cases shorten an animal's life. A white rabbit would be more surely the prey of hawk or buzzard, and the white mole, or field mouse, could not long escape from the vigilant owl. So, also, any deviation from those tints best adapted to conceal a carnivorous animal would render the pursuit of its prey mush more difficult, would place it at a disadvantage among its fellows, and in a time of scarcity would probably cause it to starve to death. On the other hand, if an animal spreads from a temperate into an arctic district, the conditions are changed. During a large portion of the year, and just when the struggle for existence is most severe, white is the prevailing tint of nature, and dark colours will be the most conspicuous. The white varieties will now have an advantage; they will escape from their enemies or will secure food, while their brown companions will be devoured or will starve; and as "like produces like" is the established rule in nature, the white race will become permanently established, and dark varieties, when they occasionally appear, will soon die out from their want of adaptation to their environment. In each case the fittest will survive, and a race will be eventually produced adapted to the conditions in which it lives.

We have here an illustration of the simple and effectual means by which animals are brought into harmony with the rest of nature. That slight amount of variability in every species, which we often look upon as something accidental or abnormal, or so insignificant as to be hardly worthy of notice, is yet the foundation of all those wonderful and harmonious resemblances which play such an important part in the economy of nature. Variation is generally very small in amount,[1] but it is all that is required, because the change in the external conditions to which an animal is subject is generally very slow and intermittent. When these changes have taken place too rapidly, the result has often been the extinction of species; but the general rule is, that climatal and geological changes go on slowly, and the slight but continual variations in the colour, form, and structure of all animals have furnished individuals adapted to these changes, and who have become the progenitors of modified races. Rapid multiplication, incessant slight variation, and survival of the fittest—these are the laws which ever keep the organic world in harmony with the inorganic, and with itself. These are the laws which we believe have produced all the cases of protective resemblance already adduced, as well as those still more curious examples we have yet to bring before our readers.

It must always be borne in mind that the more wonderful examples, in which there is not only a general but a special resemblance—as in the walking leaf, the mossy phasma, and the leaf-winged butterfly—represent those few instances in which the process of modification has been going on during an immense series of generations. They all occur in the tropics, where the conditions of existence are the most favourable, and where climatic changes have for long periods been hardly perceptible. In most of them favourable variations both of colour, form, structure, and instinct or habit, must have occurred to produce the perfect adaptation we now behold. All these are known to vary, and favourable variations, when not accompanied by others that were unfavourable, would certainly survive. At one time a little step might be made in this direction, at another time in that—a change of conditions might sometimes render useless that which it had taken ages to produce—great and sudden physical modifications might often produce the extinction of a race just as it was approaching perfection, and a hundred checks of which we can know nothing may have retarded the progress towards perfect adaptation; so that we can hardly wonder at there being so few cases in which a completely successful result has been attained as shown by the abundance and wide diffusion of the creatures so protected.

[1] Later research has shown that variation is more frequent and of greater amount than at first supposed. See *Darwinism,* chap. iii.

Objection that Colour, as being dangerous, should not exist in Nature

It is as well here to reply to an objection that will no doubt occur to many readers—that if concealment is so useful; to all animals, and so easily brought about by variation and survival of the fittest, there ought to be no conspicuously-coloured creatures; and they will perhaps ask how we account for the brilliant birds, and painted snakes, and gorgeous insects that occur abundantly all over the world. It will be advisable to answer this question rather fully, in order that we may be prepared to understand the phenomena of "mimicry," which it is the special object of this chapter to illustrate and explain.

the slightest observation of the life of animals will show us that they escape from their enemies and obtain their food in an infinite number of ways, and that their varied habits and instincts are in every case adapted to the conditions of their existence. The porcupine and the hedgehog have a defensive armour that saves them from the attacks of most animals. The tortoise in not injured by the conspicuous colours of his shell, because that shell is in most cases an effectual protection to him. The skunks of North America find safety in their power of emitting an unbearably offensive odour; the beaver in its aquatic habits and solidly constructed abode. In some cases the chief danger to an animal occurs at one particular period of its existence, and if that is guarded against its numbers can easily be maintained. This is the case with many birds, the eggs and young of which are especially obnoxious to danger, and we find accordingly a variety of curious contrivances to protect them. We have nests carefully concealed, hung from the slender extremities of grass or boughs over water; or placed in the hollow of a tree with a very small opening. When these precautions are successful, so many more individuals will be reared than can possibly find food during the least favourable seasons, that there will always be a number of weakly and inexperienced young birds who will fall a prey to the enemies of the race, and thus render necessary for the stronger and healthier individuals no other safeguard than their strength and activity. The instincts most favourable to the production and rearing of offspring will in these cases be most important, and the survival of the fittest will act so as to keep up and advance those instincts, while other causes which tend to modify colour and marking may continue their action almost unchecked.

It is perhaps in insects that we may best study the varied means by which animals are defended or concealed. One of the uses of the phosphorescence with which many insects are furnished is probably to frighten away their enemies; for Kirby any Spence state that a ground-beetle (Carabus) has been observed running round and round a luminous centipede as if afraid to attack it. An immense number of insects have stings, and some stingless ants of the genus Polyrachis are armed with strong and sharp spines on the back, which must render them unpalatable to many of the smaller insectivorous birds. Many beetles of the family Curculionidæ have the wing cases and other external parts so excessively hard, that they cannot be pinned without first drilling a hole to receive the pin, and it is probable that all such find a protection in this excessive hardness. Great numbers of insects hide themselves among the petals of flowers, or in the cracks of bark and timber; and finally, extensive groups and even whole orders have a more or less powerful and disgusting smell and taste, which they either possess permanently, or can emit at pleasure. The attitudes of some insects may also protect them, as the habit of turning up the tail by the harmless rove-beetles (Staphylindidæ) no doubt leads other animals besides children to the belief that they can sting. The curious attitude assumed by sphinx caterpillars is probably a safeguard, as well as the blood-red tentacles which can suddenly be thrown out from the neck by the caterpillars of all the true swallow-tailed butterflies.

It is among the groups that possess some of these varied kind of protection in a high degree that we find the greatest amount of conspicuous colour, or at least the most complete absence of protective imitation. The stinging Hymenoptera, wasps, bees, and hornets are, as a rule, very showy and brilliant insects, and there is not a single instance recorded in which any one of them is coloured so as to resemble a vegetable or inanimate substance. The Chrysididæ, or golden wasps, which do not sting, possess as a substitute the power of rolling themselves up into a ball, which is almost as hard and polished as if really made of metal,—and they are all adorned with the most gorgeous colours.[1] The whole order

Hemiptera (comprising the bugs) emit a powerful odour, and they present a very large proportion of gay-coloured and conspicuous insects. The lady-birds (Coccinellidæ) and their allies the Eumorphidæ, are often brightly spotted, as if to attract attention; but they can both emit fluids of a very disagreeable nature; they are certainly rejected by some birds and are probably never eaten by any.

The great family of ground-beetles (Carabidæ) almost all possess a disagreeable and some a very pungent smell, and a few, called bombardier beetles, have the peculiar faculty of emitting a jet of very volatile liquid, which appears like a puff of smoke, and is accompanied by a distinct crepitating explosion. It is probably because these insects are mostly nocturnal and predacious that they do not present more vivid hues. They are chiefly remarkable for brilliant metallic tints or dull red patches when they are not wholly black, and are therefore very conspicuous by day, when insect-eaters are kept off by their bad odour and taste, but are sufficiently invisible at night, when it is of importance that their prey should not become aware of their proximity.

It seems probable that, in some cases, that which would appear at first to be source of danger to its possessor may really be a means of protection. Many showy and weak-flying butterflies have a very broad expanse of wing, as in the brilliant blue Morphos of Brazilian forests, and the large Eastern Papilios; yet these groups are tolerably plentiful.

Now, specimens of these butterflies are often captured with pierced and broken wings, as if they had been seized by birds from whom they had escaped; but if the wings had been much smaller in proportion to the body, it seems probable that the insect would be more frequently struck or pierced in a vital part, and thus the increased expanse of the wings may have been indirectly beneficial.

In other cases the capacity of increase in a species is so great that however many of the perfect insect may be destroyed, there is always ample means for the continuance of the race. Many of the flesh-flies, gnats, ants, palm-tree weevils, and locusts are in this category. The whole family of Cetoniadæ or rose chafers, so full of gaily-coloured species, are probably saved from attack by a combination of characters. They fly very rapidly with a zigzag or waving course; they hide themselves the moment they alight, either in the corolla of flowers, or in rotten wood, or in cracks and hollows of trees, and they are generally encased in a very hard and polished coat of mail, which may render them unsatisfactory food to such birds as would be able to capture them. The causes which lead to the development of colour have been here able to act unchecked, and we see the result in a large variety of the most gorgeously-coloured insects.

[1] These colours may, however, be protective by causing the rolled-up insect to look like a piece of shining stone or mineral.

Here, then, with out very imperfect knowledge of the life-history of animals, we are able to see that there are widely varied modes by which they may obtain protection from their enemies or concealment from their prey. Some of these seem to be so complete and effectual as to answer all the wants of the race, and lead to the maintenance of the largest possible population. When this is the case, we can well understand that no further protection derived from a modification of colour can be of the slightest use, and the most brilliant hues may be developed without any prejudicial effect upon the species. On some of the laws that determine the development of colour something may be said presently. It is now merely necessary to show that concealment by obscure or imitative tints is only one out of very many ways by which animals maintain their existence; and having done this we are prepared to consider the phenomena of what has been termed "mimicry." It is to be particularly observed, however, that the word is not here used in the sense of voluntary imitation, but to imply a particular kind of resemblance—a resemblance not in internal structure but in external appearance—a resemblance in those parts only that catch the eye—a resemblance that deceives. As this kind of resemblance has the same effect as voluntary imitation or mimicry, and as we have no word that expresses the required meaning, "mimicry" was adopted by Mr. Bates (who was the first to explain the facts), and has led to some misunderstanding; but there need be none, if it is remembered that both "mimicry" and "imitation" are used in a metaphorical sense, as implying that close external likeness which causes things unlike in structure to be mistaken for each other.

Mimicry

It has been long known to entomologists that certain insects bear a strange external resemblance to others belonging to distinct genera, families, or even orders, and with which they have no real affinity whatever. The fact, however, appears to have been generally considered as dependent upon some unknown law of "analogy"—some "system of nature," or "general plan," which had guided the Creator in designing the myriads of insect forms, and which we could never hope to understand. In only one case does it appear that the resemblance was thought to be useful, and to have been designed as a means to a definite and intelligible purpose. The flies of the genus Volucella enter the nests of bees to deposit their eggs, so that their larvæ may feed upon the larvæ of the bees, and these flies are each wonderfully like the bee on which it is parasitic. Kirby and Spence believed that this resemblance or "mimicry" was for the express purpose of protecting the flies from the attacks of the bees, and the connection is so evident that it was hardly possible to avoid this conclusion. The resemblance, however, of moths to butterflies or to bees, of beetles to wasps, and of locusts to beetles, has been many times noticed by eminent writers; but scarcely ever till within the last few years does it appear to have been considered that these resemblances had any special purpose, or were of any direct benefit to the insects themselves. In this respect they were looked upon as accidental, as instances of the "curious analogies" in nature which must be wondered at but which could not be explained. Recently, however, these instances have been greatly multiplied; the nature of the resemblances has been more carefully studied, and it has been found that they are often carried out into such details as almost to imply a purpose of deceiving the observer. The phenomena, moreover, have been shown to follow certain definite laws, which again all indicate their dependence on the more general law of the "survival of the fittest," or, "the preservation of favoured races in the struggle for life." It will, perhaps, be as well here to state what these laws or general conclusions are, and then to give some account of the facts which support them.

The first law is, that in an overwhelming majority of cases of mimicry, the animals (or the groups) which resemble each other inhabit the same country, the same district, and in most cases are to be found together on the very same spot.

The second law is, that these resemblances are not

indiscriminate, but are limited to certain groups, which in every case are abundant in species and individuals, and can often be ascertained to have some special protection.

The third law is, that the species which resemble or "mimic" these dominant groups are comparatively less abundant in individuals, and are often very rare.

These laws will be found to hold good in all the cases of true mimicry among various classes of animals to which we have now to call the attention of our readers.

Mimicry among Lepidoptera

As it is among butterflies that instances of mimicry are most numerous and most striking, an account of some of the more prominent examples in this group will first be given. There is in South America an extensive family of these insects, the Heliconidæ which are in many respects very remarkable. They are so abundant and characteristic in all the woody portions of the American tropics, that in almost every locality they will be seen more frequently than any other butterflies. They are distinguished by very elongate wings, body, and antennæ, and are exceedingly beautiful and varied in their colours; spots and patches of yellow, red, or pure white upon a black, blue, or brown ground being most general. They frequent the forests chiefly, and all fly slowly and weakly; yet although they are so conspicuous, and could certainly be caught by insectivorous birds more easily than almost any other insects, their great abundance all over the wide region they inhabit shows that they are not so persecuted. It is to be especially remarked also, that they possess no adaptive colouring to protect them during repose, for the under side of their wings presents the same, or at least an equally conspicuous colouring as the upper side; and they may be observed after sunset suspended at the end of twigs and leaves, where they have taken up their station for the night, fully exposed to the attacks of enemies if they have any. These beautiful insects possess, however, a strong pungent semi-aromatic or medicinal odour, which seems to pervade all the juices of their system. When the entomologist squeezes the breast of one of them between his fingers to kill it, a yellow liquid exudes which stains the skin, and the smell of which can only be got rid of by time and repeated washings. Here we have probably the cause of their immunity from attack, since there is a great deal of evidence to show that certain insects are so disgusting to birds that they will under no circumstances touch them. Mr. Stainton has observed that a brood of young turkeys greedily devoured all the worthless moths he had amassed in a night's "sugaring," yet one after another seized and rejected a single white moth which happened to be among them. Young pheasants and partridges which eat many kinds of caterpillars seem to have an absolute dread of that of the common currant moth, which they will never touch, and tomtits as well as other small birds appear never to eat the same species. In the case of the Heliconidæ, however, we have some direct evidence to the same effect. In the Brazilian forests there are great numbers of insectivorous birds—as jacamars, trogons, and puffbirds—which catch insects on the wing, and that they destroy many butterflies is indicated by the fact that the wings of these insects are often found on the ground where their bodies have been devoured. But among these there are no wings of Heliconidæ, while those of the large showy Nymphalidæ which have a much swifter flight, are often met with. Again, a gentleman who had recently returned from Brazil stated at a meeting of the Entomological Society that he once observed a pair of puffbirds catching butterflies, which they brought to their nest to feed their young; yet during half an hour they never brought one of the Heliconidæ, which were flying lazily about in great numbers, and which they could have captured more easily than any others. It was this circumstance that led Mr. Belt to observe them so long, as he could not understand why the most common insects should be altogether passed by. Mr. Bates also tells us that he never saw them molested by lizards or predacious flies, which often pounce on other butterflies.

If, therefore, we accept it as highly probable (if not

proved) that the Heliconidæ are very greatly protected from attack by their peculiar odour and taste, we find it much more easy to understand their chief characteristics—their great abundance, their slow flight, their gaudy colours, and the entire absence of protective tints on their under surfaces. This property places them somewhat in the position of those curious wingless birds of oceanic islands, the dodo, the apteryx, and the moas, which are with great reason supposed to have lost the power of flight on account of the absence of carnivorous quadrupeds. Our butterflies have been protected in a different way, but quite as effectually; and the result has been that as there has been nothing to escape from, there has been no weeding out of slow flyers, and as there has been nothing to hide from, there has been no extermination of the bright-coloured varieties, and no preservation of such as tended to assimilate with surrounding objects.

Now let us consider how this kind of protection must act. Tropical insectivorous birds very frequently sit on dead branches of a lofty tree, or on those which overhand forest paths, gazing intently around, and darting off at intervals to seize an insect at a considerable distance, which they generally return to their station to devour. If a bird began by capturing the slow-flying conspicuous Heliconidæ, and found them always so disagreeable that it could not eat them, it would after a very few trials leave off catching them at all; and their whole appearance, form, colouring, and mode of flight are so peculiar that there can be little doubt birds would soon learn to distinguish them at a long distance, and never waste any time in pursuit of them. Under these circumstances, it is evident that any other butterfly of a group which birds were accustomed to devour would be almost equally well protected by closely resembling a Heliconian externally, as if it acquired also the disagreeable odour; always supposing that there were only a few of them among a great number of the Heliconias. If the birds could not distinguish the two kinds externally, and there were on the average only one eatable among fifty uneatable, they would soon give up seeking for the eatable ones, even if they knew them to exist. If, on the other hand, any particular butterfly of an eatable group acquired the disagreeable taste of the Heliconias while it retained the characteristic form and colouring of its own group, this would be really of no use to it whatever; for the birds would go on catching it among its eatable allies (compared with which it would rarely occur), it would be wounded and disabled, even if rejected, and its increase would thus be as effectually checked as if it were devoured. It is important, therefore, to understand that if any one genus of an extensive family of eatable butterflies were in danger of extermination from insect-eating birds, and if two kinds of variation were going on among them, some individuals possessing a slightly disagreeable taste, others a slight resemblance to the Heliconidæ, this latter quality would be much more valuable than the former. The change in flavour would not at all prevent the variety from being captured as before, and it would almost certainly be thoroughly disabled before being rejected. The approach in colour and form to the Heliconidæ, however, would be at the very first a positive, though perhaps a slight advantage; for although at short distances this variety would be easily distinguished and devoured, yet at a longer distance it might be mistaken for one of the uneatable group, and so be passed by and gain another day's life, which might in many cases be sufficient for it to lay a quantity of eggs and leave a numerous progeny, many of which would inherit the peculiarity which had been the safeguard of their parent.

Now, this hypothetical case is exactly realised in South America. Among the white butterflies forming the family Pieridæ (many of which do not greatly differ in appearance from our own cabbage butterflies) is a genus of rather small size (Leptalis), some species of which are white like their allies, while the larger number exactly resemble the Heliconidæ in the form and colouring of the wings. It must always be remembered that these two families are as absolutely distinguished from each other by structural characters as are the carnivora and the ruminants among quadrupeds, and that an entomologist can always distinguish the one from the other by the structure of the feet, just as certainly as a zoologist can tell a bear from a buffalo by the skull or by a tooth. Yet the resemblance of a species of the one family to another species in the other family was often so great, that both Mr. Bates and myself were many times deceived at the time of capture, and did not discover the distinctness of the two insects till a closer examination detected their essential differences. During his residence of eleven years in the Amazon valley, Mr. Bates found a number of species or varieties of Leptalis, each of which was a more or less exact copy of one of the Heliconidæ of the district it inhabited; and the results of his observations are embodied in a paper published in the *Linnœan Transactions,* in which he first explained the phenomena of "mimicry" as the result of natural selection, and showed its identity in casue and purpose with protective resemblance to vegetable or inorganic forms.

The imitation of the Heliconidæ by the Leptalides is carried out to a wonderful degree in form as well as in colouring. The wings have become elongated to the same extent, and the antennæ and abdomen have both become lengthened, to correspond with the unusual condition in which they exist in the former family. In coloration there are several types in the different genera of Heliconidæ. The genus Mechanitis is generally of a rich semi-transparent brown, banded with black and yellow; Methona is of large size, the wings transparent like horn, and with black transverse bands; while the delicate Ithomias are all more or less transparent, with black veins and borders, and often with marginal and transverse bands of orange red. These different forms are all copied by the various species of Leptalis, every band and spot and tint of colour, and the various degrees of transparency, being exactly reproduced. As if to derive all the benefit possible from this protective mimicry, the habits have become so modified that the Leptalides generally frequent the very same spots as their models, and have the same mode of flight; and as they are always very scarce (Mr. Bates estimating their numbers at about one to a thousand of the group they resemble), there is hardly a possibility of their being found out by their enemies. It is also very remarkable that in almost every case the particular Ithomias and other species of Heliconidæ which they resemble are noted as being very common species, swarming in individuals, and found over a wide range of country. This indicates antiquity and permanence in the species, and is exactly the condition most essential both to aid in the development of the resemblance and to increase its utility.

But the Leptalides are not the only insects who have prolonged their existence by imitating the great protected group of Heliconidæ;—a genus of quite another family of most lovely small American butterflies, the Erycinidæ, and three genera of diurnal moths, also present species which often mimic the same dominant forms, so that some, as Ithomia ilerdina of St. Paulo, for instance, have flying with them a few individuals of three widely different insects, which are yet disguised with exactly the same form, colour, and markings, so as to be quite undistinguishable when upon the wing. Again, the Heliconidæ are not the only group that are imitated, although they are the most frequent models. The black and red group of South American Papilios, and the handsome Erycinian genus Stalachtis, have also a few who copy them; but this fact offers no difficulty, since these two groups are almost as dominant as the Heliconidæ. They both fly very slowly, they are both conspicuously coloured, and they both abound in individuals; so that there is every reason to believe that they possess a protection of a similar kind to the Heliconidæ, and that it is therefore equally an advantage to other insects to be mistaken for them. There is also another extraordinary fact that we are not yet in a position clearly to comprehend: some groups of the Heliconidæ themselves mimic other groups. Species of Heliconia mimic Mechanitis, and every species of Napeogenes mimics some other Heliconideous butterfly.[1] This would seem to indicate that the distasteful secretion is not produced alike by all members of the family, and that where it is deficient protective imitation comes into play. It is this, perhaps, that has caused such a general resemblance among the Heliconidæ, such a uniformity of type with great diversity of colouring, since any aberration causing an insect to cease to look like one of the family would inevitably lead to its being attacked, wounded, and exterminated, even although it was not eatable.

In other parts of the world an exactly parallel series of facts have been observed. The Danaidæ and the Acrædæ of the Old World tropics form in fact one great group with the Heliconidæ. They have the same general form, structure, and habits; they possess the same protective odour, and are equally abundant in individuals, although not so varied in colour, blue and white spots on a black ground being the most general pattern. The insects which mimic these are chiefly Papilios and Diadema, a genus allied to our peacock and tortoiseshell butterflies. In tropical Africa there is a peculiar group of the genus Danais, characterised by dark-brown and bluish-white colours, arranged in bands or stripes. One of these, Danais niavius, is exactly imitated both by Papilio hippocoon and by Diadema anthedon; another, Danais echeria, by Papilio cenea; and in Natal a variety of the Danais is found having a white spot at the tip of wings, accompanied by a variety of the Papilio bearing a corresponding white spot. Acræa gea is copied in its very peculiar style of coloration by the female of Papilio cynorta, by Panopæa hirce, and by the female of Elymnias phegea. Acræa euryta of Calabar has a female variety of Panopea hirce from the same place which exactly copies it; and Mr. Trimen, in his paper on "Mimetic Analogies among African Butterflies," published in the *Transactions of the Linnœan*

Society for 1868, gives a list of no less than sixteen species and varieties of Diadema and its allies, and ten of Papilio, which in their colour and makings are perfect mimics of species or varieties of Danais or Acræ which inhabit the same districts.

[1] A satisfactory explanation of this phenomenon has now been found. See *Darwinism,* p. 252.

Passing on to India, we have Danais tytia, a butterfly with semi-transparent bluish wings and a border of rich reddish brown. This remarkable style of colouring is exactly reproduced in Papilio agestor and in Diadema nama, and all three insects not unfrequently come together in collections made at Darjeeling. In the Philippine Islands the large and curious Idea leuconöe, with its semi-transparent white wings, veined and spotted with black, is copied by the rare Papilio idæoides from the same islands.

In the Malay archipelago the very common and beautiful Euplæa midamus is so exactly mimicked by two rare Papilios (P. paradoxa and P. ænigma) that I generally caught them under the impression that they were the more common species; and the equally common and even more beautiful Euplæa rhadamanthus, with its pure white bands and spots on a ground of glossy blue and black, is reproduced in the Papilio caunus. Here also there are species of Diadema imitating the same group in two or three instances; but we shall have to adduce these further on in connection with another branch of the subject.

It has been already mentioned that in South America there is a group of Papilios which have all the characteristics of a protected race, and whose peculiar colours and markings are imitated by other butterflies not so protected. There is just such a group also in the East, having very similar colours and the same habits, and these also are mimicked by other species in the same genus not closely allied to them, and also by a few of other families. Papilio hector, a common Indian butterfly of a rich black colour spotted with crimson, is so closely copied by Papilio romulus that the latter insect has been thought to be its female. A close examination shows, however, that it is essentially different, and belongs to another section of the genus. Papilio antiphus and P. diphilus, black swallow-tailed butterflies with cream-coloured spotes, are so well imitated by varieties of P. theseus, that several writers have classed them as the same species. Papilio liris, found only in the island of Timor, is accompanied there by P. ænomaus, the female of which so exactly resembles it that they can hardly be separated in the cabinet, and on the wing are quite undistinguishable. But one of the most curious cases is the fine yellow-spotted Papilio cöon, which is unmistakably imitated by the female tailed from of Papilio memnon. These are both from Sumatra; but in North India

P. cöon is replaced by another species, which has been named P. doubledayi, having red spots instead of yellow; and in the same district the corresponding female tailed form of Papilio androgeus, sometimes considered a variety of P. memnon, is similarly red-spotted. Mr. Westwood has described some curious day-flying moths (Epicopeia) from North India, which have the form and colour of Papilios of this section, and two of these are very good imitations of Papilio polydorus and Papilio varuna, also from North India.

Almost all these cases of mimicry are from the tropics, where the forms of life are more abundant, and where insect development especially is of unchecked luxuriance; but there are also one or two instances in temperate regions. In North America, the large and handsome red and black butterfly Danais Archippus is very common; and the same country is inhabited by Limenitis Misippus, which closely resembles the Danais, while it differs entirely from every species of its own genus.

The only case of probable mimicry in our own country is that of the common which moth (Spilosoma menthastri), referred to at p. 56 as being rejected by young turkeys among hundreds of other moths on which they greedily fed. Each bird in succession took hold of this moth and threw it down again, as if too nasty to eat. Mr. Jenner Weir also found that this moth was refused by the Bullfinch, Chaffinch,

Yellow Hammer, and Red Bunting, but eaten after much hesitation by the Robin. We may therefore fairly conclude that this species would be disagreeable to many other birds, and would thus have an immunity from attack, which may be the cause of its great abundance and of its conspicuous white colour. Now it is a curious thing that there is another moth, Diaphora mendica, which appears about the same time, and whose female only is white. It is about the same size Spilosoma menthastri, and sufficiently resembles it in the dusk, and this moth is much less common. It seems very probable, therefore, that these species stand in the same relation to each other as the mimicking butterflies of various families do to the Heliconidæ and Danaidæ. It would be very interesting to experiment on all white moths, to ascertain if those which are most common are generally rejected by birds. It may be anticipated that they would be so, because white is the most conspicuous of all colours for nocturnal insects, and had they not some other protection would certainly be very injurious to them.

Lepidoptera mimicking other Insects

In the preceding cases we have found Lepidoptera imitating other species of the same order, and such species only as we have good reason to believe were free from the attacks of many insectivorous creatures; but there are other instances in which they altogether lose the external appearance of the order to which they belong, and take on the dress of bees of wasps—insects which have an undeniable protection in their stings. The Sesiidæ and ægeriidæ two families of day-flying moths, are particularly remarkable in this respect, and a mere inspection of the names gives to the various species shows how the resemblance has struck every one. We have apiformis, vespiforme, ichneumoniforme, scoliæforme, sphegiforme (bee-like, wasp-like, ichneumon-like, etc.) and many others, all indicating a resemblance to stinging Hymenoptera. In Britain we may particularly notice Sesia bombiliformis, which very closely resembles the male of the large and common humble bee, Bombus hortorum; Sphecia craboniforme, which is coloured like a hornet, and is (on the authority of Mr. Jenner Weir) much more like it when alive than when in the cabinet, from the way in which it carries its wings; and the currant clear-wing, Trochilium tipuliforme, which resembles a small black wasp (Odynerus sinuatus) very abundant in gardens at the same season. It has been so much the practice to look upon these resemblances as mere curious analogies playing no part in the economy of nature, that we have scarcely any observations of the habits and appearance when alive of the hundreds of species of these groups in various parts of the world, or how far they are accompanied by Hymenoptera, which they specifically resemble. There are many species in India (like those figured by Professor Westwood in his *Oriental Entomology)* which have the hind legs very broad and densely hairy, so as exactly to imitate the brush-legged bees (Scopulipedes) which abound in the same country. In this case we have more than mere resemblance of colour, for that which is an important functional structure in the one group is imitated in another whose habits render it perfectly useless.

Mimicry among Beetles

It may fairly be expected that if these imitations of one creature by another really serve as a protection to weak and decaying species, instances of the same kind will be found among other groups than the Lepidoptera; and such is the case, although they are seldom so prominent and so easily recognised as those already pointed out as occurring in that order. A few very interesting examples may, however, be pointed out in most of the other orders of insects. The Coleoptera or beetles that imitate other Coleoptera of distinct groups are very numerous in tropical countries, and they generally follow the law already laid down as regulating these phenomena. The insects which other imitate always have a special protection, which leads them to be avoided as dangerous or uneatable by small insectivorous animals; some have a disgusting taste (analogous to that of Heliconidæ); others have such a hard and stony covering that they cannot be crushed or digested; while a third set are very active, and armed with powerful jaws, as well as having some disagreeable secretion. Some species of Eumorphidæ and Hispidæ, small flat or hemispherical beetles which are exceedingly abundant, and have a disagreeable secretion, are imitated by other of the very distinct group of Longicornes (of which our common musk-beetle may be taken as an example). The extraordinary little Cyclopeplus batesii belongs to the same sub-family of this group as the Onychocerus scorpio and O. concentricus, which have already been adduced as imitating with such wonderful accuracy the bark of the trees they habitually frequent; but it differs totally in outward appearance from every one of its allies, having taken upon itself the exact shape and colouring of a globular Corynomalus, a little stinking beetle with clubbed antennæ. It is curious to see how these clubbed antennæ are imitated by an insect belonging to a group with long slender antennæ. The sub-family Anisocerinæ, to which Cyclopeplus belongs, is characterised by all its members possessing a little knob or dilatation about the middle of the antennæ. This knob is considerably enlarged in C. batesii, and the terminal portion of the antennæ beyond it is so small and slender as to be scarcely visible, and thus an excellent substitute is obtained for the short clubbed antennæ of the

Corynomalus. Erythroplatis corallifer is another curious broad flat beetle, that no one would take for a Longicorn, since it almost exactly resembles Cephalodonta spinipes, one of the commonest of the South American Hispidæ; and what is still more remarkable, another Longicorn of a distinct group, Streptolabis hispoides, was found by Mr. Bates, which resembles that same insect with equal minuteness,—a case exactly parallel to that among butterflies, where species of two or three distinct groups mimicked the same Heliconia. Many of the soft-winged beetles (Malacoderms) are excessively abundant in individuals, and it is probable that they have some similar protection, more especially as other species often strikingly resemble them. A Longicorn beetle, Pæciloderma terminale, found in Jamaica, is coloured exactly in the same way as a Lycus (one of the Malacoderms) from the same island. Eroschema poweri, a Longicorn from Australia, might certainly be taken for one of the same group, and several species from the Malay Islands are equally deceptive. In the Island of Celebes I found one of this group, having the whole body and elytra of a rich deep blue colour, with the head only orange; and in company with it an insect of a totally different family (Eucnemidæ) with identically the same coloration, and of so nearly the same size and form as to completely puzzle the collector on every fresh occasion of capturing them. I have been recently informed by Mr. Jenner Weir, who keeps a variety of small birds, that none of them will touch our common "soliders and sailors" (species of Malacoderms), thus confirming my belief that they were a protected group, founded on the fact of their being at once very abundant, of conspicuous colours, and the objects of mimicry.

There are a number of the larger tropical weevils which have the elytra and the whole covering of the body so hard as to be a great annoyance to the entomologist, because in attempting to transfix them the points of his pins are constantly turned. I have found it necessary in these cases to drill a hole very carefully with the point of a sharp penknife before attempting to insert a pin. Many of the fine long-antennæd Anthribidæ (an allied group) have to be treated in the same way. We can easily understand that after small birds have in vain attempted to eat these insects, they should get to know them by sight, and ever after leave them alone, and it will then be an advantage for other insects which are comparatively soft and eatable to be mistaken for them. We need not be surprised, therefore, to find that there are many Longicorns which strikingly resemble the "hard beetles" of their own district. In South Brazil, Acanthotritus dorsalis is strikingly like a Curculio of the hard genus Heiliplus, and Mr. Bates assures me that he found Gymnocerus cratosomoides (a Longicorn) on the same tree with a hard Cratosomus (a weevil), which it exactly mimics. Again, the pretty Longicorn, Phacellocera batesii, mimics one of the hard Anthribidæ of the genus Ptychoderes, having long slender antennæ. In the Moluccas we find Cacia anthriboides, a small Longicorn which might be easily mistaken for a very common species of Anthribidæ found in the same districts; and the very rare Capnolymma stygium closely imitates the common Mecocerus gazella, which abounded where it was taken. Doliops curculionoides and other allied Longicorns from the Philippine Islands most curiously resemble, both in form and colouring, the brilliant Pachyrhynchi,—Curculionidæ, which are almost peculiar to that group of islands. The remaining family of Coleoptera most frequently imitated is the Cicindelidæ. The rare and curious Longicorn, Collyrodes lacordairei, has exactly the form and colouring of the genus Collyris, while an undescribed species of Heteromera is exactly like a Therates, and was taken running on the trunks of trees, as is the habit of that group. There is one curious example of a Longicorn mimicking a Longicorn, like the Papilios and Heliconidæ which mimic their own allies. Agnia fasciata, belonging to the sub-family Hypselominæ, and Nemophas grayi, belonging to the Lamiinæ, were taken in Amboyna on the same fallen tree at the same time, and were supposed to be the same species till they were more carefully examined, and found to be structurally quite different. The colouring of these insects is very remarkable, being rich steel-blue black, crossed by broad hairy bands of orange buff, and out of the many thousands of known species of Longicorns they are probably the only two which are so coloured. The Nemophas grayi is the larger, stronger, and better armed insect, and belongs to a more widely spread and dominant group, very rich in species and individuals, and is therefore most probably the subject of mimicry by the other species.

Beetles mimicking other Insects

We will now adduce a few cases in which beetles imitate other insects, and insects of other orders imitate beetles.

Charis melipona, a South American Longicorn of the family Necydalidæ, has been so named from its resemblance to a small bee of the genus Melipona. It is one of the most remarkable cases of mimicry, since the beetle has the thorax and body densely hairy like the bee, and the legs are tufted in a manner most unusual in the order Coleoptera. Another Longicorn, Odontocera odyneroides, has the abdomen banded with yellow, and constricted at the base, and is altogether so exactly like a small common wasp of the genus Odynerus, that Mr. Bates informs us he was afraid to take it out of his net with his fingers for fear of being stung. Had Mr. Bates' taste for insects been less omnivorous than it was, the beetle's disguise might have saved it from his pin, as it had no doubt often done from the beak of hungry birds. A larger insect, Sphecomorpha chalybea, is exactly like one of the large metallic blue wasps, and like them has the abdomen connected with the thorax by a pedicel, rendering the deception most complete and striking. Many Eastern species of Longicorns of the genus Oberea, when on the wing, exactly resemble Tenthredinidæ, and many of the small species of Hesthesis run about on timber, and cannot be distinguished from ants. There is one genus of South American Longicorns that appears to mimic the shielded bugs of the genus Scutellera. The Gymnocerus capucinus is one of these, and is very like Pachyotris fabricii, one of the Scutelleridæ. The beautiful Gymnocerus dulcissimus is also very like the same group of insects, though there is no known species that exactly corresponds to it; but this is not to be wondered at, as the tropical Hemiptera have been comparatively so little cared for by collectors.

Insects mimicking Species of other Orders

The most remarkable case of an insect of another order mimicking a beetle is that of the Condylodera tricondyloides, one of the cricket family from the Philippine Islands, which is so exactly like a Tricondyla (one of the tiger beetles) that such an experienced entomologist as Professor Westwood placed it among them in his cabinet, and retained it there a long time before he discovered his mistake! Both insects run along the trunks of trees, and whereas Tricondylas are very plentiful, the insect that mimics it is, as in all other cases, very rare. Mr. Bates also informs us that he found at Santarem, on the Amazon, a species of locust which mimicked one of the tiger beetles of the genus Odontocheila, and was found on the same trees which they frequented.

There are a considerable number of Diptera, or two-winged flies, that closely resemble wasps and bees, and no doubt derive much benefit from the wholesome dread which those insects excite. The Midas dives, and other species of large Brazilian flies, have dark wings and metallic blue elongate bodies, resembling the large stinging Sphegidæ of the same country; and a very large fly of the genus Asilus had black-banded wings and the abdomen tipped with rich orange, so as exactly to resemble the fine bee Euglossa dimidiata, and both are found in the same parts of South America. We have also in our own country species of Bombylius which are almost exactly like bees. In these cases the end gained by the mimicry is no doubt freedom from attack, but it had sometimes an altogether difference purpose. There are a number of parasitic flies whose larvæ feed upon the larvæ of bees, such as the British genus Volucella and many of the tropical Bombylii, and most of these are exactly like the particular species of bee they prey upon, so that they can enter their nests unsuspected to deposit their eggs. There are also bees that mimic bees. The cuckoo bees of the genus Nomada are parasitic on the Andrenidæ, and they resemble either wasps or species of Andrena; and the parasitic humble bees of the genus Apathus almost exactly resemble the species of humble bees in whose nests they are reared. Mr. Bates informs us that he found numbers of these "cuckoo" bees and flies on the Amazon, which all wore the livery of working bees peculiar to the same country.

There is a genus of small spiders in the tropics which feet on ants, and they are exactly like ants themselves, which no doubt gives them more opportunity of seizing their prey; and Mr. Bates found on the Amazon a species of Mantis which exactly resembled the white ants which it fed upon, as well as several species of crickets (Scaphura), which resembled in a wonderful manner different sand-wasps of large size, which are constantly on the search for crickets with which to provision their nests.

Perhaps the most wonderful case of all is the large caterpillar mentioned by Mr. Bates, which startled him by its close resemblance to a small snake. The first three segments behind the head were dilatable at the will of the insect, and had on each side a large black pupillated spot, which resembled the eye of the reptile. Moreover, it resembled a poisonous viper, not a harmless species of snake, as was proved by the imitation of keeled scales on the crown produced by the recumbent feet, as the caterpillar threw itself backward!

The attitudes of many of the tropical spiders are most extraordinary and deceptive, but little attention has been paid to them. They often mimic other insects, and some, Mr. Bates assures us, are exactly like flower buds, and take their station in the axils of leaves, where they remain motionless waiting for their prey.

Cases of Mimicry among the Vertebrata

Having thus shown how varied and extraordinary are the modes in which mimicry occurs among insects, we have now to inquire if anything of the same kind is to be observed among vertebrated animals. When we consider all the conditions necessary to produce a good deceptive imitation, we shall see at once that such can very rarely occur in the higher animals, since they possess none of those facilities for the almost infinite modifications of external form which exist in the very nature of insect organisation. The outer covering of insects being more or less solid and horny, they are capable of almost any amount of change of form and appearance without any essential modification internally. In many groups the wings give much of the character, and these organs may be much modified both in form and colour without interfering with their special functions. Again, the number of species of insects is so great, and there is such diversity of form and proportion in every group, that the chances of an accidental approximation in size, form, and colour of one insect to another of a different group are very considerable; and it is these chance approximations that furnish the basis of mimicry, to be continually advanced and perfected by the survival of those varieties only which tend in the right direction.

In the vertebrata, on the contrary, the skeleton being internal, the external form depends almost entirely on the proportions and arrangement of that skeleton, which again is strictly adapted to the functions necessary for the well-being of the animal. The form cannot, therefore, be rapidly modified by variation, and the thin and flexible integument will not admit of the development of such strange protuberances as occur continually in insects. The number of species of each group in the same country is also comparatively small, and thus the chances of that first accidental resemblance which is necessary for natural selection to work upon are much diminished. We can hardly see the possibility of a mimicry by which the elk could escape from the wolf, or the buffalo from the tiger. There is, however, in one group of Vertebrata such a general similarity of form, that a very slight modification, if accompanied by identity of colour, would produce the necessary amount of resemblance; and at the same time there exist a number of species which it would be advantageous for other to resemble, since they are armed with the most fatal weapons of offence. We accordingly find that reptiles furnish us with a very remarkable and instructive case of true mimicry.

Mimicry among Snakes

There are in tropical America a number of venomous snakes of the genus Elaps, which are ornamented with brilliant colours disposed in a peculiar manner. The ground colour is generally bright red, on which are black bands of various widths and sometimes divided into two or three by yellow rings. Now, in the same country are found several genera of harmless snakes, having no affinity whatever with the above, but coloured exactly the same. For example, the poisonous Elaps fulvius often occurs in Guatemala with simple black bands on a coral-red ground; and in the same country is found the harmless snake Pliocerus equalis, coloured and banded in identically the same manner. A variety of Elaps corallinus has the black band narrowly bordered with yellow on the same red ground colour, and a harmless snake, Homalocranium semi-cinctum (Colubridæ), has exactly the same markings, and both are found in Mexico. The deadly Elaps lemniscatus has the black bands very broad, and each of them divided into three by narrow yellow rings; and this again is exactly copied by a harmless snake, Pliocerus elapoides, which is found along with its model in Mexico.

But, more remarkable still, there is in South America a third group of snakes, the genus Oxyrhopus (Scytalidæ), doubtfully venomous, and having no immediate affinity with either of the preceding, which has also the same curious distribution of colours, namely, variously disposed rings of red, yellow, and black; and there are some cases in which species of all three of these groups similarly marked inhabit the same district. For example, Elaps mipartitus has single black rings very close together. It inhabits the west side of the Andes, and in the same districts occur Pliocerus euryzonus and Oxyrhopus petolarius, which exactly copy its pattern. In Brazil Elaps lemniscatus is copied by Oxyrhopus trigeminus, both having black rings disposed in threes. In Elaps hemiprichii the ground colour appears to be black, with alternations of two narrow yellow bands and a broader red one; and of the pattern again we have an exact double in Oxyrhopus formosus, both being found in many localities of tropical South America.

What adds much to the extraordinary character of these resemblances is the fact, that nowhere in the world but in America are there any snakes which have this style of colouring. Dr. Gunther, of the British Museum, who has kindly furnished some of the details here referred to, assures me that this is the case; and that red, black, any yellow rings occur together on no other snakes in the world but on Elaps and the species which so closely resemble it. In all these cases, the size and form as well as the coloration are so much alike, that none but a naturalist would distinguish the harmless from the poisonous species.

Many of the small tree-frogs are no doubt also mimickers. When seen in their natural attitudes, I have been often unable to distinguish them from beetles or other insects sitting upon leaves, but regret to say I neglected to observe what species or groups they most resembled, and the subject does not yet seem to have attracted the attention of naturalists abroad.

Mimicry among Birds

In the class of birds there are a number of cases that make some approach to mimicry, such as the resemblance of the cuckoos, a weak and defenceless group of birds, to hawks and Gallinaceæ. There is, however, one example which goes much further than this, and seems to be of exactly the same nature as the many cases of insect mimicry which have been already given. In Australia and the Moluccas there is a genus of honeysuckers called Tropidorhynchus, good sized birds, very strong and active, having powerful grasping claws and long, curved, sharp beaks. They assemble together in groups and small flocks, and they have a very loud bawling note, which can be heard at a great distance, and serves to collect a number together in time of danger. They are very plentiful and very pugnacious, frequently driving away crows, and even hawks, which perch on a tree where a few of them are assembled. They are all of rather dull and obscure colours. Now in the same countries there is a group of orioles, forming the sub-genus Mimeta, much weaker birds, which have lost the gay colouring of their allies, the golden orioles, being usually olive-green or brown; and in several cases these most curiously resemble the Tropidorhynchus of the same island. For example, in the island of Bouru is found the Tropidorhynchus bouruensis, of a dull earthy colour, and the Mimeta bouruensis, which resembles it in the following particulars: The upper and under surfaces of the two birds are exactly of the same tints of dark and light brown; the Tropidorhynchus has a large, bare black patch round the eyes; this is copied in the Mimeta by a patch of black feathers. The top of the head of the Tropiodrhynchus had a scaly appearance, from the narrow scale-formed feathers, which are imitated by the broader feathers of the Mimeta having a dusky line down each. The Tropidorhynchus has a pale ruff formed of curious recurved feathers on the nape (which has given the whole genus the name of Friar birds); this is represented in the Mimeta by a pale band in the same position. Lastly, the bill of the Tropidorhynchus is raised into a protuberant keel at the base, and the Mimeta has the same character, although it is not a common one in the genus. The result is, that on a superficial examination the birds are identical, although they have important structural differences, and cannot be placed near each other in any natural arrangement. As a proof that the resemblance is really deceptive, it may be mentioned that the Mimeta is figured and described as a honeysucker in the costly *Voyage de l'Astrolabe,* under the name of Philedon bouruensis!

Passing to the island of Ceram, we find allied species of both genera. The Tropidorhynchus subcornutus is of an earthy brown colour washed with yellow ochre, with bare orbits, dusky cheeks, and the usual pale recurved nape-ruff. The Mimeta forsteni is absolutely identical in the tints of every part of the body, the details of which are imitated in the same manner as in the Bouru birds already described. In two other islands there is an approximation towards mimicry, although it is not so perfect as in the two preceding cases. In Timor the Tropidorhynchus timoriensis is of the usual earthy brown above, with the nape-ruff very prominent, the cheeks black, the throat nearly white, and the whole under surface pale whitish brown. These various tints are all well reproduced in Mimeta virescens, the chief want of exact imitation being that the throat and breast of the Tropidorhynchus has a very scaly appearance, being covered with rigid pointed feathers which are not imitated in the Mimeta, although there are signs of faint dusky spots which may easily furnish the groundwork of a more exact imitation by the continued survival of favourable variations in the same direction. There is also a large knob at the base of the bill of the Tropidorhynchs which is not at all imitated by the Mimeta. In the island of Morty (north of Gilolo) there exists the Tropidorhynchus fuscicapillus, of a dark sooty brown colour, especially on the head, while the under parts are rather lighter, and the characteristic ruff of the nape is wanting. Now it is curious that in the adjacent island of Gilolo should be found the Mimeta phæochromus, the upper surface of which is of exactly the same dark sooty tint as the Tropidorhynchus, and is the only known species that is of such a dark colour. The under side is not quite light enough, but it is a good approximation. This Mimeta is a rare bird, and may very probably exist in Morty, though not yet found there; or, on the other hand, recent changes in physical geography may have led to the restriction of the Tropidorhynchus to that island, where it is very common.

Here, then, we have two cases of perfect mimicry and two others of good approximation, occurring between species of the same two genera of birds; and in three of these cases the pairs that resemble each other are found together in the same island, and to which they are peculiar. In all these cases the Tropidorhynchus is rather larger than the Mimeta, but the difference is not beyond the limits of variation in species, and the two genera are somewhat alike in form and proportion. There are, no doubt, some special enemies by which many small birds are attacked, but which are afraid of the Tropidorhynchus (probably some of the hawks), and thus it becomes advantageous for the weak Mimeta to resemble the strong, pugnacious, noisy, and very abundant Tropidorhynchus.

My friend, Mr. Osbert Salvin, has given me another interesting case of bird mimicry. In the neighbourhood of Rio Janeiro is found an insect-eating hawk (Harpagus diodon), and in the same district a bird-eating hawk (Accipiter pileatus) which closely resembles it. Both are of the same ashy tint beneath, with the thighs and under wing-coverts reddish brown, so that when on the wing and seen from below they are undistinguishable. The curious point, however, is that the Accipiter has a much wider range than the Harpagus, and in the regions where the insect-eating species is not found it no longer resembles it, the under wing-coverts varying to white; thus indicating that the red-brown colour is kept true by its being useful to the Accipiter to be mistaken for the insect-eating species, which birds have learnt not to be afraid of.

Mimicry among Mammals

Among the Mammalia the only case which may be true mimicry is that of the insectivorous genus Cladobates, found in the Malay countries, several species of which very closely resemble squirrels. The size is about the same, the long bushy tail is carried in the same way, and the colours are very similar. In this case the use of the resemblance must be to enable the Cladobates to approach the insects or small birds on which it feeds under the disguise of the harmless fruit-eating squirrel.

Objections to Mr. Bates' Theory of Mimicry

Having now completed our survey of the most prominent and remarkable cases of mimicry that have yet been noticed, we must say something of the objections that have been made to the theory of their production given by Mr. Bates, and which we have endeavoured to illustrate and enforce in the preceding pages. Three counter explanations have been proposed. Professor Westwood admits the fact of the mimicry and its probable use to the insect, but maintains that each species was created a mimic for the purpose of the protection thus afforded it. Mr. Andrew Murray, in his paper on the "Disguises of Nature," inclines to the opinion that similar conditions of food and of surrounding circumstances have acted in some unknown way to produce the resemblances; and when the subject was discussed before the Entomological Society of London, a third objection was added—that heredity or the reversion to ancestral types of form and coloration might have produced many of the cases of mimicry.

Against the special creation of mimicking species there are all the objections and difficulties in the way of special creation in other cases, with the addition of a few that are peculiar to it. The most obvious is, that we have gradations of mimicry and of protective resemblance—a fact which is strongly suggestive of a natural process having been at work. Another very serious objection is, that a mimicry has been shown to be useful only to those species and groups which are rare and probably dying out, and would cease to have any effect should the proportionate abundance of the two species be reversed, it follows that on the special-creation theory the one species must have been created plentiful, the other rare; and, notwithstanding the many causes that continually tend to alter the proportions of species, these two species must have always been specially maintained at their respective proportions, or the very purpose for which they each received their peculiar characteristics would have completely failed. A third difficulty is, that although it is very easy to understand how mimicry may be brought about by variation and the survival of the fittest, it seems a very strange thing for a Creator to protect an animal by making it imitate another, when the very assumption of a Creator implies his power to create it so as to require no such circuitous protection. These appear to be fatal objections to the application of the special-creation theory to this particular case.

The other two supposed explanations, which may be shortly expressed as the theories of "similar conditions" and of "heredity," agree in making mimicry, where it exists, an adventitious circumstance no necessarily connected with the well-being of the mimicking species. But several of the most striking and most constant facts which have been adduced directly contradict both these hypotheses. The law that mimicry is confined to a few groups only is one of these, for "similar conditions" must act more or less on all groups in a limited region, and "heredity" must influence all groups related to each other in an equal degree. Again, the general fact that those species which mimic others are rare, while those which are imitated are abundant, is in no way explained by either of these theories, any more than is the frequent occurrence of some papable mode of protection in the imitated species. "Reversion to an ancestral type" no way explains why the imitator and the imitated always inhabit the very same district, whereas allied forms of every degree of nearness and remoteness generally inhabit different countries, and often different quarters of the globe; and neither it nor "similar conditions" will account for the likeness between species of distinct groups being superficial only—a disguise, not a true resemblance; for the imitation of bark, of leaves, of sticks, of dung; for the resemblances between species in different orders, and even different classes and sub-kingdoms; and finally, for the graduated series of the phenomena, beginning with a general harmony and adaptation of tint in autumn and winter moths and in arctic and desert animals, and ending with those complete cases of detailed mimicry which not only deceive predacious animals, but puzzle the most experienced insect collectors and the most learned entomologists.

Mimicry by Female Insects only

But there is yet another series of phenomena connected with this subject, which considerably strengthens the view here adopted, while it seems quite incompatible with either of the other hypotheses; namely, the relation of protective colouring and mimicry to the sexual differences of animals. It will be clear to every one that if two animals, which as regards "external conditions" and "hereditary descent" are exactly alike, yet differ remarkably in coloration, one resembling a protected species and the other not, the resemblance that exists in one only can hardly be imputed to the influence of external conditions or as the effect of heredity. And if, further, it can be proved that the one requires protection more than the other, and that in several cases it is that one which mimics the protected species, while the one that least requires protection never does so, it will afford very strong corroborative evidence that there is a real connection between the necessity for protection and the phenomenon of mimicry. Now the sexes of insects offer us a test of the nature here indicated, and appear to furnish one of the most conclusive arguments in favour of the theory that the phenomena termed "mimicry" are produced by natural selection.

The comparative importance of the sexes varies much in different classes of animals. In the higher vertebrates, where the number of young produced at a birth is small and the same individuals breed many years in succession, the preservation of both sexes is almost equally important. In all the numerous cases in which the male protects the female and her offspring, or helps to supply them with food, his importance in the economy of nature is proportionately increased, though it is never perhaps quite equal to that of the female. In insects the case is very different; they pair but once in their lives, and the prolonged existence of the male is in most cases quite unnecessary for the continuance of the race. The female, however, must continue to exist long enough to deposit her eggs in a place adapted for the development and growth of the progeny. Hence there is a wide difference in the need for protection in the two sexes; and we should, therefore, expect to find that in some cases the special protection given to the female was in the male less in amount or altogether wanting. The facts entirely confirm this expectation. In the spectre insects (Phasmidæ) it is often the females alone that so strikingly resemble leaves, while the males show only a rude approximation. The male Diadema misippus is a very handsome and conspicuous butterfly, without a sign of protective or imitative colouring, while the female is entirely unlike her partner, and is one of the most wonderful cases of mimicry on record, resembling most accurately the common Danais chrysippus, in whose company it is often found. So in several species of South American Pieris, the males are white and black, of a similar type of colouring to our own "cabbage" butterflies, while the females are rich yellow and buff, spotted and marked so as exactly to resemble species of Heliconidæ, with which they associate in the forest. In the Malay archipelago is found a Diadema which had always been considered a male insect on account of its glossy metallic-blue tints, while its companion of sober brown was looked upon as the female. I discovered, however, that the reverse is the case, and that the rich and glossy colours of the female are imitative and protective, since they cause her exactly to resemble the common Euplœa midamus of the same regions, a species which has been already mentioned in this essay as mimicked by another butterfly, Papilio paradoxa. I have since named this named this interesting species Daidema anomala (see the *Transactions of the Entomological Society*, 1869, p. 285). In this case, and in that of Diadema misippus, there is no difference in the habits of the two sexes, which fly in similar localities; so that the influence of "external conditions" cannot be invoked here as it has been in the case of the South American Pieris pyrrha and allies, where the white males frequent open sunny places, while the Heliconia-like females haunt the shades of the forest.

We may impute to the same general cause (the greater need of protection for the female, owing to her weaker flight, greater exposure to attack, and supreme importance)—the fact of the colours of female insect being so very generally duller and less conspicuous than those of the other sex. And that it is chiefly due to this cause rather than to what Mr. Darwin terms "sexual selection" appears to be shown by the otherwise inexplicable fact, that in the groups which have a protection of any kind independent of concealment, sexual difference of colour are either quite wanting or slightly developed. The Heliconidæ and Danaidæ, protected by a disagreeable falvour, have the females as bright and conspicuous as the males, and very rarely differing at the from them. The stinging Hymenoptera have the two sexes equally well coloured. The Carabidæ, the Coccinellidæ, Chrysomelidæ, and the Telephori have both sexes equally conspicuous, and seldom differing in colours. The brilliant Curculios, which are protected by their hardness, are brilliant in both sexes. Lastly, the glittering Cetoniadæ and Buprestidæ, which seem to be protected by their hard and polished coats, their rapid motions and peculiar habits, present few sexual differences of colour, while sexual selection has often manifested itself by structural differences, such as horns, spines, or other processes.

Cause of the dull Colours of Female Birds

The same law manifests itself in Birds. The female while sitting on her eggs requires protection by concealment to a much greater extent than the male; and we accordingly find that in a large majority of the cases in which the male birds are distinguished by unusual brilliancy of plumage, the females are much more obscure, and often remarkably plain-coloured. The exceptions are such as eminently to prove the rule, for in most cases we can see very good reason for them. In particular, there are a few instances among wading and gallinaceous birds in which the female has decidedly more brilliant colours than the male; but it is a most curious and interesting fact that in most if not all these cases the males sit upon the eggs; so that this exception to the usual rule almost demonstrates that it is because the process of incubation is at once very important and very dangerous, that the protection of obscure colouring is developed. The most striking example is that of the gray phalarope (Phalaropus fulicarius). When in winter plumage, the sexes of this bird are alike in coloration, but in summer the female is much the most conspicuous, having a black head, dark wings, and reddish-brown back, which the male is nearly uniform brown, with dusky spots. Mr. Gould in his *Birds of Great Britain* figures the two sexes in both winter and summer plumage, and remarks on the strange peculiarity of the usual colours of the two sexes being reversed, and also on the still more curious fact that the "male alone sits on the eggs," which are deposited on the bare ground. In another British bird, the dotterell, the female is also larger and more brightly coloured than the male; and it seems to be proved that the males assist in incubation even if they do not perform it entirely, for Mr. Gould tells us "that they have been shot with the breast bare of feathers, caused by sitting on the eggs." The small quail-like birds forming the genus Turnix have also generally large and bright-coloured females, and we are told by Mr. Jerdon in his *Birds of India* that "the natives report that during the breeding season the females desert their eggs and associate in flocks while the males are employed in hatching the eggs." It is also an ascertained fact that the females are more bold and pugnacious that the males. A further confirmation of this view is to be found in the fact (not hitherto noticed) that in a large majority of the cases in which bright colours exist in both sexes incubation takes place in a dark hole or in a dome-shaped nest. Female kingfishers are often equally brilliant with the male, and they build in holes in banks. Bee-eaters, trogons, motmots, and toucans all build in holes, and in none is there any difference in the sexes, although they are, without exception, showy birds. Parrots build in holes in trees, and in the majority of cases they present no marked sexual difference tending to concealment of the female. Woodpeckers are in the same category, since, though the sexes often differ in colour, the female is not generally less conspicuous than the male. Wagtails and titmice build concealed nests, and the females are nearly as gay as their mates. The female of the pretty Australian bird, Pardalotus punctatus, is very conspicuously spotted on the upper surface, and it builds in a hole in the ground. The gay-coloured hang-nests (Icterinæ) and the equally brilliant tanagers may be well contrasted; for the former, concealed in their covered nests, present little or no sexual difference of colour—while the open-nested tanagers have the females dull-coloured and sometimes with almost protective tints. No doubt there are many individual exceptions to the rule here indicated, because many and various causes have combined to determine both the coloration and the habits of birds. These have no doubt acted and reacted on each other; and when conditions have changed one of these characters may often have become modified, while the other, though useless, may continue by hereditary descent an apparent exception to what otherwise seems a very general rule. The facts presented by the sexual differences of colour in birds and their mode of nesting are on the whole in perfect harmony with that law of protective adaptation of colour and form, which appears to have checked to some extent the powerful action of sexual selection, and to have materially influenced the colouring of female birds, as it has undoubtedly done that of female insects.

Use of the gaudy Colours of many Caterpillars

Since this essay was first published a very curious difficulty has been cleared up by the application of the general principle of protective colouring. Great numbers of caterpillars are so brilliantly marked and coloured as to be very conspicuous even at a considerable distance, and it has been noticed that such caterpillars seldom hide themselves. Other species, however, are green or brown, closely resembling the colours of the substances on which they feed, while others again imitate sticks, and stretch themselves out motionless form a twig so as to look like one of its branches.

Now, as caterpillars form so large a part of the food of birds, it was not easy to understand why any of them should have such bright colours and markings as to make them specially visible. Mr. Darwin had put the case to me as a difficulty from another point of view, for he had arrived at the conclusion that brilliant coloration in the animal kingdom is mainly due to sexual selection, and this could not have acted in the case of sexless larvæ. Applying here the analogy of others insects, I reasoned that since some caterpillars were evidently protected by their imitative colouring, and others by their spiny or hairy bodies, the bright colours of the rest must also be in some way useful to them. I further thought that as some butterflies and moths were greedily eaten by birds, while others were distasteful to them, and these latter were mostly of conspicuous colours, so probably these brilliantly coloured caterpillars were distasteful, and therefore never eaten by birds. Distastefulness alone would, however, be of little service to caterpillars, because their soft and juicy bodies are so delicate that if seized and afterwards rejected by a bird, they would almost certainly be killed. Some constant and easily perceived signal was therefore necessary to serve as a warning to birds never to touch these uneatable kinds, and a very gaudy and conspicuous colouring with the habit of fully exposing themselves to view becomes such a signal, being in strong contrast with the green or brown tints and retiring habits of the eatable kinds. The subject was brought by me before the Entomological Society (see *Proceedings*, 4th March 1867), in order that those members having opportunities for making observations might do so in the following summer; and I also wrote a letter to the *Field* newspaper, begging that some of its readers would co-operate in making observations on what insects were rejected by birds, at the same time fully explaining the great interest and scientific importance of the problem. It is a curious example of how few of the country readers of that paper are at all interested in questions of simple natural history, that I only obtained one answer from a gentleman in Cumberland, who gave me some interesting observations on the general dislike and abhorrence of all birds to the "Gooseberry Caterpillar," probably that of the Magpie moth (Abraxas grossulariata). Neither young pheasants, partridges, nor wild ducks could be induced to eat it, sparrows and finches never touched it, and all birds to whom he offered it rejected it with evident dread and abhorrence. It will be seen that these observations are confirmed by those of two members of the Entomological Society, to whom we are indebted for more detailed information.

In March 1869 Mr. J. Jenner Weir communicated a valuable series of observations made during many years, but more especially in the two preceding summers, in his aviary, containing the following birds of more or less insectivorous habits:—Robin, Yellow-hammer, Reed-bunting, Bullfinch,

Chaffinch, Crossbill, Thrush, Tree-pipit, Siskin, and Redpoll. He found that hairy caterpillars were uniformly rejected; five distinct species were quite unnoticed by all his birds, and were allowed to crawl about the aviary for days with impunity. The spiny caterpillars of the Tortoiseshell and Peacock butterflies were equally rejected; but in both these cases Mr. Weir thinks it is the taste, not the hairs or spines, that is disagreeable, because some very young caterpillars of a hairy species were rejected although no hairs were developed, and the smooth pupæ of the above-named butterflies were refused as persistently as the spined larvæ. In these cases, then, both hairs and spines would seem to be mere signs of uneatableness.

His next experiments were with those smooth gaily-coloured caterpillars which never conceal themselves, but on the contrary appear to court observation. Such are those of the Magpie moth (Abraxas grossulariata), whose caterpillar is conspicuously white and black spotted—the Diloba cæruleocephala, whose larva is pale yellow with a broad blue or green lateral band—the Cucullia verbasci, whose larva is greenish white with yellow bands and black spots, and Anthrocera filipendulæ (the six spot Burnet moth), whose caterpillar is yellow with black spots. These were given to the birds at various times, sometimes mixed with other kinds of larvæ which were greedily eaten, but they were in every case rejected apparently unnoticed, and were left to crawl about till they died.

The next set of observations were on the dull-coloured and protected larvæ, and the results of numerous experiments are thus summarised by Mr. Weir. "All caterpillars whose habits are nocturnal, which are dull-coloured, with fleshy bodies and smooth skins, are eaten with the greatest avidity. Every species of green caterpillar is also much relished. All Geometræ, whose larvæ resemble twigs as they stand out from the plant on their anal prolegs, are invariably eaten."

At the same meeting Mr. A. G. Butler, of the British Museum, Communicated the results of his observations with lizards, frogs, and spiders, which strikingly corroborate those of Mr. Weir. Three green lizards (Lacerta viridis), which he kept for several years, were very voracious, eating all kinds of food, from a lemon cheesecake to a spider, and devouring flies, caterpillars, and humble bees; yet there were some caterpillars and moths which they would seize only to drop immediately. Among these the principal were the caterpillar of the Magpie moth (Abraxas grossulariata) and the perfect six spot Burnet moth (Anthrocera filipendulæ). These would be first seized but invariably dropped in disgust, and afterwards left unmolested. Subsequently frogs were kept and fed with caterpillars from the garden, but two of these—that of the before-mentioned Magpie moth, and that of the V. moth (Halia wavaria), which is green with conspicuous white or yellow stripes and black spots—were constantly rejected. When these species were first offered, the frogs sprang at them eagerly and licked them into their mouths; no sooner, however, had they done so than they seemed to be aware of the mistake that they had made, and sat with gaping mouths, rolling their tongues about until they had got quit of the nauseous morsels.

With spiders the same thing occurred. These two caterpillars were repeatedly put into the webs both of the geometrical and hunting spiders (Epeira diadema and Lycosasp.), but in the former case they were cut out and allowed to drop; in the latter, after disappearing in the jaws of their captor down his dark silken funnel, they invariably reappeared, either from below or else taking long strides up the funnel again. Mr. Butler has observed lizards fight with and finally devour humble bees, and a frog sitting on a bed of stone-crop leap up and catch the bees which flew over his head, and swallow them, in utter disregard of their stings. It is evident, therefore, that the possession of a disagreeable taste or odour is a more effectual protection to certain conspicuous caterpillars and moths than would be even the possession of a sting.

The observations of these two gentlemen supply a very remarkable confirmation of the hypothetical solution of the difficulty which I had given two years before. And as it is generally acknowledged that the best test of the truth and completeness of a theory is the power which it gives us of prevision, we may, I think, fairly claim this as a case in which the power of prevision has been successfully exerted, and therefore as furnishing a very powerful argument in favour of the truth of the theory of Natural Selection.

Summary

I have now completed a brief, and necessarily very imperfect, survey of the various ways in which the external form and colouring of animals is adapted to be useful to them, either by concealing them from their enemies or from the creatures they prey upon. It has, I hope, been shown that the subject is one of much interest, both as regards a true comprehension of the place each animal fills in the economy of nature, and the means by which it is enabled to maintain that place; and also as teaching us how important a part is played by the minutest details in the structure of animals, and how complicated and delicate is the equilibrium of the organic world.

My exposition of the subject having been necessarily somewhat lengthy and full of details, it will be as well to recapitulate its main points.

There is a general harmony in nature between the colours of an animal and those of its habitation. Arctic animals are white, desert animals are sand-coloured; dwellers among leaves and grass are green; nocturnal animals are dusky.

These colours are not universal, but are very general, and are seldom reversed. Going on a little further, we find birds, reptiles, and insects so tinted and mottled as exactly to match the rock, or bark, or leaf, or flower, they are accustomed to rest upon,—and thereby effectually concealed. Another step in advance, and we have insects which are formed as well as coloured so as exactly to resemble particular leaves, or sticks, or mossy twigs, or flowers; and in these cases very peculiar habits and instincts come into play to aid in the deception and render the concealment more complete. We now enter upon a new phase of the phenomena, and come to creatures whose colours neither conceal them nor make them like vegetable or mineral substances; on the contrary, they are conspicuous enough, but they completely resemble some other creature of a quite different group, while they differ much in outward appearance from those with which all essential parts of their organisation show them to be really closely allied. They appear like actors or masqueraders dressed up and painted for amusement, or like swindlers endeavouring to pass themselves off for well-known and respectable members of society. What is the meaning of this strange travesty? Does Nature descend to imposture or masquerade? We answer, she does not. Her principles are too severe. There is a use in every detail of her handiwork. The resemblance of one animal to another is of exactly the same essential nature as the resemblance to a leaf, or to bark, or to desert sand, and answers exactly the same purpose. In the one case the enemy will not attack the leaf or the bark, and so the disguise is a safeguard; in the other case it is found that for various reasons the creature resembled is passes over, and not attacked by the usual enemies of its order, and thus the creature that resembles it has an equally effectual safeguard. We are plainly shown that the disguise is of the same nature in the two cases, by the occurrence in the same group of one species resembling a vegetable substance, while another resembles a living animal of another group; and we know that the creatures resembled possess an immunity from attack, by their being always very abundant, by their being conspicuous and not concealing themselves, and by their having generally no visible means of escape from their enemies; while, at the same time, the particular quality that makes them disliked is often very clear, such as a nasty taste or an indigestible hardness. Further examination reveals the fact that, in several cases of both kinds of disguise, it is the female only that is thus disguised; and as it can be shown that the female needs protection much more than the male, and that her preservation for a much longer period is absolutely necessary for the continuance of the race, we have an additional indication that the resemblance is in all cases subservient to a great purpose—the preservation of the species.

In endeavouring to explain these phenomena as having

been brought about by variation and natural selection, we start with the fact that white varieties frequently occur, and when protected from enemies show no incapacity for continued existence and increase. We know, further, that varieties of many other tints occasionally occur; and as "the survival of the fittest" must inevitably weed out those whose colours are prejudicial and preserve those whose colours are a safeguard, we require no other mode of accounting for the protective tints of arctic and desert animals. But this being granted, there is such a perfectly continuous and graduated series of examples of every kind of protective imitation, up to the most wonderful cases of what is termed "mimicry," that we can find no place at which to draw the line, and say: So far variation and natural selection will account for the phenomena, but for all the rest we require a more potent cause. The counter theories that have been proposed, that of the "special creation" of each imitative form, that of the action of "similar conditions of existence" for some of the cases, and of the laws of "hereditary descent and the reversion to ancestral forms" for others,—have all been shown to be beset with difficulties, and the two latter to be directly contradicted by some of the most constant and most remarkable of the facts to be accounted for.

General deductions as to Colour in Nature

The important part that "protective resemblance" has played in determining the colours and markings of many groups of animals, will enable us to understand the meaning of one of the most striking facts in nature, the uniformity in the colours of the vegetable as compared with the wonderful diversity of the animal world. There appears no good reason why trees and shrubs should not have been adorned with as many varied hues and as strikingly designed patterns as birds and butterflies, since the gay colours of flowers show that there is no incapacity in vegetable tissues to exhibit them. But even flowers show themselves present us with none of those wonderful designs, those complicated arrangements of stripes and dots and patches of colour, that harmonious blending of hues in lines and bands and shaded spots, which are so general a feature in insects. It is the opinion of Mr. Darwin that we owe much of the beauty of flowers to the necessity of attracting insects to aid in their fertilisation, and that much of the development of colour in the animal world is due to "sexual selection," colour being universally attractive, and thus leading to its propagation and increase; but while fully admitting this, it will be evident, from the facts and arguments here brought forward, that very much of the *variety* both of colour and markings among animals is due to the supreme importance of concealment, and thus the various tints of minerals and vegetables have been directly reproduced in the animal kingdom, and again and again modified as more special protection became necessary. We shall thus have two causes for the development of colour in the animal world, and shall be better enabled to understand how, by their combined and separate action, the immense variety we now behold has been produced. Both causes, however, will come under the general law of "Utility," the advocacy of which, in its broadest sense, we owe almost entirely to Mr. Darwin. A more accurate knowledge of the varied phenomena connected with this subject may not improbably give us some information both as to the senses and the mental faculties of the lower animals. For it is evident that if colours which please us also attract them, and if the various disguises which have been here enumerated are equally deceptive to them as to ourselves, then both their powers of vision and their faculties of perception and emotion must be essentially of the same nature as our own—a fact of high philosophical importance in the study of our own nature and our true relations to the lower animals.

Conclusion

Although such a store of interesting facts has been already accumulated, the subject we have been discussing is one of which comparatively little is really known. The natural history of the tropics has never yet been studied on the spot with a full appreciation of "what to observe" in this matter. The varied ways in which the colouring and form of animals serve for their protection, their strange disguises as vegetable or mineral substances, their wonderful mimicry of other beings, offer an almost unworked and inexhaustible field of discovery for the zoologist, and will assuredly throw much light on the laws and conditions which have resulted in the wonderful variety of colour, shade, and marking which constitutes one of the most pleasing characteristics of the animal world, but the immediate causes of which it has hitherto been most difficult to explain.

If I have succeeded in showing that in this wide and picturesque domain of nature, results which have hitherto been supposed to depend either upon those incalculable combinations of laws which we term chance or upon the direct volition of the Creator, are really due to the action of comparatively well-known and simple causes, I shall have attained my present purpose, which has been to extend the interest so generally felt in the more striking facts of natural history to a large class of curious but much neglected details;

and to further, in however slight a degree, our knowledge of the subjection of the phenomena of life to the *Reign of Law.*

IV

ON INSTINCT IN MAN AND ANIMALS

THE most perfect and most striking examples of what is termed instinct—those in which reason or observation appear to have the least influence, and which seem to imply the possession of faculties farthest removed from our own—are to be found among insects. The marvelous constructive powers of bees and wasps, the social economy of ants, the careful provision for the safety of a progeny they are never to see manifested by many beetles and flies, and the curious preparations for the pupa state by the larvæ of butterflies and moths, are typical examples of this faculty, and are supposed to be conclusive as to the existence of some power or intelligence very different from that which we derive from our senses or from our reason.

How Instinct may be best Studied

Whatever we may define instinct to be, it is evidently some form of mental manifestation, and as we can only judge of mind by the analogy of our own mental functions and by observation of the results of mental action in other men and in animals, it is incumbent on us, first, to study and endeavour to comprehend the minds of infants, of savage men, and of animals not very far removed from ourselves, before we pronounce positively as to the nature of the mental operations in creatures so radically different from us as insects. We have not yet even been able to ascertain what are the senses they possess, or what relation their powers of seeing, hearing, and feeling have to ours. Their sight may far exceed ours both in delicacy and in range, and may possibly give them knowledge of the internal constitution of bodies analogous to that which we obtain by the spectroscope; and that their visual organs do possess some powers which ours do not, is indicated by the extraordinary crystalline rods radiating from the optic ganglion to the facets of the compound eye, which rods vary in form and thickness in different parts of their length, and possess distinctive characters in each group of insects. This complex apparatus, so different from anything in the eyes of vertebrates, may subserve some function quite inconceivable by us, as well as that which we know as vision. There is reason to believe that insects appreciate sounds of extreme delicacy, and it is supposed that certain minute organs, plentifully supplied with nerves, and situated in the subcostal vein of the wing in most insects, are the organs of hearing. But besides these, the Orthoptera (such as grasshoppers, etc.) have what are supposed to the ears on their fore legs, and Mr. Lowne believes that the little stalked balls, which are the sole remnants of the hind wings in flies, are also organs of hearing or of some analogous sense. In flies, too, the third joint of the antennæ contains thousands of nerve-fibres, which terminate in small open cells, and this Mr. Lowne believes to be the organ of smell, or of some other, perhaps new, sense. It is quite evident, therefore, that insects may possess senses which give them a knowledge of that which we can never perceive, and enable them to perform acts which to us are incomprehensible. In the midst of this complete ignorance of their faculties and inner nature, is it wise for us to judge so boldly of their powers by a comparison with our own? How can we pretend to fathom the profound mystery of their mental nature, and decide what, and how much, they can perceive or remember, reason or reflect! To leap at one bound from our own consciousness to that of an insect's is as unreasonable and absurd as if, with a pretty good knowledge of the multiplication table, we were to go straight to the study of the calculus of functions, or as if our comparative anatomists should pass from the study of man's bony structure to that of the fish, and, without any knowledge of the numerous intermediate forms, were to attempt to determine the homologies between these distant types of vertebrata. In such a case would not error be inevitable, and would not continued study in the same direction only render the erroneous conclusions more ingrained and more irremovable.

Definition of Instinct

Before going further into this subject we must determine what we mean by the term instinct. It has been variously defined as—"disposition operating without the aid of instruction or experience," "a mental power totally independent of organisation," or "a power enabling an animal to do that which, in those things man can do, results from a chain of reasoning, and in things which man cannot do, is not to be explained by any efforts of the intellectual faculties." We find, too, that the word instinct is very frequently applied to acts which are evidently the result either of organisation or of habit. The colt or calf is said to walk instinctively, almost as soon as it is born; but this is solely due to its organisation, which renders walking both possible and pleasurable to it. So we are said instinctively to hold out our hands to save ourselves from falling, but this is an acquired habits, which the infant does not possess. It appears to me that instinct should be defined as—"the performance by an animal of complex acts, absolutely without instruction or previously acquired knowledge." Thus, acts are said to the performed by birds in building their nests by bees in constructing their cells, and by many insects in providing for the future wants of themselves or their progeny, without ever having seen such acts performed by others, and without any knowledge of why they perform them themselves. This is expressed by the very common term “blind instinct.” But we have here a number of assertions of matters of fact, which, strange to say, have never been proved to be facts at all. They are thought to be so self-evident that they may be taken for granted. No one has ever yet obtained the eggs of some bird which builds an elaborate nest, hatched these eggs by steam or under a quite distinct parent, placed them afterwards in an extensive aviary or covered garden, where the situation and the materials of a nest similar to that of the parent birds may be found, and then seen what kind of nest these birds would build. If under these rigorous conditions they choose the same materials, the same situation, and construct the nest in the same way and as perfectly as their parents did, instinct would be proved in their case; now it is only assumed, and assumed, as I shall show further on, without any sufficient reason. So, no one has ever carefully taken the pupæ of a hive of bees out of the comb, removed them from the presence of other bees, and loosed them in a large conservatory with plenty of flowers and food, and observed what kind of cells they would construct. But till this is done, no one can say that bees build without instruction, no one can say that with every new swarm there are no bees older than those of the last brood, who may be the teachers in forming the new comb. Now, in a scientific inquiry, a point which can be proved should not be assumed, and a totally unknown power should not be brought in to explain facts, when known powers may be sufficient. For both these reasons I decline to accept the theory of instinct in any case where all other possible modes of explanation have not been exhausted.

Does Man possess Instincts.

Many of the upholders of the instinctive theory maintain that man has instinct exactly of the same nature as those of animals, but more or less liable to be obscured by his reasoning powers; and as this is case more open to our observation than any other, I will devote a few pages to its consideration. Infants are said to suck by instinct, and afterwards to walk by the same power, while in adult man the most prominent case of instinct is supposed to be the powers possessed by savage races to find their ways across a trackless and previously unknown wilderness. Let us take first the case of the infant's sucking. It is sometimes absurdly stated that the new-born infant "seeks the breast," and this is held to be a wonderful proof of instinct. No doubt it would be if true, but unfortunately for the theory it is totally false, as every nurse and medical man can testify. Still, the child undoubtedly sucks without teaching, but this is one of those *simple* acts dependent upon organisation, which cannot properly be termed instinct, any more than breathing or muscular motion. Any object of suitable size in the mouth of an infant excites the nerves and muscles so as to produce the act of suction, and when, at a little later period, the will comes into play, the pleasurable sensations consequent on the act lead to its continuance. So walking is evidently dependent on the arrangement of the bones and joints, and the pleasurable exertion of the muscles, which lead to the vertical posture becoming gradually the most agreeable one; and there can be little doubt that an infant would learn of itself to walk, even if suckled by a wild beast.

How Indians travel through unknown and trackless Forests

Let us now consider the fact of Indians finding their way through forests they have never traversed before. This is much misunderstood, for I believe it is only performed under such special conditions as at once to show that instinct has nothing to do with it. A savage, it is true, can find his way through his native forests in a direction in which he has never traversed them before; but this is because from infancy he has been used to wander in them, and to find his way by indications which he has observed himself or learnt from others. Savages make long journeys in many directions, and, their whole faculties being directed to the subject, they gain a wide and accurate knowledge of the topography, not only of their own district, but of all the regions round about. Every one who has travelled in a new direction communicates his knowledge to those who have travelled less, and descriptions of routes and localities, and minute incidents of travel, form one of the main staples of conversation round the evening fire. Every wanderer or captive from another tribe adds to the store of information, and as the very existence of individuals and of whole families and tribes depends upon the completeness of this knowledge, all the acute perceptive faculties of the adult savage are devoted to acquiring and perfecting it. The good hunter or warrior thus comes to know the bearing of every hill and mountain range, the directions and junctions of all the streams, the situation of each tract characterised by peculiar vegetation, not only within the area he has himself traversed, but for perhaps a hundred miles around it. His acute observation enables him to detect the slightest undulations of the surface, the various changes of subsoil and alterations in the character of the vegetation, that would be imperceptible or meaningless to a stranger. His eye is always open to the direction in which he is going; the mossy side of trees, the presence of certain plants under the shade of rocks, the morning and evening flight of birds, are to him indications of direction almost as sure as the sun in the heavens. Now, if such a savage is required to find his way across this country in a direction in which he has never been before, he is quite equal to the task. By however circuitous a route he has come to the point he is to start from, he has observed all the bearings and distances so well, that he knows pretty nearly where he is, the direction of his own home and that of the place he is required to go to. He starts towards it, and knows that by a certain time he must cross an upland or a river, that the streams should flow in a certain direction, and that he should cross some of them at a certain distance from their sources. The nature of the soil throughout the whole region is known to him, as well as all the great features of the vegetation. As he approaches any tract of country he has been in or near before, many minute indications guide him, but he observes them so cautiously that his white companions cannot perceive by what he has directed his course. Every now and then he slightly changes his direction, but he is never confused, never loses himself, for he always feels at home; till at last he arrives at a well-known country, and directs his course so as to reach the exact spot desired. To the Europeans whom he guides he seems to have come without trouble, without any special observation, and in a nearly straight unchanging course. They are astonished, and ask if he has ever been the same route before, and when he answers "No," conclude that some unerring instinct could alone have guided him. But take his same man into another country very similar to his own, but with other streams and hills, another kind of soil, with a somewhat different vegetation and animal life; and after bringing him by a circuitous route to a given point, ask him to return to his starting-place, by a straight line of fifty miles through the forest, and he will certainly decline to attempt it, or, attempting it, will more or less completely fail. His supposed instinct does not act out of his own country.

A savage, even in a new country, has, however, undoubted advantage from his familiarity with forest life, his entire fearlessness of being lost, his accurate perception of direction and of distance, and he is thus able very soon to acquire a knowledge of the district that seems marvelous to a civilised man; but my own observation of savages in forest countries has convinced me that they find their way by the use of no other faculties than those which we ourselves possess. It appears to me, therefore, that to call in the aid of a new and mysterious power to account for savages being able to do that which, under similar conditions, we could almost all of us perform, although perhaps less perfectly, is almost ludicrously unnecessary.

In the next essay I shall attempt to show that much of what has been attributed to instinct in birds can be also very well explained by crediting them with those faculties of observations, memory, and imitation, and with that limited amount of reason, which they undoubtedly exhibit.

V

THE PHILOSOPHY OF BIRDS' NESTS[1]

Instinct or Reason in the Construction of Birds' Nests

BIRDS, we are told, build their nests by *instinct,* while man constructs his dwelling by the exercise of *reason.* Birds never change, but continue to build for ever on the self-same plan; man alters and improves his houses continually. Reason advance; instinct is stationary.

This doctrine is so very general that it may almost be said to be universally adopted. Men who agree on nothing else accept this as good explanation of the facts. Philosophers and poets, metaphysicians and divines, naturalists and the general public, not only agree in believing this to be probable, but even adopt it as a sort of axiom that is so self-evident as to need no proof, and use it as the very foundation of their speculations on instinct and reason. A belief so general, one would think, must rest on indisputable facts, and be a logical deduction from them. Yet I have come to the conclusion that not only is it very doubtful, but absolutely erroneous; that it not only deviates widely from the truth, but is in almost every particular exactly opposed to it. I believe, in short, that birds do *not* build their nests by instinct; that man does *not* construct his dwelling by reason; that birds do change and improve when affected by the same causes that make men do so; and that mankind neither alter nor improve when they exist under conditions similar to those which are almost universal among birds.

[1] First published in the *Intellectual Observer,* July 1867; reprinted in *Contributions,* etc., with considerable alterations and additions; and with further additions in the present volume.

Do Men build by Reason or by Imitation?

Let us first consider the theory of reason, as alone determining the domestic architecture of the human race. Man, as a reasonable animal, it is said, continually alters and improves his dwelling. This I entirely deny. As a rule, he neither alters nor improves, any more than the birds do. What have the houses of most savage tribes improved from, each as invariable as the nest of a species of bird? The tents of the Arab are the same now as they were two or three thousand years ago, and the mud villages of Egypt can scarcely have improved since the time of the Pharaohs.

The palm-leaf huts and hovels of the various tribes of South America and the Malay Archipelago, what have they improved from since those regions were first inhabited? The Patagonian's rude shelter of leaves, the hollowed bank of the South African Earthmen, we cannot even conceive to have been ever inferior to what they now are. Even nearer home, the Irish turf cabin and the Highland stone shelty can hardly have advanced much during the last two thousand years. Now, no one imputes this stationary condition of domestic architecture among these savage tribes to instinct, but to simple imitation from one generation to another, and the absence of any sufficiently powerful stimulus to change or improvement. No one imagines that if an infant Arab could be transferred to Patagonia or to the Highlands, it would, when it grew up, astonish its foster-parents by constructing a tent of skins. On the other hand, it is quite clear that physical conditions, combined with the degree of civilisation arrived at, almost necessitate certain types of structure. The turf, or stones, or snow—the palm-leaves, bamboo, or branches—which are the materials of houses in various countries, are used because nothing else is so readily to be obtained. The Egyptian peasant has none of these, not even wood. What, then, can he use but mud? In tropical forest-countries, the bamboo and the broad palm-leaves are the natural material for houses, and the form and mode of structure will be decided in part by the nature of the country, whether hot or cool, whether swampy or dry, whether rocky or plain, whether frequented by wild beasts, or whether subject to the attacks of enemies.

When once a particular mode of building has been adopted, and has become confirmed by habit and by hereditary custom, it will be long retained, even when its utility has been lost through changed conditions, or through migration into a very different region. As a general rule, throughout the whole continent of America, native houses, when permanent, are built directly upon the ground—strength and security being given by thickening the low walls and the roof. In almost the whole of the Malay Islands, on the contrary, the houses are raised on posts, often to a great height, with an open bamboo floor; and the whole structure is exceedingly slight and thin. Now, what can be the reason of this remarkable difference between countries, many parts of which are strikingly similar in physical conditions, natural productions, and the state of civilisation of their inhabitants? We appear to have some clue to it in the supposed origin and migrations of their respective populations. The indigenes of tropical America are believed to have immigrated from the north—from a country where the winters are sever, and raised houses with open floors would be hardly habitable. They moved southwards by land along the mountain ranges and uplands, and in an altered climate continued the mode of construction of their forefathers, modified only by the new materials they met with. By minute observations of the Indians of the Amazon Valley, Mr. Bates arrived at the conclusion that they were comparatively recent immigrants from a colder climate. He says: "No one could live long among the Indians of the Upper Amazon without being struck with their constitutional dislike to the heat. … Their skin is hot to the touch, and they perspire little. … They are restless and discontented in hot, dry weather, but cheerful on cool days, when the rain is pouring down their naked backs." And, after giving many other details, he concludes, "How different all this is with the Negro, the true child of tropical climes! The impression gradually forced itself on my mind that the Red Indians lives as an immigrant or stranger in these hot regions, and that his constitution was not originally adapted, and has not since become perfectly adapted, to the climate."

The Malay races, on the other hand, are no doubt very ancient inhabitants of the hottest regions, and are particularly addicted to forming their first settlements at the mouths of rivers or creeks, or in land-locked bays and inlets. They are a pre-eminently maritime or semi-aquatic people, to whom a canoe is a necessary of life, and who will never travel by land if they can do so by water. In accordance with these tastes, they have built their houses on posts in the water, after the manner of the lake-dwellers of ancient Europe; and this mode of construction has become so confirmed, that even those tribes which have spread far into the interior, on dry plains and rocky mountains, continue to build in exactly the same manner, and find safety in the height to which they elevate their dwellings above the ground.

Why does each Bird build a peculiar kind of Nest?

These general characteristics of the abode of savage man will be found to be exactly paralleled by the nest of birds. Each species uses the materials it can most readily obtain, and builds in situations most congenial to its habits. The wren, for example, frequenting hedgerows and low thickets, builds its nest generally of moss, a material always found where it lives, and among which it probably obtains much of its insect food; but it varies sometimes, using hay or feathers when these are at hand. Rooks dig in pastures and ploughed fields for grubs, and in doing so must continually encounter roots and fibres. These are used to line its nest. What more natural! The cow feeding on carrion, dead rabbits, and lambs, and frequenting sheep-walks and warrens, chooses fur and wool to line its nest. The lark frequents cultivated fields, and makes its nest, on the ground, of dry grass-stems lined with finer grass and rootlets—materials the most easy to meet with, and the best adapted to its needs. The kingfisher makes its nest of the bones of the fish which it has eaten. Swallows use clay and mud from the margins of the ponds and rivers over which they find their insect food. The materials of birds' nest, like those used by savage man for his house, are, then, those which come first to hand; and it certainly requires no more special instinct to select them in one case than in the other.

But, it will be said, it is not so much the materials as the form and structure of nests, that vary so much, and are so wonderfully adapted to the wants and habits of each species; how are these to be accounted for except by instinct? I reply: They may be in a great measure explained by the general habits of the species, the nature of the tools they have to work with, and the materials they can most easily obtain, with the very simplest adaptations of means to an end, quite within the mental capacities of birds. The delicacy and perfection of the nest will bear a direct relation to the size of the bird, its structure and habits. That of the wren or the humming-bird is perhaps not finer or more beautiful in proportion than that of the blackbird, the magpie, or the crow. The wren, having a slender beak, long legs, and great activity, is able with great ease to form a well-woven nest of the finest materials, and places it in thickets and hedgerows which it frequents in its search for food. The titmouse, haunting fruit-trees and walls, and searching in cracks and crannies for insects, is naturally led to build in holes where it has shelter and security; while its great activity, and the perfection of its tools (bill and feet) enable it readily to form a beautiful receptacle for its eggs and young. Pigeons having heavy bodies and weak feet and bills (imperfect tools for forming a delicate structure) build rude, flat nests of sticks, laid across strong branches, which will bear their weight and that of their bulky young. They can do no better. The Caprimulgidæ have the most imperfect tools of all, feet that will not support them except on a flat surface (for they cannot truly perch) and a bill excessively broad, short, and weak, and almost hidden by feathers and bristles. They cannot build a nest of twigs or fibres, hair or moss, like other birds, and they therefore generally dispense with one altogether, laying their eggs on the bare ground, or on the stump or flat limb of a tree. The clumsy hooked bills, short necks and feet, and heavy bodies of parrots, render them quite incapable of building a nest like most other birds. They cannot climb up a branch without using both bill and feet; they cannot even turn round on a perch without holding on with their bill. How, then, could they inlay, or weave, or twist the materials of a nest? Consequently they all lay in holes of trees, the top of rotten stumps, or in deserted ants' nests, the soft materials of which they can easily hollow out.

Many terns and sandpipers lay their eggs on the bare sand of the sea-shore, and no doubt the Duke of Argyll is correct when he says that the cause of this habit is not that they are unable to form a nest, but that, in such situations, any nest would be conspicuous and lead to the discovery of the eggs. The choice of *place* is, however, evidently determined by the habits of the birds, who, in their daily search for food, are continually roaming over extensive tide-washed flats. Gulls vary considerably in their mode of nesting, but it is always in accordance with their structure and habits. The situation is either on bare rock or on ledges of sea-cliffs, in marshes or on weedy shores. The materials are sea-weed, tuffs of grass or rushes, or the *débris* of the shore, heaped together with as little order and constructive art as might be excepted from the webbed feet and clumsy bill of these birds, the latter better adapted for seizing fish than for forming a delicate nest. The long-legged broad-billed flamingo, who is continually stalking over muddy flats in search of food, heaps up the mud into a conical stool, on the top of which it lays its eggs. The bird can thus sit upon them conveniently, and they are kept dry, out of reach of the tides.

Now I believe that throughout the whole class of birds the same general principles will be found to hold good, sometimes distinctly, sometimes more obscurely apparent, according as the habits of the species are more marked, or their structure more peculiar. It is true that, among birds differing but little in structure or habits, we see considerable diversity in the mode of nesting, but we are now so well assured that important changes of climate and of the earth's surface have occurred within the period of existing species, that it is by no means difficult to see how such differences have arisen. Simple habits are known to be hereditary, and as the area now occupied by each species is different from that of every other, we may be sure that such changes would act differently upon each, and would often bring together species which had acquired their peculiar habits in distinct regions and under different conditions.

How do Young Birds learn to Build their first Nest?

But it is objected, bird do not *learn* to make their nest as man does to build, for all birds will make exactly the same nest as the rest of their species, even if they have never seen one, and it is instinct alone that can enable them to do this. No doubt this would be instinct if it were true, and I simply ask for proof of the fact. This point, although so important to the question at issue, is always assumed without proof, and even against proof, for what facts there are, are opposed to it. Birds brought up from the egg in cages do not make the characteristic nest of their species, even though the proper materials are supplied them, and often make no nest at all, but rudely heap together a quantity of materials; and the experiment has never been fairly tried of turning out a pair of birds so brought up into an enclosure covered with netting, and watching the result of their untaught attempts at nest-making. With regard to the songs of birds, however, which is thought to be equally instinctive, the experiment has been tried, and it is found that young birds never have the song peculiar to their species if they have not heard it, whereas they acquire very easily the song of almost any other bird with which they are associated.

Do Birds sing by Instinct or by Imitation?

The Hon. Daines Barrington was of opinion that "notes in birds are no more innate than language is in man, and depend entirely on the master under which they are bred, *as far as their organs will enable them to imitate* the sounds which they have frequent opportunities of hearing." He has given an account of his experiments in the *Philosophical Transactions* for 1773 (vol. lxiii.) He says: "I have educated nestling linnets under the three best singing larks—the skylark, woodlark, and titlark, every one of which, instead of the linnet's song, adhered entirely to that of their respective instructors. When the note of the titlark linnet was thoroughly fixed, I hung the bird in a room with two common linnets for a quarter of a year, which were full in song; the titlark linnet, however, did not borrow any passage from the linnet's song, but adhered steadfastly to that of the titlark,"

He then goes on to say that birds taken from the nest at two or three weeks old have already learnt the call-note of their species. To prevent this the birds must be taken from the nest when a day or two old, and he gives an account of a goldfinch which he saw a Knighton in Radnorshire, and which sang exactly like a wren, without any portion of the proper note of its species. This bird had been taken from the nest at two or three days old, and had been hung at a window opposite a small garden, where it had undoubtedly acquired the notes of the wren without having any opportunity of learning even the call of the goldfinch.

He also saw a linnet, which had been taken from the nest when only two or three days old, and which, not having any other sounds to imitate, had learnt almost to articulate, and could repeat the words, "Pretty Boy," and some other short sentences.

Another linnet was educated by himself under a *vengolina* (a small African finch, which he says sings better than any foreign bird but the American mocking bird), and it imitated its African master so exactly that it was impossible to distinguish the one from the other.

Still more extraordinary was the case of a common house sparrow, which only chirps in a wild state, but which learnt the song of the linnet and goldfinch by being brought up near those birds.

The Rev. W. H. Herbert made similar observations, and

states that the young whinchat and wheatear, which have naturally little variety of song, are ready in confinement to learn from other species, and become much better songsters. The bullfinch, whose natural notes are weak, harsh, and insignificant, has nevertheless a wonderful musical faculty, since it can be taught to whistle complete tunes. The nightingale, on the other hand, whose natural song is so beautiful, is exceedingly apt in confinement to learn that of other birds instead. Bechstein gives an account of a redstart which has built under the eaves of his houses, which imitated the song of a caged chaffinch in a window underneath, while another in his neighbour's garden repeated some of the notes of a blackcap, which had a nest close by.

These facts, and many others which might be quoted, render it certain that the peculiar notes of birds are acquired by imitation, a surely as a child learns English or French, not by instinct, but by hearing the language spoken by its parents.

It is especially worthy of remark that, for young birds to acquire a new song correctly, they must be taken out of hearing of their parents very soon, for in the first three or four days they have already acquired some knowledge of the parent notes, which they will afterwards imitate. This shows that very young birds can both hear and remember, and it would be very extraordinary if, after they could see, they could neither observe nor recollect, and could live for days and weeks in a nest and know nothing of its materials and the manner of its construction. During the time they are learning to fly and return often to the nest, they must be able to examine it inside and out in every detail, and as we have seen that their daily search of food invariably leads them among the materials of which it is constructed, and among places similar to that in which it is placed, is it so very wonderful that when they want one themselves they should make one like it? How else, in fact, should they make it? Would it not be much more remarkable if they went out of their way to get materials quite different from those used in the parent nest, if they arranged them in a way they had seen no example of, and formed the whole structure differently from that in which they themselves were reared, and which we may fairly presume is that which their whole organisation is best adapted to put together with celerity and ease? It has, however, been objected that observation, imitation, or memory can have nothing to do with a bird's architectural powers, because the young birds, which in England are born in May or June, will proceed in the following April or May to build a nest as perfect and as beautiful as that in which it was hatched, although it could never have seen one built. But surely the young birds *before* they left the nest had ample opportunities of observing its *form,* its *size,* its *position,* the *materials* of which it was constructed, and the manner in which those materials were arranged. Memory would retain these observations till the following spring, when the materials would come in their ways during their daily search for food, and it seems highly probable that the older birds would begin building first, and that those born the preceding summer would follow their example, learning from them how the foundations of the nest are laid and the material put together.[1]

Again, we have no right to assume that young birds generally pair together. It seems probable that in each pair there is more frequently only one bird born the preceding summer, who would be guided, to some extent, by its partner.

[1] It has been very pertinently remarked by a friend that, if young birds did observe the nest they were reared in, they would consider it to be a natural production, like the leaves and branches and matted twigs that surrounded it, and could not possibly conclude that their parents had constructed the one and not the other. This may be a valid objection, and if so, we shall have to depend on the mode of instruction described in the succeeding paragraphs, but the question can only be finally decided by a careful set of experiments.

My friend, Dr. Richard Spruce, the well-known traveller and botanist, thinks this is the case, and has kindly allowed me to publish the following observations, which he sent me after reading my book.

How young Birds may learn to build Nests

"Among the Indians of Peru and Ecuador, many of whose customs are relics of the semi-civilisation that prevailed before the Spanish conquest, it is usual for the young men to marry old women, and the young women old men. A young man, they say, accustomed to be tended by his mother, would fare ill if he had only an ignorant young girl to take care of him; and the girl herself would be better off with a man of mature years, capable of supplying the place of a father to her.

"Something like this custom prevails among many animals. A stout old buck can generally fight his way to the doe of his choice, and indeed of as many does as he can manage; but a young buck 'of his fist horns' must either content himself with celibacy, or with some dame well-stricken in years.

"Compare the nearly parallel case of the domestic cock and of many other bird. Then consider the consequences amongst birds that pair, if an old cock sorts with a young hen and an old hen with a young cock, as I think is certainly the case with blackbirds and others that are known to fight for the youngest and handsomest females. One of each pair being already an 'old bird,' will be competent to instruct its younger partner (not only in the futility of 'chaff,' but) in the selection of a site for a nest and how to build it; then, how eggs are hatched and young birds reared.

"Such, in brief, is my idea of how a bird on its first espousals may be taught the Whole Duty of the married state."

On this difficult point I have sought for information from some of our best field ornithologists, but without success, as it is in most cases impossible to distinguish old from young birds after the first year. I am informed, however, that the males of blackbirds, sparrows, and many other kinds fight furiously, and the conqueror of course has the choice of a mate. Dr. Spruce's view is at least as probable as the contrary one (that young birds, *as a rule,* pair together), and it is to some extent supported by the celebrated American observer, Wilson, who strongly insists on the variety in the nests of birds of the same species, some being so much better finished than others; and he believes *that the less perfect nests are built by the younger, the more perfect by the older, birds.*

Nearly a century ago the Swiss naturalist, Leroy, made a similar observation. He maintained that there is a distinctly perceptible inferiority in the nests built by young birds; and he further remarks that the best constructed nests are made by birds whose young remain a long time in them, and thus have more opportunity of learning how they are made. He says that the nests of young birds are ill made and badly situated, and that these defects are remedied in time, when their builders have been instructed by a sense of the inconveniences they have endured. He maintains that nests of the same species of bird differ as much as human dwellings, and that of a hundred swallows' nests no two are exactly alike; and he imputed to want of long-continued observation our failure to discover improvement in them.[1]

At all events, till the crucial experiment is made, and a pair of wild bird, raised from the egg without ever seeing a nest, are shown to be capable of making one exactly of the parental type, I do not think we are justified in calling in the aid of an unknown and mysterious faculty to do that which is so strictly analogous to the house-building of savage man.

[1] *The Intelligence and Perfectibility of Animals from a Philosophic Point of View.* By Charles Georges Leroy.

The observations and experiments of the late Mr. Spalding may seem opposed to this view, as they undoubtedly prove some very remarkable instinctive actions on the part of young chickens hatched in an incubator. These birds appear to recognise the call of a hen; and one chick walked or ran straight towards her, leaping over or running round small obstacles; and this only twenty minutes after its eyes had been allowed to see the light and the first time it had ever moved its legs. A young chicken, ten minutes after its eyes had been unveiled for the first time seized and swallowed a fly at the first stroke.[1]

In subsequent papers Mr. Spalding showed that young

swallows could fly well and avoid obstacles on the first attempt; that young pigs a few minutes old could hear and run to their mother, though out of sight; and that most young animals give indications of fear at the voice or presence of their natural enemies.

But in all these cases we have comparatively simple motions or acts induced by feeling of liking or disliking; and we can see that they may be due of definite nervous and muscular co-ordinations which are essential to the existence of the species. That a chicken should feel pleasure at the sound of a hen's voice and pain or fear at that of a hawk, and should move towards the one and away from the other, is a fact of the same nature as the liking of an infant for milk and its dislike of beer with the motion of the head towards the one and away from the other when offered to it. But when, at a much later period, with all its senses and powers of motion fully developed by use and exercise, and with the results of the experiences of a year's eventful life, the bird proceeds to perform the highly complex operation of building a nest, we have no right to assume without direct proof that it will be guided throughout by instinct alone; and we have seen that not only is there no evidence to support this theory, but that all the facts we possess are directly opposed to it.

[1] "On Instinct." Paper read at British Association, sect. D., 1872; *Nature*, vol. vi. p. 485.

Since this essay was published, however, some amount of experiment to illustrate the question at issue has become available. Mr. B. T. Lowne, F.R.C.S., had three of the small ring-doves (Turtur risoria) which had been hatched in the breeding box of an ordinary dove's cage. They were kept at first in a similar cage, with some hay, on which the two hen birds laid eggs and hatched some young. In the following April these birds were put into an aviary in the open air, in which was a large branch of a tree with numerous twigs and buds, and there was also a breeding box with hay and straw. Noticing that the older birds perched on the branch with small pieces of stick in their bills, Mr. Lowne supplied them with a quantity of twigs and small sticks, and the very curious and interesting result was that they built a nest on the branch and laid their eggs in it. But this was not effected without much difficult, and only after they had received assistance. They first seemed to try to fix the twigs against the wall of the aviary or its roof, and waved them about above their heads till they dropped them. Mr. Lowne then fixed some perches for them lower down, and wove some small branches together to afford an additional resting-place. They took possession of this and again carried up twigs and dropped them, and Mr. Lowne then observed that while the straight smooth twigs fell to the ground those that were forked often lodged in the branches. He therefore supplied them with plenty of forked or branched twigs, and by carrying these up and dropping them (and I presume standing on them, or other-wise rendering them compact, though this is not mentioned) they at length (in three days) formed a nest "exactly like that of a wood-pigeon." This "they lined neatly with straw," and each dove laid two eggs in it.[1]

1 *Popular Science Review,* New Series, vol. iii. p. 274.

This experiment, though very interesting, is by no means satisfactory or conclusive. In the first place, pigeons are the very rudest of nest-builders, and will sometimes lay their eggs on a dense flat bough without any nest at all. Then it is clear that these birds had no notion how to begin to build; they required to be assisted, and, as Mr. Lowne says, "as soon as a few branches had lodged below them, they finished the nest *which accident had commenced for them.*" Then they lined it with straw, which is *not* their habit in a state of nature, but appears to have been the result of their having been used to such a nest. The one thing that remains, and which Mr. Lowne thinks proves instinct, is their *not* forming their nest in the box they had been accustomed to, and their using sticks and twigs instead of straw only. But they evidently preferred the light and air and movement of the branch. That was all in harmony with their special organisation, and was a return to the habits which were at once the result and the cause of that organisation. They preferred to make the nest in this pleasant place, but they did not know how to begin. As soon as the sticks, lodged by accident, furnished a sufficient base, they carried up more sticks and soon obtained a rude nest. They saw that smooth straight twigs dropped to the ground, whereas branched twigs kept in the branches, and they had quite sense and observations enough to choose the branched twigs for the purpose. In all this there seems to me to be no proof of the operation of instinct as usually understood, and the experiment yet requires trying with some of our native birds that build elaborate and very distinctive nests, such as the song-thrush, the gold-crest, the wren or the long-tailed tit. If several of these could be brought up in strange nests, and then be turned out into a large wired enclosure containing shrubs and bushes, and if under these circumstances each built an unmistakable nest of its own species, the nest-building instinct would have to be admitted.

The nearest approach to such a test experiment has been recently furnished by Mr. Charles Dixon. He states that some young chaffinches (Fringilla Cœlebs) were taken to New Zealand and there turned out. They throve well, and a nest built by a pair of them was photographed, and from this photograph the nest is thus described by Mr. Dixon: “It is evidently built in the fork of a branch, and shows very little of that neatness of fabrication for which this bird is noted in England. The cup of the nest is small, loosely put together, apparently lined with feathers, and the walls of the structure are prolonged about eighteen inches and hang loosely down the side of the supporting branch. The whole structure bears some resemblance to the nests of the hangnests, with the exception that the cavity containing the eggs is situated on the top. Clearly these New Zealand chaffinches were at a loss for a design when fabricating their nest. They had no standard to work by, no nests of their own kind to copy, no older birds to give them any instruction, and the result is the abnormal structure I have just described. Perhaps these chaffinches imitated in some degree the nest of some New Zealand species; or it may be that the few resemblances to the typical nest of the Palæarctic chaffinch are the results of memory—the dim remembrance of the nest in which they had been reared, but which had almost been effaced by novel surroundings and changed conditions of life. Any way we have here, at least, a most interesting and convincing proof that birds do not make their nests by blind instinct, but by imitating the nest in which they were reared, aided largely by rudimentary reason and by memory."[1]

This experiment also leaves much to be desired, but it undoubtedly shows that instinct alone does not determine the form and structure of a bird's nest, or we should not see so great a departure from the type in the case of the New Zealand chaffinches.

The Skill exhibited in Nest-building Exaggerated

We are too apt to assume that because a nest appears to us delicately and artfully built, it therefore requires much special knowledge and acquired skill (or their substitute, instinct) in the bird who builds it. We forget that it is formed twig by twig and fibre by fibre, rudely enough at first, but crevices and irregularities, which must seem huge gaps and chasms in the eyes of the little builders, are filled up by twigs and stalks pushed in by slender beak and active foot, and that the wool, feathers, or horsehair are laid thread by thread, so that the result seems a marvel of ingenuity to us, just as would the rudest Indian hut to a native of Brobdignag.

[1] *Nature,* vol. xxxi. p. 533 (April 1885).

Levaillant has given an account of the process of nest-building by a little African warbler, which sufficiently shows that a very beautiful structure may be produced with very little art. The foundation was laid of moss and flax interwoven with grass and tufts of cotton, and presented a rude mass, five or six inches in diameter, and four inches thick. This was pressed and trampled down repeatedly, so as at last to make it into a kind of felt. The birds pressed it with their bodies, turning round upon them in every direction, so as to get it quite firm and smooth before raising the sides. These were added bit by bit, trimmed and beaten with the wings and feet, so as to felt the whole together, projecting fibres being now and them worked in with the bill. By these simple and apparently inefficient means, the inner surface of the nest was rendered almost as smooth and compact as a piece of cloth.

Man's Works mainly Imitative

But look at civilised man! It is said; look at Grecian, and Egyptian, and Roman, and Gothic, and modern architecture! What advance! what improvement! what refinements! This is what reason leads to, whereas birds remain for ever stationary. If, however, such advances as these are required to prove the effects of reason as contrasted with instinct, then all savage and many half-civilised tribes have no reason, but build instinctively quite as much as birds do.

Man ranges over the whole earth, and exists under the most varied conditions, leading necessarily to equally varied habits. He migrates—he makes wars and conquests—one race mingles with another—different customs are brought into contact—the habits of a migrating or conquering race are modified by the different circumstances of a new country. The civilised race which conquered Egypt must have developed its mode of building in a forest country where timber was abundant, for it is not probable that the idea of cylindrical columns originated in a country destitute of trees. The pyramids might have been built by an indigenous race, but not the temples of Luxor and Karnak. In Grecian architecture almost every characteristic feature can be traced to an origin in wooden buildings. The columns, the architrave, the frieze, the fillets, the cantilevers, the form of the roof, all point to an origin in some southern forest-clad country, and strikingly corroborate the view derived from philology, that Greece was colonised from north-western India. But to erect columns and span them with huge blocks of stone, or marble, is not an act of reason, but one of pure unreasoning imitation. The arch is the only true and reasonable mode of covering over wide spaces with stone, and, therefore, Grecian architecture, however exquisitely beautiful, is false in principle, and is by no means a good example of the application of reason to the art of building. And what do most of us do at the present day but imitate the buildings of those that have gone before us? We have not even been able to discover or develop any definite style of building best suited for us. We have no characteristic national style of architecture, and to that extent are even below the birds, who have each their characteristic form of nest, exactly adapted to their wants and habits.

Birds do Alter and Improve their Nests when altered Conditions require it

The great uniformity in the architecture of each species of bird which has been supposed to prove a nest-building instinct, may, therefore, fairly be imputed to the uniformity of the conditions under which each species lives. Their range is often limited, and they very seldom permanently change their country, so as to be placed in new conditions. When, however, new conditions to occur, they take advantage of them just as freely and wisely as man could do. The chimney and house-swallows are a standing proof of a change of habit since chimneys and houses were built, and in America this change has taken place within about three hundred years. Thread and worsted are now used in many nests instead of wool and horsehair, and the jackdaw shows an affection for the church steeple, which can hardly be explained by instinct. In the more thickly populated parts of the United States the Baltimore Oriole uses all sorts of pieces of string, skeins of silk, or the gardener's bass, to weave into its fine pensile nest, instead of the single hairs and vegetable fibres it has painfully to seek in wilder regions; and, as already stated, Wilson, a most careful observer, believes that it improves in nest-building by practice—the older birds making the best nests. More recently, Dr. Abbott, the well-known American naturalist, has studies the nests of the Baltimore

Oriole. He found that, away from the habitations of man, the orioles built concealing nests; but in villages and cities, on the other hand, where they were in no special danger from predatory hawks (or more probably from snakes) the nests were built comparatively open, so that the bird within was not concealed.[1] The purple martin takes possession of empty gourds or small boxes, stuck up for its reception in almost every village and farm in America; and several of the American wrens will also build in cigar boxes, with a small hole cut in them, if placed in a suitable situation. The orchard oriole of the United States offers us an excellent example of a bird which modifies its nest according to circumstances. When built among firm and stiff branches the nest is very shallow, but if, as is often the case, it is suspended from the slender twigs of the weeping willow, it is made much deeper, so that when swayed about violently by the wind the young may not tumble out. It has been observed also that the nests built in the warm Southern States are much slighter and more open in texture than those in the colder regions of the north. Our own house-sparrow equally well adapts himself to circumstances. When he builds in trees, as he, no doubt, always did originally, he constructs a well-made domed nest, perfectly fitted to protect his young ones; but when he can find a convenient hole in a building or among thatch, or in any well-sheltered place, he takes much less trouble, and forms a very loosely-built nest.

Professor Jeitteles of Vienna has described various forms of nests of Hirundo urbica adapted to different situations, some having the form of a semi-ellipsoid placed vertically, with the entrance at one side, others being three-quarters of a sphere, with the entrance in the centre. A nest of Hirundo rustica was also observed supported on an iron hook in a wall, but not itself touching the wall. It was quite hemispherical, like that of a blackbird, a form common in England, whereas the usual form on the continent is that of a quarter of a sphere.[2]

The following case of a recent change of habit in nest-building was communicated to me by Mr. Henry Reeks in 1870: "Thirty years ago, and perhaps less, the herring-gulls used to breed on some inland rocks in a large lake called

[1] *Popular Science Monthly*, vol. vi. p. 481. Quoted by Vice-President E. S. Morse, in Address to American Association for Advancement of Science at Buffalo, N.Y., August 1876.

[2] *Ornithologischer Verein in Wien. Mitthelungen des Ausschusses,* No. 3, 12 Juli 1876. See also Seebohm's *British Birds,* vol. ii. p. 174.

'Parson Pond,' in Newfoundland, which is separated from the sea only by a high pebbly beach. Within the period above stated high tides and heavy seas have shifted the course of the brook flowing from the lake into the sea, and caused a greater, and consequently a more rapid fall of fresh water, which has so shallowed that part of the lake where the gulls were in the habit of breeding that it was no longer safe to build on rocks easily accessible to their common enemy, the fox. They therefore betook themselves to some neighbouring spruce and balsam firs not much over a hundred yards distant from their old breeding station." Audubon also notes a similar change of habit, some herring-gulls building their nests in spruce-trees on an island in the Bay of Fundy, where they had formerly built on the ground.

A curious example of a recent change of habits has occurred in Jamaica. Previous to 1854 the palm swift (Tachornis phœnicobea) inhabited exclusively the palm trees in a few districts in the island. A colony then established themselves in two cocoa-nut palms in Spanish Town, and remained there till 1857, when one tree was blown down and the other stripped of its foliage. Instead of now seeking out other palm trees the swifts drove out the swallows who built in the piazza of the House of Assembly, and took possession of it, building their nests on the tops of the end walls and at the angles formed by the beams and joists, a place which they continue to occupy in considerable numbers. It is remarked that here they form their nest with much less elaboration than when built in the palms, probably from being less exposed.

But perfection of structure and adaptation to purpose are not universal characteristics of birds' nests, since there are decided imperfections in the nesting of many birds which are quite compatible with our present theory, but are hardly so with that of instinct, which is supposed to be infallible. The passenger pigeon of America often crowds the branches with its nests till they break, and the ground is strewn with shattered nests, eggs, and young birds. Rooks' nests are often so imperfect that during high winds the eggs fall out; but the window-swallow is the most unfortunate in this respect, for White, of Selborne, informs us that he has seen them build, year after year, in places where their nests are liable to be washed away by a heavy rain and their young ones destroyed.

Conclusion

A fair consideration of all these facts will, I think, fully support the statement with which I commenced, and show that the chief mental faculties exhibited by birds in the construction of their nests are the same in kind as those manifested by mankind in the formation of their dwellings. These are, essentially, imitation, and a slow and partial adaptation to new conditions. To compare the work of birds with the highest manifestations of human art and science is totally beside the question. I do not maintain that birds are gifted with reasoning faculties at all approaching in variety and extent to those of man. I simply hold that the phenomena presented by their mode of building their nests, when fairly compared with those exhibited by the great mass of mankind in building their houses, indicate no essential difference in the kind or nature of the mental faculties employed. If instinct means anything, it means the capacity to perform some complex act without teaching or experience. It implies not only innate ideas but innate knowledge of a very definite kind, and, if established, would overthrow Mr. Mill's sensationalism and all the modern philosophy of experience. That the existence of true instinct may be established in other cases is not impossible; but in the particular instance of birds' nests, which is usually considered one of its strongholds, I cannot find a particle of evidence to show the existence of anything beyond those lower reasoning and imitative powers which animals are universally admitted to possess.

VI

A THEORY OF BIRDS' NESTS, SHOWING THE RELATION OF CERTAIN DIFFERENCES OF COLOUR IN FEMALE BIRDS TO THEIR MODE OF NIDIFICATION[1]

THE habit of forming a more or less elaborate structure for the reception of their eggs and young must undoubtedly be looked upon as one of the most remarkable and interesting characteristics of the class of birds. In other classes of vertebrate animals, such structures are few and exceptional, and never attain to the same degree of completeness and beauty. Birds' nests have, accordingly, attracted much attention, and have furnished one of the stock arguments to prove the existence of a blind but unerring instinct in the lower animals. The very general belief that every bird is enabled to build its nest, not by the ordinary faculties of observation, memory, and imitation, but by means of some innate and mysterious impulse, has had the bad effect of withdrawing attention from the very evident relation that exists between the structure, habits, and intelligence of birds, and the kind of nests they construct.

In the preceding essay I have detailed several of these relations, and they teach us that a consideration of the structure, the food, and other specialities of a bird's existence will give a clue, and sometimes a very complete one, to the reason why it builds its nest of certain materials, in a definite situation, and in a more or less elaborate manner.

I now propose to consider the question from a more general point of view, and to discuss its application to some important problems in the natural history of birds.

[1] Published in the *Journal of Travel and Natural History*, No. 2; reprinted in *Contributions*, etc., with considerable additions and corrections.

Changed Conditions and persistent Habits as influencing Nidification

Besides the causes above alluded to, there are two other factors whose effect in any particular case we can only vaguely guess at, but which must have had an important influence in determining the existing details of nidification. These are—changed conditions of existence, whether internal or external, and the influence of hereditary or imitative habit; the first inducing alterations in accordance with changes of organic structure, of climate, or of the surrounding fauna and flora; the other preserving the peculiarities so produced, even when changed conditions render them no longer necessary. Many facts have been already given which show that birds do adapt their nests to the situations in which they place them, and the adoption of eaves, chimneys, and boxes by swallows, wrens, and many other birds, shows that they are always ready to take advantage of changed conditions. It is probable, therefore, that a permanent change of climate would cause many birds to modify the form or materials of their abodes, so as better to protect their young. The introduction of new enemies to eggs or young birds might introduce many alterations tending to their better concealment. A change in the vegetation of a country would often necessitate the use of new materials. So, also, we may be sure, that as a species slowly became modified in any external or internal characters, it would necessarily change in some degree it mode of building. This effect would be produced by modifications of the most varied nature; such as the power and rapidity of flight, which must often determine the distance to which a bird will go to obtain materials for its nest; the capacity of sustaining itself almost motionless in the air, which must sometimes determine the position in which a nest can be built; the strength and grasping power of the foot in relation to the weight of the bird, a power absolutely essential to the constructor of a delicately-woven and well-finished nest; the length and fineness of the beak, which has to be used like a needle in building the best textile nests; the length and mobility of the neck, which is needful for the same purpose; the possession of a salivary secretion like that used in the nests of many of the swifts and swallows, as well as that of the song-thrush—peculiarities of habits which ultimately depend on structure, and which often determine the material most frequently met with or most easily to be obtained. Modifications in any of these characters would necessarily lead either to a change in the materials of the nest, or in the mode of combining them in the finished structure, or in the form or position of that structure.

During all these changes, however, certain specialities of nest-building would continue for a shorter or a longer time after the causes which had necessitated them had passed away. Such records of a vanished past meet us everywhere, even in man's works, notwithstanding his boasted reason. Not only are the main features of Greek architecture mere reproductions in stone of what were originally parts of a wooden building, but our modern copyists of Gothic architecture often build solid buttresses capped with weighty pinnacles to support a wooden roof which has no outward thrust to render them necessary; and even think they ornament their buildings by adding sham spouts of carved stone, while modern waterpipes, stuck on without any attempt at harmony, do the real duty. So, when railways superseded coaches, it was thought necessary to build the first-class carriages to imitate a number of coach-bodies joined together; and the arm-loops for each passenger to hold on by, which were useful when bad roads made every journey a succession of jolts and lurches, were continued on our smooth macadamised mail-routes, and, still more absurdly, remain to this day[1] in our railway carriages, the relic of a kind of locomotion we can now hardly realise. Another good example is to be seen in our boots. When elastic sides came into fashion we had been so long used to fasten them with buttons or laces, that a boot without either looked bare and unfinished, and accordingly the makers often put on a row of useless buttons or imitation laces, because habit rendered the appearance of them necessary to us. It is universally admitted that the habits of children and of savages give us the best clue to the habits and mode of thought of animals; and every one most have observed how children at first imitate the actions of their elders, without any regard to the use or applicability of the particular acts. So, in savages, many customs peculiar to each tribe are handed down from father to son merely by the force of habit, and are continued long after the purpose which they originally served has ceased to exist. With these and a hundred similar facts everywhere around us, we may fairly impute much of what we cannot understand in the details of Bird-Architecture to an analogous cause. If we do not do so, we must assume either that birds are guided in every action by pure reason to a far greater extent than men are, or that an infallible instinct leads them to the same result by a different road. The first theory has never, that I am aware of, been maintained by any author, and I have already shown that the second, although constantly assumed, has never been proved, and that a large body of facts is entirely opposed to it. One of my critics has, indeed, maintained that I admit "instinct" under the term "hereditary habit"; but the whole course of my argument shows that I do not do so. Hereditary habit is, indeed, the same as instinct when the term is applied to some simple action dependent upon a peculiarity of structure which is hereditary; as when the descendants of tumbler pigeons tumble, and the descendants of pouter pigeons pout. In the present case, however, I compare it strictly to the hereditary, or more properly, persistent or imitative, habits of savages, in building their houses as their fathers did. Imitation is a lower faculty than invention. Children and savages imitate before they originate; birds, as well as all other animals, do the same.

[1] Since this was written they have generally been disused.

The preceding observations are intended to show that the exact mode of nidification of each species of bird is probably the result of a variety of causes, which have been continually inducing changes in accordance with changed organic or physical conditions. The most important of these causes seem to be, in the first place, the structure of the species, and, in the second, its environment or conditions of existence. Now, we know that every one of the characters or conditions included under these two heads is variable. We have seen that, on the large scale, the main features of the nest built by each group of birds bears a relation to the organic structure of that group, and we have, therefore, a right to infer that as structure varies, the nest will vary also in some particular corresponding to the changes of structure. We have seen also that birds change the position, the form, and the construction of their nest whenever the available materials or the available situations vary naturally or have been altered by man; and we have, therefore, a right to infer that similar changes have taken place when, by a natural process, external conditions have become in any way permanently altered. We must remember, however, that all these factors are very stable during many generations, and only change at a rate commensurate with those of the great physical features of the earth as revealed to us by geology; and we may, therefore, infer that the form and construction of nests, which we have shown to be dependent on them, are equally stable. If, therefore, we find less important and more easily modified characters than these so correlated with peculiarities of nidification as to indicate that one is probably the cause of the other, we shall be justified in concluding that these variable characters are dependent on the mode of nidification, and not that the form of the nest has been determined by these variable characters. Such a correlation I am now about to point out.

Classification of Nests

For the purpose of this inquiry it is necessary to group nests into two great classes, without any regard to their most obvious differences or resemblances, but solely looking to the fact of whether the contents (eggs, young, or sitting bird) are hidden or exposed to view. In the first class we place all those in which the eggs and young are completely hidden, no matter whether this is effected by an claborate covered structure, or by depositing the eggs in some hollow tree or burrow underground. In the second, we group all in which the eggs, young, and sitting bird are exposed to view, no matter whether there is the most beautifully formed nest or none at all. Kingfishers, which build almost invariably in holes in banks; woodpeckers and parrots, which build in hollow trees; the Icteridæ of America, which all make beautiful covered and suspended nests; and our own wren, which builds a domed nest—are examples of the former; while our thrushes, warblers, and finches, as well as the crow-shrikes, chatterers, and tanagers of the tropics, together with all raptorial birds and pigeons, and a vast number of others in every part of the world, all adopt the latter mode of building.

It will be seen that this division to birds, according to their nidification, bears little relation to the character of the nest itself. It is a functional not a structural classification. The most rude and the most perfect specimens of bird-architecture are to be found in both sections. It has, however, a certain relation to natural affinities, for large groups of birds, undoubtedly allied, fall into one or the other division exclusively. The species of a genus or of a family are rarely divided between the two primary classes, although they are frequently divided between the two very distinct modes of nidification that exist in the first of them.

All the Scansorial or climbing, and most of the Fissirostral or wide-gaped birds, for example, build concealed nests; and in the latter group the two families which build open nests, the swifts and the goatsuckers, are undoubtedly very widely separated from the other families with which they are associated in our classifications.[1] The tits vary much in their mode of nesting, some making open nests concealed in a hole, while others build domed or even pendulous covered nests, but they all come under the same class. Starlings vary in a similar way. The talking mynahs, like our own starlings, build in holes, the glossy starlings of the East (of the genus Calornis) form a hanging covered nest, while the genus Sturnopastor builds in a hollow tree. One of the most striking cases in which one family of birds is divided between the two classes is that of the finches; for while most of the European species build exposed nests, many of the Australian finches make them dome-shaped.

Sexual differences of Colour in Birds

Turning now from the nests to the creatures who make them, let us consider birds themselves from a somewhat unusual point of view, and form them into separate groups, according as both sexes, or the males only, are adorned with conspicuous colours.

[1] Recent research places the goatsuckers nearest to (though still far from) the owls, while swifts are again brought nearer to the swallows. Dr. R. W. Shufeldt in *Journ. of the Linn. Soc.*, vol. xx. *Zoology*, p. 383.

The sexual differences of colour and plumage in birds are very remarkable, and have attracted much attention; and, in the case of polygamous birds, have been explained by Mr. Darwin's principle of sexual selection. We may, perhaps, understand how male pheasants and grouse have acquired their more brilliant plumage and greater size by the continual rivalry of the males both in strength and beauty; but this theory does not throw any light on the causes which have made the female toucan, bee-eater, parroquet, macaw, and tit in almost every case as gay and brilliant as the male, while the gorgeous chatterers, manakins, tanagers, and birds of paradise, as well as our own blackbird, have mates so dull and inconspicuous that they can hardly be recognised as belonging to the same species.

The Law which connects the Colours of Female Birds with the mode of Nidification

The above-stated anomaly can, however, now be explained by the influence of the mode of nidification, since, with very few exceptions, I find it to be the rule—*that when both sexes are of strikingly gay and conspicuous colours the nest is of the first class, or such as to conceal the sitting birds; while, whenever the male is gay and conspicuous and the nest is open so as to expose the sitting bird to view, the female is of dull or obscure colours.* I will now proceed to indicate the chief facts that support this statement, and will afterwards explain the manner in which I conceive the relation has been brought about.

We will first consider those groups of birds in which the female is gaily or at least conspicuously coloured, and is in most cases exactly like the male.

1. Kingfishers (Alcedinidæ). In some of the most brilliant

species of this family the female exactly resembles the male; in others there is a sexual difference, but it rarely tends to make the female less conspicuous. In some the female has a coloured band across the breast, which is wanting in the male, as in the beautiful blue and white Haleyon diops of Ternate. In others the band is rufous in the female, as in several of the American species; while in Dacelo gaudichaudii, and others of the same genus, the tail of the female is rufous, while that of the male is blue. In most kingfishers the nest is in a deep hole in the ground; in Tanysiptera it is said to be a hole in the nests of termites, or sometimes in crevices under overhanging rocks.

2. Motmots (Momotidæ). In these showy birds the sexes are exactly alike, and the nest in a hole under ground.

3. Puff-birds (Bucconidæ). These birds are often gaily coloured; some have coral-red bills; the sexes are exactly alike, and the nest is in a hole in sloping ground.

4. Trogons (Trogonidæ). In these magnificent birds the females are generally less brightly coloured than the males, but are yet often gay and conspicuous. The nest is in a hole of a tree.

5. Hoopoes (Upupidæ). The barred plumage and long crests of these birds render them conspicuous. The sexes are exactly alike, and the nest is in a hollow tree.

6. Hornbills (Bucerotidæ) These large birds have enormous coloured bills, which are generally quite as well

coloured and conspicuous in the females. Their nests are always in hollow trees, where the female is entirely concealed.

7. Barbets (Capitonidæ). These birds are all very gaily-coloured, and, what is remarkable, the most brilliant patches of colour are disposed about the head and neck, and are very conspicuous. The sexes are exactly alike, and the nest is in a hole of a tree.

8. Toucans (Rhamphastidæ). These fine birds are coloured in the most conspicuous parts of their body, especially on the large bill, and on the upper and lower tail coverts, which are crimson, white, or yellow. The sexes are exactly alike, and they always build in a hollow tree.

9. Plaintain-eaters (Musophagidæ). Here again the head and bill are most brilliantly coloured in both sexes, and the nest is in a hole of a tree.

10. Ground cuckoos (Centropus). These birds are often of conspicuous colours, and are alike in both sexes. They build a domed nest.

11. Woodpeckers (Picidæ). In this family the females often differ from the males in having a yellow or white, instead of a crimson crest, but are almost as conspicuous. They all nest in holes in trees.

12. Parrots (Psittaci). In this great tribe, adorned with the most brilliant and varied colours, the rule is that the sexes are precisely alike, and this is the case in the most gorgeous families, the lories, the cockatoos, and the macaws; but in

some there is a sexual difference of colour to a slight extent. All build in holes, mostly in trees, but sometimes in the ground, or in while ants' nests. In the single case in which the nest is exposed, that of the Australian ground parrot, Pezoporus formosus, the bird has lost the gay colouring of its allies, and is clothed in sombre and completely protective tints of dusky green and black.

13. Gapers (Eurylæmidæ). In these beautiful Eastern birds, somewhat allied to the American chatterers, the sexes are exactly alike, and are adorned with the most gay and conspicuous markings. The nest is a woven structure, *covered over,* and suspended from the extremities of branches over water.

14. Pardalotus (Ampelidæ). In these Australian birds the females differ from the males, but are often very conspicuous, having brightly-spotted heads. Their nests are sometimes dome shaped, sometimes in holes of trees, or in burrows in the ground.

15. Tits (Paridæ). These little birds are always pretty, and many (especially among the Indian species) are very conspicuous. They always have the sexes alike, a circumstance very unusual among the smaller gaily-coloured birds of our own country. The nest is always covered over or concealed in a hole.

16. Nuthatches (Sitta). Often very pretty birds, the sexes alike, and the nest in a hole.

17. —(Sittella). The female of these Australian nuthatches is often the most conspicuous, being white and black marked. The nest is, according to Gould, "completely concealed among upright twigs connected together."

18. Creepers (Climacteris). In these Australian creepers the sexes are alike, or the female most conspicuous, and the nest is in a hole of a tree.

19. Estrelda, Amadina. In these genera of Eastern and Australian finches the females, although more or less different from them males, are still very conspicuous, having a red rump, or being white spotted. They differ from most others of the family in building domed nests.

20. Certhiola. In these pretty little American creepers the sexes are alike, and they build a domed nest.

21. Mynahs (Sturnidæ). These showy Eastern starlings have the sexes exactly alike. They build in holes of trees.

22. Calornis (Sturnidæ). These brilliant metallic starlings have no sexual differences. They build a pensile covered nest.

23. Hangnests (Icteridæ). The red or yellow and black plumage of most of these birds is very conspicuous, and is exactly alike in both sexes. They are celebrated for their fine purse-shaped pensile nests.

It will be seen that this list comprehends six important families of Fissirostres, four of Scansores, the Psittaci, and several genera, with three entire families of Passeres, comprising about twelve hundred species, or about one-seventh of all known birds.

The cases in which, whenever the male is gaily coloured, the female is much less gay or quite inconspicuous are exceedingly numerous, comprising, in fact, almost all the bright-coloured Passeres, except those enumerated in the preceding class. The following are the most remarkable:—

1. Chatterers (Cotingidæ). These comprise some of the most gorgeous birds in the world, vivid blues, rich purples, and bright reds being the most characteristic colours. The females are always obscurely tinted, and are often of a greenish hue, not easily visible among the foliage.

2. Manakins (Pipridæ). These elegant birds, whose caps or crests are of the most brilliant colours, are usually of a somber green in the female sex.

3. Tanagers (Tanagridæ). These rival the chatterers in the brilliancy of their colours, and are even more varied. The females are generally of plain and somber hues, and always less conspicuous than the males.

4. Sugar-birds (Cœrebidæ). The males are a beautiful blue; the females green.

5. Pheasants (Phasianidæ). These include some of the most brilliant and gorgeously coloured birds in the world, such as the peacock, gold and silver pheasants, fire-backed pheasants, and many others; but the females are always comparatively dull coloured, and generally of highly protective tints.

In the extensive families of the warblers (Sylviadæ), thrushes (Turdidæ), flycatchers (Muscicapidæ), and shrikes (Laniadæ), a considerable proportion of the species are beautifully marked with gay and conspicuous tints, but in every case the females are less gay, and are most frequently of the very plainest and least conspicuous hues. Now, throughout *the whole of these families the nest is open,* and I am not aware of a single instance in which any one of these birds builds a *domed nest,* or places it in a *hole of a tree,* or *under ground,* or in any place where it is effectually concealed.

In considering the question we are now investigating, it is not necessary to take into account the larger and more powerful birds, because they seldom depend much on concealment to secure their safety. In the raptorial birds bright colours are as a rule absent; and their structure and habits are such as not to require any special protection for the female. The larger waders are sometimes very brightly coloured in both sexes; but they are probably little subject to the attacks of enemies, since the scarlet ibis, the most conspicuous of birds, exist in immense quantities in South America. In game birds and water-fowl, however, the females are often very plainly coloured, when the males are adorned with brilliant hues; and the abnormal family of the Megapodidæ offers us the interesting fact of an identity in the colours of the sexes (which in Megacephalon and Talegalla are somewhat conspicuous), in conjunction with the habit of not sitting on the eggs at all.

What the Facts Teach us

Taking the whole body of evidence here brought forward, embracing as it does almost every group of bright-coloured birds, it will, I think, be admitted that the relation between the two series of facts in the colouring and nidification of birds has been sufficiently established. There are, it is true, a few apparent and some real exceptions, which I shall consider presently; but they are too few and unimportant to weigh much against the mass of evidence on the other side, and may for the present be neglected. Let us then consider what we are to do with this unexpected set of correspondences between groups of phenomena which, at first sight, appear so disconnected. Do they fall in with any other groups of natural phenomena? Do they teach us anything of the way in which nature works, and give us any insight into the causes which have brought about the marvellous variety, and beauty, and harmony of living things? I believe we can answer these questions in the affirmative; and I may mention, as a sufficient proof that these are not isolated facts, that I was first led to see their relation to each other by the study of an analogous though distinct set of phenomena among insects, that of protective resemblance and "mimicry."

On considering this remarkable series of corresponding facts, the first thing we are taught by them seems to be, that there is no incapacity in the female sex among birds to receive the same bright hues and strongly contrasted tints with which their partners are so often decorated, since whenever they are *protected and concealed* during the period of incubation *they are similarly adorned.* The fair inference is, that it is chiefly due to the absence of protection or concealment during this important epoch, that gay and conspicuous tints are withheld or left undeveloped. The mode in which this has been effected is very intelligible, if we admit the action of natural and sexual selection. It would appear from the numerous cases in which both sexes are adorned with equally brilliant colours (while both sexes are rarely armed with equally developed offensive and defensive weapons when not required for individual safety), that the normal action of "sexual selection" or of other unknown causes, is to develop colour and beauty in both sexes, by the preservation and multiplication of all varieties of colour in either sex which are pleasing to the other. Several very close observers of the habits of animals have assured me that male birds and quadrupeds do often take very strong likes and dislikes to individual females, and we can hardly believe that the one sex (the female) can have a general taste for colour while the other has no such taste. However this may be, the fact remains, that in a vast number of cases the female acquires as brilliant and as varied colours as the male, and therefore most probably acquires them in the same way as the male does—that is, either because the colour is useful to it, or is correlated with some useful variation, or is pleasing to the other sex. The only remaining supposition is that it is transmitted from the other sex, without being of any use. From the number of examples above adduced of bright colours in the female, this would imply that colour-characters acquired by one sex are generally (but not necessarily) transmitted to the other. If this be the case it will, I think, enable us to explain the phenomena, even if we do not admit that the male bird is ever influenced in the choice of a mate by her more gay or perfect plumage.

The female bird, while sitting on her eggs in an uncovered nest, is much exposed to the attacks of enemies, and any modification of colour which rendered her more conspicuous would often lead to her destruction and that of her offspring. All variations of colour in this direction in the female would therefore sooner or later be eliminated, while such modifications as rendered her inconspicuous, by assimilating her to surrounding objects, as the earth or the foliage, would, on the whole, survive the longest, and thus lead to the attainment of those brown or green and inconspicuous tints, which form the colouring (of the upper surface at least) of the vast majority of female birds which sit upon open nests.

This does not imply, as some have thought, that all female birds were once as brilliant as the males. The change has been a very gradual one, generally dating from the origin of genera or of larger groups, but there can be no doubt that the remote ancestry of birds having great sexual differences of colour were nearly or quite alike, sometimes (perhaps in most cases) more nearly resembling the female, but occasionally perhaps being nearer what the male is now. The young birds (which usually resemble the females) will probably give some idea of this ancestral type, and it is well known that the young of allied species and of different sexes are often undistinguishable.

Colour more variable than Structure or Habits, and therefore the Character which has generally been Modified

At the commencement of this essay I have endeavoured to prove that the characteristic differences and the essential features of birds' nests are dependent on the structure of the species and upon the present and past conditions of their existence. Both these factors are more important and less variable than colour; and we must therefore conclude that in most cases the mode of nidificaiton (dependent on structure and environment) has been the cause, and not the effect, of the similarity or differences of the sexes as regards colour. When the confirmed habit of a group of birds was to build their nests in holes of trees like the toucans, or in holes in the ground like the kingfishers, the protection the female thus obtained, during the important and dangerous time of incubation, placed the two sexes on an equality as regards exposure to attack, and allowed "sexual selection," or any other cause, to act unchecked in the development of gay colours and conspicuous markings in both sexes.

When, on the other hand (as in the tanagers and flycatchers), the habit of the whole group was to build open cup-shaped nests in more or less exposed situations, the production of colour and marking in the female, by whatever cause, was continually checked by its rendering her too conspicuous, while in the male it had free play, and developed in him the most gorgeous hues. This, however, was not perhaps universally the case; for where there was more than usual intelligence and capacity for change of habits, the danger the female was exposed to be a partial brightness of colour or marking might lead to the construction of a concealed or covered nest, as in the case of the tits and hangnests. When this occurred, a special protection to the female would be no longer necessary; so that the acquisition of colour and the modification of the nest might in some cases act and react on each other and attain their full development together.

Exceptional Cases confirmatory of the above Explanation

There exist a few very curious and anomalous facts in the natural history of birds, which fortunately serve as crucial tests of the truth of this mode of explaining the inequalities of sexual coloration. It has been long known that in some species the males either assisted in, or wholly performed, the act of incubation. It has also been often noticed that in certain birds the usual sexual differences were reversed, the male being the more plainly coloured, the female more gay and often larger. I am not, however, aware that these two anomalies had ever been supposed to stand to each other in the relation of cause and effect, till I adduced them in support of my views of the general theory of protective adaptation. Yet it is undoubtedly the fact that in the best known cases in which the female bird is more conspicuously coloured than the male, it is either positively ascertained that the latter performs the duties of incubation, or there are good reasons for believing such to be the case. The most satisfactory example is that of the Gray Phalarope (Phalaropus fulicarius), the sexes of which are alike in winter, while in summer the female instead of the male takes on a gay and conspicuous nuptial plumage; but the male performs the duties of incubation, sitting upon the eggs, which are laid upon the bare ground.

In the dotterel (Eudromias morinellus) the female is larger and more brightly coloured than the male; and here, also, it is almost certain that the latter sits upon the eggs. The turnices of India also have the female larger and often more brightly coloured; and Mr. Jerdon states, in his *Birds of India,* that the natives report that, during the breeding season, the females desert their eggs and associate in flocks, while the males are employed in hatching the eggs. In the few other cases in which the females are more brightly coloured, the habits are not accurately known. The case of the ostriches and emeus will occur to many as a difficulty, for here the male incubates, but is not less conspicuous than the female; but there are two reasons why the case does not apply: the birds are too large to derive any safety from concealment; from enemies which would devour the eggs they can defend themselves by force, while to escape from their personal foes they trust to speed.

We find, therefore, that a very large mass of facts relating to the sexual coloration and the mode of nidification of birds, including some of the most extraordinary anomalies to be found in their natural history, can be shown to have an interdependent relation to each other, on the simple principle of the need of greater protection to that parent which performs the duties of incubation. Considering the very imperfect knowledge we possess of the habits of most extra-European birds, the exceptions to the prevalent rule are few, and generally occur in isolated species or in small groups; while several apparent exceptions can be shown to be really confirmations of the law.

Real or apparent Exceptions to the Law stated at page 124

The only marked exceptions I have been able to discover are the following—

1. King crows (Dicrourus). These birds are of a glossy black colour, with long forked tails. The sexes present no difference, and they build open nests. This apparent exception may probably be accounted for by the fact that these birds do not need the protection of a less conspicuous colour. They are very pugnacious, and often attack and drive away crows, hawks, and kites; and as they are semi-gregarious in their habits, the females are not likely to be attacked while incubation.

2. Orioles (Oriolidæ). The true orioles are very gay

birds; the sexes are, in many Eastern species, either nearly or quite alike, and the nests are open. This is one of the most serious exceptions, but it is one that to some extent proves the rule; for in this case it has been noticed that the parent birds display excessive care and solicitude in concealing the nest among thick foliage, and in protecting their offspring by incessant and anxious watching. This indicates that the want of protection consequent on the bright colour of the female make itself felt, and is obviated by an increased development of the mental faculties.

3. Ground thrushes (Pittidæ). These elegant and brilliantly-coloured birds are generally alike in both sexes, and build an open nest. It is curious, however, that this is only an apparent exception, for almost all the bright colours are on the under surface, the back being usually olive-green on brown, and the head black, with brown or whitish stripes, all which colours would harmonise with the foliage, sticks, and roots which surround the nest, built on or near the ground, and thus serve as a protection to the female bird.

4. Grallina Australis. This Australian bird is of strongly contrasted black and white colours. The sexes are exactly alike, and it builds an open clay nest in an exposed situation on a tree. This appears to be a most striking exception, but I am by no mean sure that it is so. We require to know

what tree it usually builds on, the colour of the bark or of the lichens that grow upon it, the tints of the ground, or of other surrounding objects, before we can say that the bird, when sitting on its nest, it really conspicuous. It has been remarked that small patches of white and black blend at a short distance to form gray, one of the commonest tints of natural objects.

5. Sunbirds (Nectariniidæ). In these beautiful little birds the males only are adorned with brilliant colours, the females being quite plain, yet they build covered nests in all the cases in which the nidification is known. This is a negative rather than a positive exception to the rule, since there may be other causes besides the need for protection which prevent the female acquiring the gay colours of her mate, and there is one curious circumstance which tends to elucidate it. The male of Leptocoma zeylanica is said to assist in incubation. It is possible, therefore, that the group may originally have used open nests, and some change of conditions, leading the male bird to sit, may have been followed by the adoption of a domed nest. This is, however, the most serious exception I have yet found to the general rule.

6. Superb warblers (Maluridæ). The males of these little birds are adorned with the most gorgeous colours, while the females are very plain, yet they make domed nests. It is to be

observed, however, that the male plumage is nuptial merely, and is retained for a very short time; the rest of the year both sexes are plain alike. It is probable, therefore, that the domed nest is for the protection of these delicate little birds against the rain, and that there is some unknown cause which has led to the development of colour in the males only.

There is one other case which at first sight looks like an exception, but which is far from being one is reality, and deserves to be mentioned. In the beautiful waxwing (Bombycilla garrula) the sexes are very nearly alike, and the elegant red wax tips to the wing-feathers are nearly, and sometimes quite, as conspicuous in the female as in the male. Yet it builds an open nest, and a person looking at the bird would say it ought, according to my theory, to cover its nest. But it is, in reality, as completely protected by its coloration as the most plainly coloured bird that flies. It breeds only in very high latitudes, and the nest, placed in fir-trees, is formed chiefly of fir-twigs and lichens. Now the delicate gray and ashy and purplish hues of the head and back, together with the yellow of the wings and tail, are tints that exactly harmonise with the colours of fir leaves, bark, and lichens, while the brilliant red wax tips exactly represent the crimson fructification of the common lichen, Cladonia coccifera. When sitting on its nest, therefore, the female bird will exhibit no colours that are not common to the materials by which it is surrounded; and the several tints are distributed in about the same proportions as they occur in nature. At a short distance the bird would be undistinguishable from the nest it is sitting on, or from a natural clump of lichens, and will thus be completely protected.

I think I have now noticed all exceptions of any importance to the law of dependence of sexual colour on nidification. It will be seen that they are very few in number, compared with those which support the generalisation; and in several cases there are circumstances in the habits or structure of the species that sufficiently explain them. It is remarkable also that I have found scarcely any *positive* exceptions—that is, cases of very brilliant or conspicuous female birds in which the nest was not concealed. Much less can there be shown any group of birds in which the females are all of decidedly conspicuous colours on the upper surface, and yet sit in open nests. The many cases in which birds of dull colours in both sexes make domed or concealed nests do not, of course, affect this theory one way or the other; since its purpose is only to account for the fact that brilliant females of brilliant males are *always* found to have covered or hidden nests, while obscure females of brilliant males *almost always* have open and exposed nests. The fact that all classes of nests occur with birds which are dull coloured in both sexes merely shows that these dull colours serve to protect the parents at other times than when sitting on the nest, the structure of which is determined by the requirements of the offspring.

If the views here advocated are correct, as to the various influences that have determined the specialities of every bird's nest, and the general coloration of female birds, with their action and reaction on each other, we can hardly expect to find evidence more complete than that here set forth. Nature is such a tangled web of complex relations, that a series of correspondences running through hundreds of species, genera, and families, in every part of the system, can hardly fail to indicate a true casual connection; and when, of the two factors in the problem, one can be shown to be dependent on the most deeply seated and the most stable facts of structure and conditions of life, while the other is a character universally admitted to be superficial and easily modified, there can be little doubt as to which is cause and which effect.

Various modes of Protection of Animals

But the explanation of the phenomenon here attempted does not rest alone in the facts I have been able now to adduce. In the essay on "Mimicry" it is shown how important a part the necessity for protection has played, in determining the external form and coloration, and sometimes even the internal structure of animals.

As illustrating this latter point, I may refer to the remarkable hooked, branched, or star-like spiculæ in many sponges, which are believed to have the function chiefly of rendering them unpalatable to other creatures. The Holothuridæ or sea-cucumbers possess a similar protection, may of them having anchor-shaped spicules embedded in their skin, as the Synapta; while others (Cuviera squamata) are covered with a hard calcareous pavement. Many of these are of a bright red or purple colour, and are very conspicuous, while the allied Trepang, or Beche-de-mer (Holothuria edulis), which is not armed with any such defensive weapons, is of a dull sand or mud colour, so as hardly to be distinguished from the sea-bed on which it reposes. Many of the smaller marine animals are protected by their almost invisible transparency, while those that are most brightly coloured will be often found to have a special protection, either in stinging tentacles like Physalia, or in a hard calcareous crust, as in the star-fishes.

Females of some Groups require and obtain more Protection than the Males

In the struggle for existence incessantly going on, protection or concealment is one of the most general and most effectual means of maintaining life; and it is by modifications of colour that this protection can be most readily obtained, since no other character is subject to such numerous and rapid variations. The case I have now endeavoured to illustrate is exactly analogous to what occurs among butterflies. As a general rule, the female butterfly is of dull and inconspicuous colours, even when the male is most gorgeously arrayed; but when the species is protected from attack by a disagreeable odour and taste, as in the Heliconidæ, Danaidæ and Acræidæ, both sexes display the same or equally brilliant hues. Among the species which gain in protection by imitating these, the very weak and slow-flying Leptalides resemble them in both sexes, because both sexes alike require protection, while in the more active and strong-winged genera—Papilio, Pieris, and Diadema—it is generally the females only that mimic the protected groups, and in doing so often become actually more gay and more conspicuous than the males, thus reversing the usual and in fact almost universal characters of the sexes. So, in the wonderful Eastern leaf-insects of the genus Phyllium, it is the female only that so marvellously imitates a green leaf; and in all these cases the difference can be traced to the greater need of protection for the female, on whose continued existence, while depositing her eggs, the safety of the race depends. In Mammalia and in reptiles, however brilliant the colour may be, there is rarely any difference between that of the sexes, because the female is not necessarily more exposed to attack than the male. It may, I think, be looked upon as a confirmation of this view, that no single case is known either in the above-named genera—Papilio, Pieris, and Diadema—or in any other butterfly, of a male *alone* mimicking one of the Danaidæ or Heliconidæ Yet the necessary colour is far more abundant in the males, and variations always seem ready for any useful purpose. This seems to depend on the general law that each species and each sex can only be modified just as far as is absolutely necessary for it to maintain itself in the struggle for existence, not a step further. A male insect by its structure and habits is less exposed to danger, and also requires less protection, than the female. It cannot, therefore, along acquire any further protection through the agency of natural selection. But the female requires some extra protection, to balance the greater danger to which she is exposed and her greater importance to the existence of the species; and this she always acquires, in one way or another, through the action of natural selection.

In this *Origin of Species,* fourth edition, p. 241, Mr. Darwin recognises the necessity for protection as sometimes being a cause of the obscure colours of female birds;[1] but he does not seem to consider it so very important an agent in modifying colour as I am disposed to do. In the same paragraph (p. 240) he alludes to the fact of female birds and butterflies being sometimes very plain, sometimes as gay as the males; but, apparently, considers this mainly due to peculiar laws of inheritance, which sometimes continue acquired colour in the line of one sex only, sometimes in both. Without denying the action of such a law (which Mr. Darwin informs me he has facts to support), I impute the difference, in the great majority of cases, to the greater or less need of protection in the female sex in these groups of animals.

This need was seen to exist a century ago by the Hon. Daines Barrington, who, in the article already quoted (see p. 104), after alluding to the fact that singing birds are all small, and suggesting (but I think erroncously) that this may have arisen from the difficulty larger birds would have in concealing themselves if they called the attention of their enemies by loud notes, goes on thus: "I should rather conceive it is for the same reason no hen bird sings, because this talent would be still more dangerous during incubation, which *may possibly also account for the inferiority in point of plumage.*" This is a curious anticipation of the main idea on which this essay is founded. It has been unnoticed for near a century, and my attention was only recently called to it by Mr. Darwin himself.

[1] This passage is omitted in the sixth edition.

Conclusion

To some persons it will perhaps appear that the causes to which I impute so much of the external aspect of nature are too simple, too insignificant, and too unimportant for such a mighty work. But I would ask them to consider that the great object of all the peculiarities of animal structure is to preserve the life of the individual, and to maintain the existence of the species. Colour has hitherto been too often looked upon as something adventitious and superficial, something given to an animal not to be useful to itself, but solely to gratify man or even superior beings—to add to the beauty and ideal harmony of nature. If this were the case, then, it is evident that the colours of organised beings would be an exception to most other natural phenomena. They would not be the product of general laws, or determined by ever-changing external conditions; and we must give up all inquiry into their origin and causes, since (by the hypothesis) they are dependent on a Will whose motives must ever be unknown to us. But, strange to say, no sooner do we begin to examine and classify the colours of natural objects, than we find that they are intimately related to a variety of other phenomena, and are, like them, strictly subordinated to general laws. I have here attempted to elucidate some of these laws in the case of birds, and have shown how the mode of nidification has affected the colouring of the female sex in this group. I have before shown to how great an extent, and in how many ways, the need of protection has determined the colours of insects, and of some groups of reptiles and mammalia, and I would now call particular attention to the fact that the gay tints of flowers, so long supposed to be a convincing proof that colour has been bestowed for other purposes than the good of its possessor, have been shown by Mr. Darwin to follow the same great law of utility. Flowers do not often need protection, but very often require the aid of insects to fertilise them, and maintain their reproductive powers in the greatest vigour. Their gay colours attract insects, as do also their sweet odours and honeyed secretions; and that this is the main function of colour in flowers is shown by the striking fact that those flowers which can be perfectly fertilised by the wind, and do not need the aid of insects, *rarely or never have gaily-coloured flowers.*

This wide extension of the general principle of utility to the colours of such varied groups, both in the animal and vegetable kingdoms, compels us to acknowledge that the "reign of law" has been fairly traced into this stronghold of the advocated of special creation. And to those who oppose the explanation now given of the various facts bearing upon this subject, I would again respectfully urge that they must grapple with the whole of the facts, not one or two of them only. It will be admitted that, on the theory of evolution and natural selection, a wide range of facts with regard to colour in nature have been co-ordinated and explained. Until at least an equally wide range of facts can be shown to be in harmony with any other theory, we can hardly be expected to abandon that which has already done such good service, and which has led to the discovery of so many interesting and unexpected harmonies among the most common (but hitherto most neglected and least understood) of the phenomena presented by organised beings.

VII

CREATION BY LAW[1]

AMONG the various criticisms that have appeared on Mr. Darwin's celebrated *Origin of Species,* there is, perhaps, none that will appeal to so large a number of well educated and intelligent persons as that contained in the Duke of Argyll's *Reign of Law.* The noble author represents the feelings and expresses the ideas of that large class of persons who take a keen interest in the progress of science in general, and especially that of Natural History, but have never themselves studied nature in detail, or acquired that personal knowledge of the structure of closely allied forms,—the wonderful gradations from species to species and from group to group, and the infinite variety of the phenomena of "variation" in organic beings,—which is absolutely necessary for a full appreciation of the facts and reasonings contained in Mr. Darwin's great work.

Nearly half of the Duke's book is devoted to an exposition of his idea of "Creation by Law," and he expresses so clearly what are his difficulties and objections as regards the theory of "Natural Selection," that I think it advisable that they should be fairly answered, and that his own views should be shown to lead to conclusions as hard to accept as any which he imputes to Mr. Darwin.

The point on which the Duke of Argyll lays most stress is, that proofs of Mind everywhere meet us in Nature, and are more especially manifest wherever we find "contrivance" or "beauty." He maintains that this indicates the constant supervision and direct interference of the Creator, and cannot possibly be explained by the unassisted action of any combination of laws. Now, Mr. Darwin's work has for its main object to show that all the phenomena of living things,—all their wonderful organs and complicated structures, their infinite variety of form, size, and colour, their intricate and involved relations to each other,—may have been produced by the action of a few general laws of the simplest kind, laws which are in most cases mere statements of admitted facts. The chief of these laws or facts are the following:—

[1] First published in the *Quarterly Journal of Science,* October 1868; reprinted in *Contribution,* etc., with a few alterations and additions.

1. *The Law of Multiplication in Geometrical Progression.*—All organized beings have enormous powers of multiplication. Even man, who increases slower than all other animals, could under the most favourable circumstances double his numbers every fifteen years, or a hundredfold in a century.

Many animals and plants could increase their numbers from ten to a thousandfold every year.

2. *The Law of Limited Population.*—The number of living individuals of each species in any country, or in the whole globe, is practically stationary; whence it follows that the whole of this enormous increase must die off almost as fast as produced, except only those individuals for whom room is made by the death of parents. As a simple but striking example, take an oak forest. Every oak will drop annually many thousands of acorns, but till an old tree falls not one of the millions of acorns produced can grow up into an oak. They must die at various stages of growth.

3. *The Law of Heredity, or Likeness of Offspring to their Parents.*—This is a universal, but not an absolute law. All creatures resemble their parents in a high degree, and in the majority of cases very accurately; so that even individual peculiarities, of whatever kind, in the parents, are almost always transmitted to some of the offspring.

4. *The Law of Variation.*—This is fully expressed by the lines:— "No being on this earthly ball, Is like another, all in all."

Offspring resemble their parents very much, but not wholly—each being possesses its individuality. This "variation" itself varies in amount, but it is always present, not only in the whole organism, but in every part of each organism. Every organ, every character, every feeling, is individual; that is to say, *varies* from the same organ, character, or feeling in every other individual.

5. *The Law of unceasing Change of Physical Conditions upon the Surface of the Earth.*—Geology shows us that this change has always gone on in times past, and we also know that it is now everywhere going on.

6. *The Equilibrium of Harmony of Nature.*—When a species is well adapted to the conditions which environ it, it flourishes; when imperfectly adapted it decays; when ill-adapted it becomes extinct. If *all* the conditions which determine an organism's wellbeing are taken into consideration, this statement can hardly be disputed.

This series of facts or laws are mere statements of what is the condition of nature. They are facts or inferences which are generally known, generally admitted—but, in discussing the subject of the "Origin of Species," as generally forgotten. It is from these universally admitted facts that the origin of all the varied forms of nature may be deduced by a logical chain of reasoning, which, however, is at every step verified and shown to be in strict accord with facts; and, at the same time, many curious phenomena which can be no other means be understood are explained and accounted for. It is probable that these primary facts or laws are but results of the very nature of life, and of the essential properties of organised and unorganised matter. Mr. Herbert Spencer, in his *First Principles* and his *Biology*, has, I think, made us able to understand how this may be; but at present we may accept these simple laws without going further back, and the question then is—whether the variety, the harmony, the contrivance, and the beauty we perceive in organic beings can have been produced by the action of these laws alone, or whether we are required to believe in the incessant interference and direct action of the mind and will of the Creator. It is simply a question of how the Creator has worked. The Duke (and I quote him as having well expressed the views of the more intelligent of Mr. Darwin's opponents) maintains that He has personally applied general laws to produce effects which those laws are not in themselves capable of producing; that the universe alone, with all its laws intact, would be a sort of chaos, without variety, without harmony, without design, without beauty; that there is not (and therefore we may presume that there could not be) any self-developing power in the universe. I believe, on the contrary, that the universe is so constituted as to be self-regulating; that as long as it contains Life, the forms under which that life is manifested have an inherent power of adjustment to each other and to surrounding nature; and that this adjustment necessarily leads to the greatest amount of variety and beauty and enjoyment, because it does depend on general laws, and not on a continual supervision and rearrangement of details. As a matter of feeling and religion, I hold this to be a far higher conception of the Creator and of the Universe than that which may be called the "continual interference" hypothesis; but it is not a question to be decided by our feelings or convictions—it is a question of facts and of reason. Could the change which geology shows us has continually taken place in the forms of life, have been produced by general laws, or does it imperatively require the incessant supervision of a creative mind? This is the question for us to consider, and our opponents have the difficult task of proving a negative, if we show that there are both facts and analogies in our favour.[1]

Mr. Darwin's Metaphors liable to Misconception

Mr. Darwin has laid himself open to much misconception, and has given to his opponents a powerful weapon against himself, by his continual use of metaphor in describing the wonderful co-adaptations of organic beings.

[1] In addition to the laws referred to above, there are of course the fundamental laws and properties of organised

matter and the mysterious powers of Life, which we shall probably never be able to explain, but which must be taken as the basis of all attempts to account for the details of form and structure in organised beings.

"It is curious," says the Duke of Argyll, "to observe the language which this most advanced disciple of pure naturalism instinctively uses, when he has to describe the complicated structure of this curious order of plants (the Orchids). 'Caution in ascribing intentions to nature' does not seem to occur to him as possible. Intention is the one thing which he does see, and which, when he does not see, he seeks for diligently until he finds it. He exhausts every form of words and of illustration, by which intention or mental purpose can be described. 'Contrivance'—'curious contrivance.'—'beautiful contrivance,'—these are expressions which occur over and over again. Here is one sentence describing the parts of a particular species: 'The labellum is developed into a long nectary, *in order to* to attract Lepidoptera, and we shall presently give reason for suspecting that the nectar is *purposely* so lodged that it can be sucked only slowly *in order* to give time for the curious chemical quality of the viscid matter setting hard and dry.'" Many other examples of similar expressions are quoted by the Duke, who maintains that no explanation of these "contrivances" has been or can be given, except on the supposition of a personal contriver, specially arranging the details of each case, although causing them to be produced by the ordinary processes of growth and reproduction.

Now there is a difficulty in this view of the origin of the structure of Orchids which the Duke does not allude to. The majority of flowering plants are fertilised, either without the agency of insects or, when insects are required, without any very important modification of the structure of the flower. It is evident, therefore, that flowers might have been formed as varied, fantastic, and beautiful as the orchids, and yet have been fertilised without more complexity of structure than is found in violets, or clover, or primroses, or a thousand other flowers. The strange springs and traps and pitfalls found in the flowers of orchids cannot be necessary *per se,* since exactly the same end is gained in ten thousand other flowers which do not possess them. Is it not then an extraordinary idea, to imagine the Creator of the universe *contriving* the various complicated parts of these flowers, as a mechanic might contrive an ingenious toy or a difficult puzzle? Is it not a more worth conception that they are some of the results of those general laws which were so co-ordinated at the first introduction of life upon the earth as to result necessarily in the utmost possible development of varied forms?

But let us take one of the simpler cases adduced and see if our general laws are unable to account for it.

A Case of Orchid-structure explained by Natural Selection

There is a Madagascar orchid—the Angræcum sesquipedale—with an immensely long and deep nectary. How did such an extraordinary organ come to be developed? Mr. Darwin's explanation is this. The pollen of this flower can only be removed by the base of the proboscis of some very large moths, when trying to get at the nectar at the bottom of the vessel. The moths with the longest probosces would do this most effectually; they would be rewarded for their long tongues by getting the most nectar; whilst on the other hand, the flowers with the deepest nectaries would be the best fertilised by the largest moths preferring them. Consequently, the deepest nectaried orchids and the longest tongued moths would each confer on the other an advantage in the battle of life. This would tend to their respective perpetuation, and to the constant lengthening of nectaries and probosces. Now let it be remembered that what we have to account for is only the unusual length of this organ. A nectary is found in many orders of plants and is especially common in the orchids, but in this one case only is it sometimes more than a foot long. How did this arise? We begin with the fact, proved experimentally by Mr. Darwin, that moths do visit orchids, do thrust their spiral trunks into the nectaries, and do fertilise them by carrying the pollinia of one flower to the stigma of another. He has further explained the exact mechanism by which this is effected, and the Duke of Argyll admits the accuracy of his observations. In our British species, such as Orchis pyramidalis, it is not necessary that there should be any exact adjustment between the length of the nectary and that of the proboscis of the insect; and thus a number of insects of various sizes are found to carry away the pollinia and aid in the fertilisation. In the Angræcum sesquipedale, however, it is necessary that the proboscis should be forced into a particular part of the flower, and this would only be done by a large moth burying its proboscis to the very base, and straining to drain the nectar from the bottom of the long tube, in which it occupies a depth of one or two inches only. Now let us start from the time when the nectary was only half its present length or about six inches, and was chiefly fertilised by a species of moth which appeared at the time of the plant's flowering, and whose proboscis was of the same length. Among the millions of flowers of the Angræcum produced every year, some would always be shorter than the average, some longer. The former, owing to the structure of the flower, would not get fertilised, because the moths could get all the nectar without forcing their trunks down to the very base. The latter would be well fertilised, and the longest would on the average be the best fertilised of all. By this process alone the average length of the nectary would annually increase, because, the short-nectaried flowers being sterile and the long ones having abundant offspring, exactly the same effect would be produced as if a gardener destroyed the short ones and sowed the seed of the long ones only; and this we know by experience would produce a regular increase of length, since it is this very process which has increased the size and changed the form of our cultivated fruits and flowers.

But this would lead in time to such an increased length of the nectary that many of the moths could only just reach the surface of the nectar, and only the few with exceptionally long trunks be able to suck up a considerable portion.

This would cause many moths to neglect these flowers because they could not get a satisfying supply of nectar, and if these were the only moths in the country the flowers would undoubtedly suffer, and the further growth of the nectary be checked by exactly the same process which had led to its increase. But there are an immense variety of moths, of various lengths of proboscis, and as the nectary became longer, other and larger species would become the fertilisers, and would carry on the process till the largest moths became the sole agents. Now, if not before, the moth would also be affected, for those with the longest probosces would get most food, would be the strongest and most vigorous, would visit and fertilise the greatest number of flowers, and would leave the largest number of descendants. The flowers most completely fertilised by these moths being those which had the longest nectaries, there would in each generation be on the average an increase in the length of the nectaries, and also an average increase in the length of the probosces of the moths; and this would be a *necessary result* from the fact that nature ever fluctuates about a mean, or that in every generation there would be flowers with longer and shorter nectaries, and moths with longer and shorter probosces than the average. No doubt there are a hundred causes that might have checked this process before it had reached the point of development at which we find it. If, for instance, the variation in the quantity of nectar had been at any stage greater than the variation in the length of the nectary, then smaller moths could have reached it and have effected the fertilization. Or if the growth of the probosces of the moths had from other causes increased quicker than that of the nectary, or if the increased length of proboscis had been injurious to them in any way, or if the species of moth with the longest proboscis had become much diminished by some enemy or other unfavourable conditions, then, in any of these cases, the shorter nectaried flowers, which would have attracted and could have been fertilised by the smaller kinds of moths, would have had the advantage. And checks of a similar nature to these no doubt have acted in other parts of the world, and have prevented such an extraordinary development of nectary as has been produced by favourable conditions in Madagascar only, and in one single species of orchid. I may here mention that some of the larger Sphinx moths of the tropics have proboscces nearly as long as the nectary of Angræcum sesquipedale. I have carefully measured the proboscis of a specimen of Macrosila cluentius from South America, in the collection of the British Museum, and find it to be nine inches and a quarter long! One from tropical Africa (Macrosila morganii) is seven inches and a half. A species having a proboscis two or three inches longer could reach the nectar in the largest flowers of Angræcum sesquipedale, whose nectaries vary in length from ten to fourteen inches. That such a moth exists in Madagascar may be safely predicted; and naturalists who visit that island should search for it with as much confidence as astronomers searched for the planet Neptune,—and I venture to predict they will be equally successful!

Now, instead of this beautiful self-acting adjustment, the opposing theory is, that the Creator of the universe, by a direct act of His will, so disposed the natural forces influencing the growth of this one species of plant as to cause its nectary to increase to this enormous length; and at the same time, by an equally special act, determined the flow of nourishment in the organisation of the moth, so as to cause its proboscis to increase in exactly the same proportion, having previously so constructed the Angræcum that it could only be maintained in existence by the agency of this moth. But what proof is given or suggested that this was the mode by which the adjustment took place? None whatever, except a feeling that there is an adjustment of a delicate kind, and an inability to see how known causes could have produced such an adjustment. I believe I have shown, however, that such an adjustment is not only possible but inevitable, unless at some point or other we deny the action of those simple laws which we have already admitted to be but the expressions of existing facts.

Adaptation brought about by General Laws

It is difficult to find anything like parallel cases in inorganic nature, but that of a river may perhaps illustrate the subject in some degree. Let us suppose a person totally ignorant of modern geology to study carefully a great river system. He finds in its lower part a deep broad channel filled to the brim, flowing slowly through a flat country and carrying out to the sea a quantity of fine sediment. Higher up it branches into a number of smaller channels, flowing alternately through flat valleys and between high banks; sometimes he finds a deep rocky bed with perpendicular walls, carrying the water through a chain of hills; where the stream is narrow he finds it deep, where wide shallow. Farther up still, he comes to a mountainous region, with hundreds of streams and rivulets, each with its tributary rills and gullies, collecting the water from every square mile of surface, and every channel adapted to the water that it has to carry. He finds that the bed of every branch and stream and rivulet has a steeper and steeper slope as it approaches it sources, and is thus enabled to carry off the water from heavy rains, and to bear away the stones and pebbles and gravel that would otherwise block up its course. In every part of this system he would see exact adaptation of means to an end. He would say that this system of channels must have been designed, it answers its purpose so effectually. Nothing but a mind could have so exactly adapted the slopes of the channels, their capacity, and frequency, to the nature of the soil and the quantity of the rainfall. Again, he would see special adaptation to the wants of man, in broad, quiet, navigable rivers flowing through fertile plains that support a large population, while the rocky streams and mountain torrents were confined to those sterile regions suitable only for a small population of shepherds and herdsmen. He would listen with incredulity to the geologist who assured him that the adaptation and adjustment he so admired was an inevitable result of the action of general laws; that the rains and rivers, aided by subterranean forces, had modelled the country, had formed the hills and valleys, had scooped out the river beds and levelled the plains; and it would only be after much patient observation and study, after having watched the minute changes produced year by year, and multiplying them by thousands and ten thousands,—after visiting the various regions of the earth and seeing the changes everywhere going on, and the unmistakable signs of greater changes in past times,—that he could be made to understand that the surface of the earth, however beautiful and harmonious it may appear, is strictly due in every detail to the action of forces which are demonstrably self-adjusting.

Moreover, when he had sufficiently extended his inquires, he would find that every evil effect which he would imagine must be the result of non-adjustment does somewhere or other occur, only it is not always evil. Looking on a fertile valley, he would perhaps say: "If the channel of this river were not well adjusted—if for a few miles it sloped the wrong way—the water could not escape, and all this luxuriant valley, full of human beings, would become a waste of waters." Well, there are hundreds of such cases. Every lake is a valley "wasted by water," and in some cases (as the Dead Sea) it is a positive evil, a blot upon the harmony and adaptation of the surface of the earth. Again, he might say—"If rain did not fall here, but the clouds passed over us to some other regions, this verdant and highly cultivated plain would become a desert." And there are such deserts over large portions of the earth, which abundant rains would convert into pleasant dwelling-places for man. Or he might observe some great navigable river, and reflect how easily rocks, or a steeper channel in places, might render it useless to man;—and a little inquiry would show him hundreds of rivers in every part of the world, which are thus rendered useless for navigation.

Exactly the same thing occurs in organic nature. We see some one wonderful case of adjustment, some unusual development of an organ, but we pass over the hundreds of cases in which that adjustment and development do not occur. No doubt when one adjustment is absent another takes it place, because no organism can continue to exist that is not adjusted to its environment; and unceasing variation, with unlimited powers of multiplication, in most cases, furnishes the means of self-adjustment. The world is so constituted that by the action of general laws there is produced the greatest possible variety of surface and of climate; and by the action of laws equally general, the greatest possible variety of organisms has been produced, adapted to the varied conditions of every part of the earth. The objector would probably himself admit that the varied surface of the earth—the plains and valleys, the hills and mountains, the deserts and volcanoes, the winds and currents, the seas and lakes and rivers, and the various climates of the earth—are all the results of general laws acting and reacting during countless ages; and that the Creator does not appear to guide and control the action of these laws—here determining the height of a mountain, there altering the channel of a river—here making the rains more abundant, there changing the direction of a current. He would probably admit that the forces of inorganic nature are self-adjusting, and that the result necessarily fluctuates about a given mean condition (which is itself slowly changing), while within certain limits the greatest possible amount of variety is produced. If then a "contriving mind" is not necessary at every step of the process of change eternally going on in the inorganic world, why are we required to believe in the continual action of such a mind in the region of organic nature? True, the laws at work are more complex, the adjustments more delicate, the appearance of special adaptation more remarkable; but why should we measure the creative mind by our own? Why should we suppose the machine too complicated to have been designed by the Creator so complete that it would necessarily work out harmonious results? The theory of "continual interference" is a limitation of the Creator's power. It assumes that He could not work by pure law in the organic, as He has done in the inorganic world; it assumes that He could not foresee the consequences of the laws of matter and mind combined—that results would continually arise which are contrary to what is best—and that He has to change what would otherwise be the course of nature in order to produce that beauty, and variety, and harmony which even we, with our limited intellects, can conceive to be the result of self-adjustment in a universe governed by unvarying law. If we could not conceive the world of nature to be self-adjusting and capable of endless development, it would even then be an unworthy idea of a Creator to impute the incapacity of our minds to Him; but when many human minds can conceive, and can even trace out in detail, some of the adaptations in nature as the necessary results of unvarying law, it seems strange that, in the interests of religion, any one should seek to prove that the System of Nature, instead of being above, is far below our highest conceptions of it. I, for one, cannot believe that the world would come to chaos if left to law alone. I cannot believe that there is in it no inherent power of developing beauty or variety, and that the direct action of the Deity is required to produce each spot or streak on every insect, each detail of structure in every one of the millions of organisms that live or have lived upon the earth. For it is impossible to draw a line. If any modifications of structure could be the result of law, why not all? If some self-adaptations could arise, why not others? If any varieties of colour, why not all the varieties we see? No attempt is made to explain this, except by reference to the fact that "purpose" and "contrivance" are everywhere visible, and by the illogical deduction that they could only have arisen from the direct action of some mind, because the direct action of our minds produces similar "contrivances"; but it is forgotten that adaptation, however produced, must have the appearance of design. The channel of a river looks as if made *for* the river, although it is made *by* it; the fine layers and beds in a deposit of sand often look as if they had been sorted, and sifted, and levelled designedly; the sides and angles of a crystal exactly resemble similar forms designed by man; but we do not therefore conclude that these effects have, in each individual case, required the directing action of a creative mind, or see any difficulty in the being produced by natural law.

Beauty in Nature

Let us, however, leave this general argument for a while, and turn to another special case, which has been appealed to as conclusive against Mr. Darwin's views. "Beauty" is, to some persons, as great a stumbling-block as "contrivance." They cannot conceive a system of the universe so perfect as necessarily to develop every form of beauty, but suppose that when anything specially beautiful occurs, it is a step beyond what that system could have produced—something which the Creator has added for his own delectation.

Speaking of the humming birds, the Duke of Argyll says: "In the first place it is to be observed of the whole group, that there is no connection which can be traced or conceived, between the splendour of the humming birds and any function essential to their life. If there were any such connection, that splendour could not be confined, as it almost exclusively is, to only one sex. The female birds are, of course, not placed at any disadvantage in the struggle for existence by their more somber colouring." And after describing the various ornaments of these birds, he says: "Mere ornament and variety of form, and these for their own sake, is the only principle or rule with reference to which Creative Power seems to have worked in these wonderful and beautiful birds. … A crest of topaz is no better in the struggle for existence than a crest of sapphire. A frill ending in spangles of the emerald is no better in the battle of life than a frill ending in spangles of the ruby. A tail is not affected for the purposes of flight, whether its marginal or its central feathers are decorated with white. … Mere beauty and mere variety, for their own sake, are objects which we ourselves seek when we can make the forces of nature subordinate to the attainment of them. There seems to be no conceivable reason why we should doubt or question that these are ends and aims also in the forms given to living organisms" (*Reign of Law,* p. 248).

Here the statement that "no connection can be conceived between the splendour of the humming birds and any function essential to their life," is met by the fact that Mr. Darwin has not only conceived but has shown, both by observation and reasoning, how beauty of colour and form may have a direct influence on the most important of all the functions of life, that of reproduction. In the variations to which birds are subject, any more brilliant colour than usual is believed to be attractive to the females, and would therefore lead to the individuals so adorned leaving more than the average number of offspring. There are some indications that this kind of sexual selection does actually take place, and the laws of inheritance would necessarily lead to the further development of any individual peculiarity that was attractive, and thus the splendour of the humming birds is directly connected with their very existence. It is true that "a crest of topaz may be no better than a crest of sapphire," but either of these may be much better than no crest at all; and the different conditions under which the parent form must have existed in different parts of its range will have determined different variations of tint, either of which were advantageous.[1] The reason why female birds are not adorned with equally brilliant plumes is sufficiently clear; they would be injurious by rendering their possessors too conspicuous during incubation. Survival of the fittest has therefore favoured the development of those dark green tints on the upper surface of so many female humming birds, which are most conducive to their protection while the important functions of hatching and rearing the young are being carried on. Keeping in mind the laws of multiplication, variation, and survival of the fittest, which are for ever in action, these varied developments of beauty and harmonious adjustments to conditions are not only conceivable but demonstrable results.

[1] Since writing this essay I have come to the conclusion that mere diversity of colouring between species is an important factor in their differentiation, serving as a means of recognition, and thus preventing crossunions. See *Darwinism,* p. 217. I have also been led to doubt the reality of the fact of female selection of slight differences of colour on which Mr. Darwin relied, but it has not been thought advisable to alter the passages which seem to admit it, as they represent my belief at the time they were written.

The objection I am now combating is solely founded on the supposed analogy of the Creator's mind to ours as regards the love of beauty for its own sake; but if this analogy is to be trusted, then there ought to be no natural objects which are disagreeable or ungraceful in our eyes. And yet it is undoubtedly the fact that there are many such. Just as surely as the horse and deer are beautiful and graceful, the elephant, rhinoceros, hippopotamus, and camel are the reverse. The majority of monkeys and apes are not beautiful; the majority of birds have no beauty of colour; a vast number of insects and reptiles are positively ugly. Now, if the Creator's mind is like ours, whence this ugliness? It is useless to say "that is a mystery we cannot explain," because we have attempted to explain one-half of creation by a method that will not apply to the other half. We know that a man with the highest taste and with unlimited wealth practically does abolish all ungraceful and disagreeable forms and colours form his own domains. If the beauty of creation is to be explained by the Creator's love of beauty, we are bound to ask why He has not banished deformity from the earth, as the wealthy and enlightened man does from his estate and from his dwelling; and if we can get no satisfactory answer, we shall do well to reject the explanation offered. Again, in the case of flowers, which are always especially referred to as the surest evidence of beauty being an end of itself in creation, the whole of the facts are never fairly met. At least half the plants in the world have not bright-coloured or beautiful flowers; and Mr. Darwin has lately arrived at the wonderful generalisation that flowers have become beautiful solely to attract insects to assist in their fertilisation. He adds, "I have come to this conclusion from finding it an invariable rule, that when a flower is fertilised by the wind it never has gaily-coloured corolla." Here is a most wonderful case of beauty being *useful,* When it might be least expected. But much more is proved; for when beauty is of no use to the plant it is not given. It cannot be imagined to do any harm. It is simply not necessary, and is therefore withheld! We ought surely to have been told how this fact is consistent with beauty being "an end in itself," and with the statement of its being given to natural objects "for its own sake."

How New Forms are produced by Variation and Selection

Let us now consider another of the popular objections which the Duke of Argyll thus sets forth:—

"Mr. Darwin does not pretend to have discovered any law or rule, according to which new forms have been born from old forms. He does not hold that outward conditions, however changed, are sufficient to account for them. ... His theory seems to be far better than a mere theory—to be an established scientific truth—in so far as it accounts, in part at least, for the success and establishment and spread of new forms *when they have arisen.* But it does not even suggest the law under which, or by or according to which, such new forms are introduced. Natural Selection can do nothing, except with the materials presented to it hands. It cannot select except among the things open to selection. ... Strictly speaking, therefore, Mr. Darwin's theory is not a theory on the Origin of Species at all, but only a theory on the causes which lead to the relative success or failure of such new forms as may be born into the world" (*Reign of Law,* p. 230).

In this and many other passages in his work the Duke of Argyll sets forth his idea of creation as a "creation by birth," but maintains that each birth of a new form from parents differing from itself has been produced by a special interference of the Creator, in order to direct the process of development into certain channels; that each new species is in fact a "special creation," although brought into existence through the ordinary laws of reproduction. He maintains, therefore, that the laws of multiplication and variation cannot furnish the right kinds of materials at the right times for natural selection to work on. I believe, on the contrary, that it can be logically *proved* from the six axiomatic laws before laid down, that such materials would be furnished; but I prefer to show there are abundance of *facts* which demonstrate that they are furnished.

The experience of all cultivators of plants and breeders of animals shows that, when a sufficient number of individuals are examined, variations of any required kind can always be met with. On this depends the possibility of obtaining breeds, races, and fixed varieties of animals and plants; and it is found that any one form of variation may be accumulated by selection, without materially affecting the other characters of the species; each *seems* to vary in the one required direction only. For example, in turnips, radishes, potatoes, and carrots the root or tuber varies in size, colour, form, and flavour, while the foliage and flowers seem to remain almost stationary; in the cabbage and lettuce, on the contrary, the foliage can be modified into various forms and modes of growth, the root, flower, and fruit remaining little altered; in the cauliflower and broccoli the flower heads vary; in the garden pea the pod only changes. We get innumerable forms of fruit in the apple and pear, while the leaves and flowers remain almost undistinguishable; the same occurs in the gooseberry and garden currant. Directly, however (in the very same genus), we want the flower to vary in the Ribes sanguineum, it does so, although mere cultivation for hundreds of years has not produced marked differences in the flowers of Ribes grossularia. When fashion demands any particular change in the form, or size, or colour of a flower, sufficient variation always occurs in the right direction, as is shown by our roses, auriculas, and geraniums; when, as recently, ornamental leaves come into fashion, sufficient variation is found to meet the demand, and we have zoned pelargoniums and variegated ivy, and it is discovered that a host of our commonest shrubs and herbaceous plants have taken to vary in this direction just when we want them to do so! This rapid variation is not confined to old and well-known plants subjected for a long series of generations to cultivation, but the Sikkim rhododendrons, the fuchsias, and calceolarias from the Andes, and the pelargoniums from the Cape, are equally accommondating, and vary just when and where and how we require them.

Turning to animals we find equally striking examples. It we want any special quality in any animal we have only to breed it in sufficient quantities and watch carefully, and the required variety is *always* found, and can be increased to almost any desired extent. In sheep, we get flesh, fat, and wool; in cows, milk; in horses, colour, strength, size, and speed; in poultry, we have got almost any variety of colour, curious modifications of plumage, and the capacity of perpetual egg-laying. In pigeons we have a still more remarkable proof of the universality of variation, for it has been at one time or another the fancy of breeders to change the form of every part of these birds, and they have never found the required variations absent. The form, size, and shape of bill and feet have been changed to such a degree as is found only in distinct genera of wild birds; the number of tail feathers has been increased, a character which is generally one of the most permanent nature, and is of high importance in the classification of birds; and the size, the colour, and the habits have been also changed to a marvellous extent. In dogs, the degree of modification and the facility with which it is effected is almost equally apparent. Look at the constant amount of variation in opposite directions that must have been going on to develop the poodle and the greyhound from the same original stock! Instincts, habits, intelligence, size, speed, form, and colour have always varied, so as to produce the very races which the wants or fancies or passions of men may have led them to desire. Whether they wanted a bull-dog to torture another animal, a greyhound to catch a hare, or a bloodhound to hunt down their oppressed fellow-creatures, the required variations have always appeared.

Now this great mass of facts, of which a mere sketch

has been here given, are fully accounted for by the "Law of Variation" as laid down at the commencement of this paper. Universal variability—small in amount, but in every direction, ever fluctuating about a mean condition until made to advance in a given direction by "selection," natural or artificial—is the simple basis for the indefinite modification of the forms of life; partial, unbalanced, and consequently unstable modifications being produced by man, while those developed under the unrestrained action of natural laws are at every step self-adjusted to external conditions by the dying out of all unadjusted forms, and are therefore stable and comparatively permanent.[1] To be consistent in their views, our opponents must maintain that every one of the variations that have rendered possible the changes produced by man have been determined at the right time and place by the will of the Creator. Every race produced by the florist or the breeder, the dog or the pigeon fancier, the ratcatcher, the sporting man, or the slave-hunter, must have been provided for by varieties occurring when wanted; and as these variations were never withheld, it would prove that the sanction of an all-wise and all-powerful Being has been given to that which the highest human minds consider to be trivial, mean, or debasing.

[1] That the variations occurring among wild animals are ample both in number and amount is proved in *Darwinism,* chap. Iii.

This appears to be a complete answer to the theory that variation sufficient in amount to be accumulated in a given direction must be the direct act of the creative mind, but it is also sufficiently condemned by being so entirely unnecessary. The facility with which man obtains new races depends chiefly upon the number of individuals he can procure to select from. When hundreds of florists or breeders are all aiming at the same object, the work of change goes on rapidly. But a common species in nature contains a thousand or a million-fold more individuals than any domestic race; and survival of the fittest must unerringly preserve all that vary in the right direction, not only in obvious characters but in minute details—not only in external but in internal organs; so that if the materials are sufficient for the needs of man, there can be no want of them to fulfil the grand purpose of keeping up a supply of modified organisms, exactly adapted to the changed conditions that are always occurring in the inorganic world.

The Objection that there are Limits to Variation

Having now, I believe, fairly answered the chief objections of the Duke of Argyll, I proceed to notice one or two of those adduced in an able and argumentative essay on the "Origin of Species" in the *North British Review* for July 1867. The writer first attempts to prove that there are strict limits to variation. When we begin to select variations in any one direction, the process is comparatively rapid, but after a considerable amount of change has been effected it becomes slower and slower, till at length its limits are reached and no care in breeding and selection can produce any further advance. The racehorse is chosen as an example. It is admitted that, with any ordinary lot of horses to being with, careful selection would in a few years make a great improvement, and in a comparatively short time the standard of our best racers might be reached. But that standard has not for many years been materially raised, although unlimited wealth and energy are expended in the attempt. This is held to prove that there are definite limits to variation in any special direction, and that we have no reason to suppose that mere time, and the selective process being carried on by natural law, could make any material difference. But the writer does not perceive that this argument fails to meet the real question, which is, not whether indefinite and unlimited change in any or all directions is possible, but whether such differences as do occur in nature could have been produced by the accumulation of variations by selection. In the matter of speed, a limit of a definite kind as regards land animals does exist in nature. All the swiftest animals—deer, antelopes, hares, foxes, lions, leopards, horses, zebras, and many others—have reached very nearly the same degree of speed. Although the swiftest of each must have been for ages preserved, and the slowest must have perished, we have no reason to believe there is any advance of speed. The possible limit under existing conditions, and perhaps under possible terrestrial conditions, has been long ago reached. In cases, however, where this limit had not been so nearly reached as in the horse, we have been enabled to make a more marked advance and to produce a greater difference of form. The wild dog is an animal that hunts much in company, and trusts more to endurance than to speed. Man has produced the greyhound, which differs much more from the wolf or the dingo than the racer does from the wild Arabian. Domestic dogs, again, have varied more in size and in form than the whole family of Canidæ in a state of nature. No wild dog, fox, or wolf is either so small as some of the smallest terriers and spaniels, or so large as the largest varieties of hound or Newfoundland dog. And, certainly, no two wild animals of the family differ so widely in form and proportions as the Chinese pug and the Italian greyhound, or the bulldog and the common greyhound. The known range of variation is, therefore, more than enough for the derivation of all the forms of dogs, wolves, and foxes from a common ancestor.

Again, it is objected that the pouter or the fan-tail pigeon cannot be further developed in the same direction. Variation seems to have reached its limits in these birds. But so it has in nature. The fantail has not only more tail feathers than any of the three hundred and sixty existing species of pigeons, but more than any of the ten thousand known species of birds. There is, of course, some limit to the number of feathers of which a tail useful for flight can consist, and in the fantail we have probably reached that limit. Many birds have the œsophagus or the skin of the neck more or less dilatable, but in no known bird is it so dilatable as in the pouter pigeon. Here again the possible limit, compatible with a healthy existence, has probably been reached. In like manner the difference in the size and form of the beak in the various breeds of the domestic pigeon is greater than that between the extreme forms of beak in the various genera and sub-families of the whole pigeon tribe. Form these facts, and many others of the same nature, we may fairly infer that if rigid selection were applied to any organ, we could in a comparatively short time produce a much greater amount of change than that which occurs between species and species in a state of nature, since the differences which we do produce are often comparable with those which exist between distinct genera or distinct families. The facts adduced by the writer of the article referred to, of the definite limits to variability in certain directions in domesticated animals, are, therefore, no objection whatever to the view that all the modifications which exist in nature have been produced by the accumulation, by natural selection, of small and useful variations, since those very modifications have equally definite and very similar limits.

Objection to the Argument from Classification

To another of this writer's objections—that by Professor Thomson's calculations the sun can only have existed in a solid state 500,000,000 of years, and that therefore *time* would not suffice for the slow process of development of all living organisms—it is hardly necessary to reply, as it cannot be seriously contended, even if this calculation has claims to approximate accuracy, that the process of change and development may not have been sufficiently rapid to have occurred within that period. His objection to the classification argument is, however, more plausible. The uncertainty of opinion among naturalists as to which are species and which varieties, is one of Mr. Darwin's very strong arguments that these two names cannot belong to things quite distinct in nature and origin. The reviewer says that this argument is of no weight, because the works of man present exactly the same phenomena; and he instances patent inventions, and the excessive difficulty of determining whether they are new or old. I accept the analogy, though it is a very imperfect one, and maintain that, such as it is, it is all in favour of Mr. Darwin's views. For are not all inventions of the same kind directly affiliated to a common ancestor? Are not improved steam-engines or clocks the lineal descendants of some existing steam-engine or clock? Is there ever a new creation in art or science any more than in nature? Did ever patentee absolutely originate any complete and entire invention, no portion of which was derived from anything that had been made or described before? It is therefore clear that the difficulty of distinguishing the various classes of inventions which claim to be new, is of the same nature as the difficulty of distinguishing varieties and species, because neither are absolutely new creations, but both are alike descendants of pre-existing forms, from which and from each other they differ by varying and often imperceptible degrees. It appears, then, that however plausible this writer's objections may seem, whenever he descends from generalities to any specific statement, his supposed difficulties turn out to be in reality strongly confirmatory of Mr. Darwin's view.

The **TIMES** *on Natural Selection*

The extraordinary misconception of the whole subject by popular writers and reviewers is well shown by an article which appeared in the *Times* newspaper on "The Reign of Law." Alluding to the supposed economy of nature, in the adaptation of each species to its own place and its special use, the reviewer remarks: "To this universal law of the greatest economy, the law of natural selection stands in direct antagonism as the law of 'greatest possible waste' of time and of creative power. To conceive a duck with webbed feet and a spoon-shaped bill, living by suction, to pass naturally into a gull with webbed feet and a knife-like bill, living on flesh, in the longest possible time and in the most laborious possible way, we may conceive it to pass from the one to the other state by natural selection. The battle of life the ducks will have to fight will increase in peril continually as they cease (with the change of their bill) to be ducks, and attain a *maximum* of danger in the condition in which they begin to be gulls; and ages must elapse and whole generations must perish, and countless generations of the one species be created and sacrificed, to arrive at one single pair of the other."

In this passage the theory of natural selection is so absurdly misrepresented that it would be amusing, did we not consider the misleading effect likely to be produced by this kind of teaching in so popular a journal. It is assumed that the duck and the gull are essential parts of nature, each well fitted for its place, and that if one had been produced from the other by a gradual metamorphosis, the intermediate forms would have been useless, unmeaning, and unfitted for any place in the system of the universe. Now, this idea can only exist in a mind ignorant of the very foundation and essence of the theory of natural selection, which is, the preservation of *useful* variations only, or, as has been well expressed, in other words, the "survival of the fittest." Every intermediate form which could possibly have arisen during the transition from the duck to the gull, so far from having an unusually severe battle to fight for existence, or incurring any "*maximum* of danger," would necessarily have been as accurately adjusted to the rest of nature, and as well fitted to maintain and to enjoy its existence, as the duck or the gull actually are. If it were not so, it never could have been produced under the law of natural selection.

Intermediate or generalised Forms of extinct Animals, an indication of Transmutation or Development

The misconception of this writer illustrates another point very frequently overlooked. It is an essential part of Mr. Darwin's theory that one existing animal has not been derived from any other existing animal, but that both are the descendants of a common ancestor, which was at once different from either, but, in essential characters, to some extent intermediate between them both. The illustration of the duck and the gull is therefore misleading; one of these birds has not been derived from the other, but both from a common ancestor. This is not a mere supposition invented to support the theory of natural selection, but is founded on a variety of indisputable facts. As we go back into past time, and meet with the fossil remains of more and more ancient races of extinct animals, we find that many of them actually are intermediate between distinct groups of existing animals. Professor Owen continually dwells on this fact: he says in his *Palætology*, p. 284: "A more generalised vertebrate structure is illustrated, in the extinct reptiles, by the affinities to ganoid fishes, shown by Ganocephala, Labyrinthodontia, and Ichthyopterygia; by the affinities of the Pterosauria to birds, and by the approximation of the Dinosauria to mammals. (These have been recently shown by Professor Huxley to have more affinity to birds.) It is manifested by the combination of modern crocodilian, chelonian, and lacertian characters in the Cryptodontia and the Dicynodontia, and by the combined lacertian and crocodilian characters in the Thecodontia and Sauropterygia." In the same work he tells us that "the Anoplotherium, in several important characters, resembled the embryo Ruminant, but retained throughout life those marks of adhesion to a generalised mammalian type;" and assures us that he has "never omitted a proper opportunity for impressing the results of observations showing the more generalised structures of extinct as compared with the more specialised forms of recent animals." Modern palæontologists have discovered hundreds examples of these more generalised or ancestral types. In the time of

Cuvier, the Ruminants and the Pachyderms were looked upon as two of the most distinct orders of animals: but it is now demonstrated that there once existed a variety of general and species, connecting by almost imperceptible grades such widely different animals as the pig and the camel. Among living quadrupeds we can scarcely find a more isolated group than the genus Equus, comprising the horses, asses, and zebras; but through many species of Paloplotherium, Hippotherium, and Hipparion, and number of extinct forms of Equus found in Europe, India, and America, an almost complete transition is established with the Eocene Anoplotherium and Paleotherium, which are also generalised or ancestral types of the tapir and rhinoceros. The recent researches of M. Gaudry in Greece have furnished much new evidence of the same character. In the Miocene (or Pliocene) beds of Pikermi he has discovered the group of the Simocyonidæ intermediate between bears and wolves; the genus Hyænictis which connects the hyænas with the civets; the Ancylotherium, which is allied both to the extinct mastodon and to the living pangolin or scaly ant-eater; and the Helladotherium, which connects the now isolated giraffe with the deer and antelops.

Between reptiles and fishes an intermediate type has been found in the Archegosaurus of the Coal formation; while the Labyrinthodon of the Trias combined characters of the Batrachia with those of crocodiles, lizards, and ganoid fishes. Even birds, the most apparently isolated of all living forms, and the most rarely preserved in a fossil state, have been shown to possess undoubted affinities with reptiles; and in the Oolitic Archæopteryx, with its lengthened tail, feathered on each side, we have one of the connecting links from the side of birds; while Professor Huxley has recently shown that the entire order of Dinosaurians have remarkable affinities to birds, and that one of them, the Compsognathus, makes a nearer approach to bird organisation than does Archæopteryx to that of reptiles.

Analogous facts to these occur in other classes of animals, as an example of which we have the authority of a distinguished paleontologist, M. Barande, quoted by Mr. Darwin, for the statement that although the Palæozoic Invertebrata can certainly be classed under existing groups, yet at this ancient period the groups were not so distinctly separated from each other as they are now; while Mr. Scudder tells us that some of the fossil insects discovered in the Coal formation of America offer characters intermediate between those of existing orders. Agassiz, again, insists strongly that the more ancient animals resemble the embryonic forms of existing species; but as the embryos of distinct groups are known to resemble each other more than the adult animals (and in fact to be undistinguishable at a very early age), this is the same as saying that the ancient animals are exactly what, on Darwin's theory, the ancestors of existing animals ought to be; and this, it must be remembered, is the evidence of one of the strongest opponents of the theory of natural selection.

Conclusion

I have thus endeavoured to meet fairly, and to answer plainly, a few of the most common objections to the theory of natural selection, and I have done so in every case by referring to admitted facts and to logical deductions from those facts.

As an indication and general summary of the line of argument I have adopted, I here give a brief demonstration in a tabular form of the Origin of Species by means of Natural Selection, referring for the *facts* to Mr. Darwin's works, and to the pages in this volume, where they are more or less fully treated.

A Demonstration of the Origin of Species by Natural Selection

PROVED FACTS	*NECESSARY CONSEQUENCES (afterwards taken as Proved Facts)*

RAPID INCREASE OF ORGANISMS, pp. 23, 142 (*Origin of Species,* p. 75, 5th ed.) TOTAL NUMBER OF INDIVIDUALS STATIONARY, p. 23. STRUGGLE FOR EXISTENCE.	SURVIVIAL OF THE FITTEST, or Natural Selection; meaning simply, that on the whole those die who are least fitted to maintain their existence (*Origin of Species,* chap. iv.)
HEREDITY WITH VARIATION, or general likeness with individual differences of parents and off-springs, pp. 142, 156, 179 (*Origin of Species,* chaps. i. ii. v.)	SURVIVIAL OF THE FITTEST, or Natural Selection; meaning simply, that on the whole those die who are least fitted to maintain their existence (*Origin of Species,* chap. iv.)
SURVIVAL OF THE FITTEST. CHANGE OF EXTERNAL CONDITIONS, universal and unceasing.—See Lyell's *Principles of Geology.*	CHANGES OF ORGANIC FORMS, to keep them in harmony with the Changed Conditions; and as the changes of conditions are permanent changes, in the sense of not reverting back to identical previous conditions, the changes of organic forms must be in the same sense permanent, and thus originate SPECIES.

VIII

THE DEVELOPMENT OF HUMAN RACES UNDER THE LAW OF NATURAL SELECTION[1]

AMONG the most advanced students of man there exists a wide difference of opinion on some of the most vital questions respecting his nature and origin. Anthropologists are now, indeed, pretty well agreed that man is not a recent introduction into the earth. All who have studied the question now admit that his antiquity is very great; and that, though we have to some extent ascertained the minimum of time during which he *must* have existed, we have made no approximation towards determining that far greater period during which he *may* have, and probably *has* existed. We can with tolerable certainty affirm that man must have inhabited the earth a thousand centuries ago, but we cannot assert that he positively did not exist, or that there is any good evidence against his having existed, for a period of ten thousand centuries. We know positively that he was contemporaneous with many now extinct animals, and has survived changes of the earth's surface fifty or a hundred times greater than any that have occurred during the historical period; but we cannot place any definite limit to the number of species he may have outlived, or to the amount of terrestrial change he may have witnessed.

[1] First published in the Anthropological Review, May 1864; reprinted in *contributions,* etc., with some alterations and additions.

Wide differences of opinion as to Man's Origin

But while on this question of man's antiquity there is a very general agreement,—and all are waiting eagerly for fresh evidence to clear up those points which all admit to be full of doubt,—on other and not less obscure and difficult questions a considerable amount of dogmatism is exhibited; doctrines are put forward as established truths, no doubt or hesitation is admitted, and it seems to be supposed that no further evidence is required, or that any new facts can modify our convictions. This is especially the case when we inquire,—Are the various forms under which man now exists primitive, or derived from pre-existing forms; in other words, is man of one or many species? To this question we immediately obtain distinct answers diametrically opposed to each other: the one party positively maintaining that man is a *species* and is essentially *one*—that all differences are but local and temporary variations, produced by the different physical and moral conditions by which he is surrounded; the other party maintaining with equal confidence that man is a genus of *many species,* each of which is practically unchangeable, and has ever been as distinct, or even more distinct, than we now behold them. This difference of opinion is somewhat remarkable, when we consider that both parties are well acquainted with the subject; both use the same vast accumulations of facts; both reject those early traditions of mankind which profess to give an account of his origin; and both declare that they are seeking fearlessly after truth alone; yet each will persist in looking only at the portion of truth on his own side of the question, and at the error which is mingled with his opponent's doctrine. It is my wish to show how the two opposing views can be combined, so as to eliminate the error and retain the truth in each, and it is by means of Mr. Darwin's celebrated theory of Natural Selection that I hope to do this, and thus to harmonise the conflicting theories of modern anthropologists.

Let us first see what each party has to say for itself. In favour of the unity of mankind it is argued that there are no races without transitions to others; that every race exhibits within itself variations of colour, of hair, of feature, and of form, to such a degree as to bridge over, to a large extent, the gap that separates it from other races. It is asserted that no race is homogeneous; that there is a tendency to vary; that climate, food, and habits produce, and render permanent, physical peculiarities, which, though slight in the limited periods allowed to our observation, would, in the long ages during which the human race has existed, have sufficed to produce all the differences that now appear. It is further asserted that the advocates of the opposite theory do not agree among themselves; that some would make three, some five, some fifty or a hundred and fifty species of man; some would have had each species created in pairs, while others require nations to have at once sprung into existence, and that there is no stability or consistency in any doctrine but that of one primitive stock.

The advocates of the original diversity of man, on the other hand, have much to say for themselves. They argue that proofs of change in man have never been brought forward except to the most trifling amount, while evidence of his permanence meets us everywhere. The Portuguese and Spaniards, settled for two or three centuries in South America, retain their chief physical, mental, and moral characteristics; the Dutch boers at the Cape, and the descendants of the early Dutch settlers in the Moluccas, have not lost the features or the colour of the Germanic races; the Jews, scattered over the world in the most diverse climates, retain the same characteristic lineaments everywhere; the Egyptian sculptures and paintings show us that, for at least 4000 or 5000 years, the strongly contrasted features of the Negro and the Semitic races have remained altogether unchanged; while more recent discoveries prove that the mound-builders of the Mississippi valley, and the dwellers on Brazilian mountains, had, even in the very infancy of the human race, some traces of the same peculiar and characteristic type of cranial formation that now distinguishes them.

If we endeavour to decide impartially on the merits of this difficult controversy, judging solely by the evidence that each party has brought forward, it certainly seems that the best of the argument is on the side of those who maintain the primitive diversity of man. Their opponents have not been able to refute the permanence of existing races as far back as we can trace them, and have failed to show, in a single case, that at any former epoch the well marked varieties of mankind approximated more closely than they do at the present day. At the same time this is but negative evidence. A condition of immobility for four or five thousand years does not preclude an advance at an earlier epoch, and—if we can show that there are causes in nature which would check any further physical change when certain conditions were fulfilled—does not even render such an advance improbable, if there are any general arguments to be adduced in its favour. Such a cause, I believe, does exist; and I shall now endeavour to point out its nature and its mode of operation.

Outline of the Theory of Natural Selection

In order to make my argument intelligible, it is necessary for me to explain very briefly the theory of natural selection promulgated by Mr. Darwin, and the power which it possesses of modifying the forms of animals and plants. The grand feature in the multiplication of organic life is, that close general resemblance is combined with more or less individual variation. The child resembles its parents or ancestors more or less closely in all its peculiarities, deformities, or beauties; it resembles them in general more than it does any other individuals; yet children of the same parents are not all alike, and it often happens that they differ very considerably from their parents and from each other. This is equally true of man, of all animals, and of all plants. Moreover, it is found that individuals do not differ from their parents in certain particulars only, while in all others they are exact duplicates of them. They differ from them and from each other in every particular: in form, in size, in colour; in the structure of internal as well as of external organs; in those subtle peculiarities which produce differences of constitution, as well as in those still more subtle ones which lead to modifications of mind and character. In other words, in every possible way, in every organ, and in every function, individuals of the same stock vary.

Now, health, strength, and long life are the results of

a harmony between the individual and the universe that surrounds it. Let us suppose that at any given moment this harmony is perfect. A certain animal is exactly fitted to secure its prey, to escape from its enemies, to resist the inclemencies of the seasons, and to rear a numerous and healthy offspring. But a change now takes place. A series of cold winters, for instance, come on, making food scarce, and bringing an immigration of some other animals to compete with the former inhabitants of the district. The new immigrant is swift of foot, and surpasses its rivals in the pursuit of game; the winter nights are colder, and require a thicker fur as a protection, and more nourishing food to keep up the heat of the system. Our supposed perfect animal is no longer in harmony with its universe; it is in danger of dying of cold or of starvation. But the animal varies in its offspring. Some of these are swifter than others—they still manage to catch food enough; some are hardier and more thickly furred—they manage in the cold nights to keep warm enough; the slow, the weak, and the thinly clad soon die off. Again and again, in each succeeding generation, the same thing takes place. By this natural process, which is so inevitable that it cannot be conceived not to act, those best adapted to live, live; those least adapted, die. It is sometimes said that we have no direct evidence of the action of this selecting power in nature. But it seems to me we have better evidence than even direct observation would be, because it is more universal, viz., the evidence of necessity. It must be so; for, as all wild animals increase in a geometrical ratio, while their actual numbers remain on the average stationary, it follows that as many die annually as are born. If, therefore, we deny natural selection, it can only be by asserting that, in such a case as I have supposed, the strong, the healthy, the swift, the well clad, the well organised animals in every respect, have no advantage over—do not on the average live longer than—the weak, the unhealthy, the slow, the ill-clad, and the imperfectly organised individuals; and this no sane man has yet been found hardy enough to assert. But this is not all; for the offspring on the average resemble their parents, and the selected portion of each succeeding generation will therefore be stronger, swifter, and more thickly furred than the last; and if this process goes on for thousands of generations, our animal will have again become thoroughly in harmony with the new conditions in which it is placed. But it will now be a different creature. It will be not only swifter and stronger, and more furry—it will also probably have changed in colour, in form, perhaps have acquired a longer tail, or differently shaped ears; for it is an ascertained fact that when one part of an animal is modified, some other parts almost always change, as it were in sympathy, with it. Mr. Darwin calls this "correlation of growth," and gives as instances that hairless dogs have imperfect teeth; white cats, when blue-eyed, are deaf; small feet accompany short beaks in pigeons; and other equally interesting cases.

Grant, therefore, the premises: 1st, That peculiarities of every kind are more or less hereditary; 2d, That the offspring of every animal vary more or less in all parts of their organisation; 3d, That the universe in which these animals live is not absolutely invariable;—none of which propositions can be denied; and then consider that the animals in any country (those at least which are not dying out) must at each successive period be brought into harmony with the surrounding conditions; and we have all the elements for a change of form and structure in the animals, keeping exact pace with changes of whatever nature in the surrounding universe. Such changes must be slow, for the changes in the universe are very slow; but just as these slow changes become important, when we look at results after long periods of action,—as we do when we perceive the alterations of the earth's surface during geological epochs,—so the parallel changes in animal form become more and more striking, in proportion as the time they have been going on is great; as we see when we compare our living animals with those which we disentomb from each successively older geological formation.

This, is briefly, the theory of natural selection, which explains the changes in the organic world as being parallel with, and in part dependent on, those in the inorganic. What we now have to inquire is, Can this theory be applied in any way to the question of the origin of the races of man? or is there anything in human nature that takes him out of the category of those organic existences over whose successive mutations it has had such powerful sway?

Different Effects of Natural Selection on Animals and on Man

In order to answer these questions, we must consider why it is that natural selection acts so powerfully upon animals, and we shall, I believe, find that its effect depends mainly upon their self-dependence and individual isolation. A slight injury, a temporary illness, will often end in death, because it leaves the individual powerless against its enemies. If an herbivorous animal is a little sick and has not fed well for a day or two, and the herd is then pursued by a beast of prey, our poor invalid inevitably falls a victim. So, in carnivorous animal, the least deficiency of vigour prevents its capturing food, and it soon dies of starvation. There is, as a general rule, no mutual assistance between adults, which enables them to tide over a period of sickness. Neither is there any division of labour; each must fulfil *all* the conditions of its existence, and, therefore, natural selection keeps all up to a pretty uniform standard.

But in man, as we now behold him, this is different.

He is social and sympathetic. In the rudest tribes the sick are assisted, at least with food; less robust health and vigour than the average does not entail death. Neither does the want of perfect limbs or other organs produce the same effects as among animals. Some division of labour takes place; the swiftest hunt, the less active fish, or gather fruits; food is, to some extent, exchanged or divided. The action of natural selection is therefore checked; the weaker, the dwarfish, those of less active limbs, or less piercing eyesight, do not suffer the extreme penalty which falls upon animals so defective.

In proportion as these physical characteristics became of less importance, mental and moral qualities will have increasing influence on the well-being of the race. Capacity for acting in concert for protection, and for the acquisition of food and shelter; sympathy, which leads all in turn to assist each other; the sense of right, which checks depredations upon our fellows; the smaller development of the combative and destructive propensities; self-restraint in present appetites, and that intelligent foresight which prepares for the future, are all qualities that from their earliest appearance must have been for the benefit of each community, and would, therefore, have become the subjects of natural selection. For it is evident that such qualities would be for the well-being of man, would guard him against external enemies, against internal dissensions, and against the effects of inclement seasons and impending famine, more surely than could any merely physical modification. Tribes in which such mental and moral qualities were predominant would therefore have an advantage in the struggle for existence over other tribes in which they were less developed—would live and maintain their numbers, while the others would decrease and finally succumb.

Again, when any slow changes of physical geography or of climate make it necessary for an animal to alter its food, its clothing, or its weapons, it can only do so by the occurrence of a corresponding change in its own bodily structure and internal organisation. If a larger or more powerful beast is to be captured and devoured, as when a carnivorous animal which has hitherto preyed on antelopes is obliged from their decreasing numbers to attack buffaloes, it is only the strongest who can hold,—those with most powerful claws and formidable canine teeth that can struggle with and overcome such an animal. Natural selection immediately comes into play, and by its action these organs gradually become adapted to their new requirements. But man, under similar circumstances, does not require longer nails or teeth, greater bodily strength or swiftness. He makes sharper spears, or a better bow, or he constructs a cunning pitfall, or combines in a hunting party to circumvent his new prey. The capacities which enable him to do this are what he requires to be strengthened, and these will, therefore, be gradually modified by natural selection, while the form and structure of his body will remain unchanged. So, when a glacial epoch comes on, some animals must acquire warmer fur, or a covering of fat, or else die of cold. Those best clothed by nature are, therefore, preserved by natural selection. Man, under the same circumstances, will make himself warmer clothing, and build better houses, and the necessity of doing this will react upon his mental organisation and social condition—will advance them while his natural body remains naked as before.

When the accustomed food of some animal becomes scarce or totally fails, it can only exist by becoming adapted to a new kind of food, a food perhaps less nourishing and less digestible. Natural selection will now act upon the stomach and intestines, and all their individual variations will be taken advantage of, to modify the race into harmony with its new food. In many cases, however, it is probable that this cannot be done. The internal organs may not vary quick enough, and then the animal will decrease in numbers and finally become extinct. But man guards himself from such accidents by superintending and guiding the operations of nature. He plants the seed of his most agreeable food, and thus procures a supply, independent of the accidents of varying seasons or natural extinction. He domesticates animals, which serve him either to capture food or for food itself, and thus changes of any great extent in his teeth or digestive organs are rendered unnecessary. Man, too, has everywhere the use of fire, and by its means can render palatable a variety of animal and vegetable substances, which he could hardly otherwise make use of, and thus obtains for himself a supply of food far more varied and abundant than that which any animal can command.

Thus man, by the mere capacity of clothing himself, and making weapons and tools, has taken away from nature that power of slowly but permanently changing the external form and structure in accordance with changes in the external world, which she exercises over all other animals. As the competing races by which they are surrounded—the climate, the vegetation, or the animals which serve them for food—are slowly changing, they must undergo a corresponding change in their structure, habits, and constitution to keep them in harmony with the new conditions—to enable them to live and maintain their numbers. But man does this by means of his intellect alone, the variations of which enable him, with an unchanged body, still to keep in harmony with the changing universe.

There is one point, however, in which nature will still act upon him as it does on animals, and, to some extent, modify his external characters. Mr. Darwin has shown that the colour of the skin is correlated with constitutional peculiarities both in vegetables and animals, so that liability to certain diseases or freedom from them is often accompanied by marked external characters. Now, there is ever reason to believe that this had acted, and, so some extent, may still continue to act on man. In localities where certain diseases are prevalent, those individuals of savage races which were subject to them would rapidly die off, while those who were constitutionally free from the disease would survive, and become the progenitors of a new race. These favoured individuals would probably be distinguished by peculiarities of *colour,* with which again peculiarities in the texture or the abundance of *hair* seem to be correlated, and thus may have been brought about those racial differences of colour which seem to have little those relation to mere temperature or other obvious peculiarities of climate.

From the time, therefore, when the social and sympathetic feelings came into active operation, and the intellectual and moral faculties became fairly developed, man would cease to be influenced by natural selection in his physical from and structure. As an animal he would remain almost stationary, the changes of the surrounding universe ceasing to produce in him that powerful modifying effect which they exercise over other parts of the organic world. But from the moment that the form of his body became stationary, his mind would become subject to those very influences from which his body had escaped; every slight variation in his mental and moral nature which should enable him better to guard against adverse circumstances, and combine for mutual comfort and protection, would be preserved and accumulated; the better and higher specimens of our race would therefore increase and spread, the lower and more brutal would give way and successively die out, and that rapid advancement of mental orangisation would occur which has raised the very lowest races of man so far above the brutes (although differing so little from some of them in physical structure), and, in conjunction with scarcely perceptible modifications of form, has developed the wonderful intellect of the European races.

Influence of external Nature in the development of the Human Mind

But from the time when this mental and moral advance commenced, and man's physical character became fixed and almost immutable, a new series of causes would come into action and take part in his mental growth. The diverse aspects of nature would now make themselves felt, and profoundly influence the character of the primitive man.

When the power that had hitherto modified the body had its action transferred to the mind, then races would advance and become improved, merely by the harsh discipline of a sterile soil and inclement seasons. Under their influence a hardier, a more provident, and a more social race would be developed than in those regions where the earth produces a perennial supply of vegetable food, and where neither foresight nor ingenuity are required to prepare for the rigours of winter. And is it not the fact that in all ages, and in every quarter of the globe, the inhabitants of temperate have been superior to those of hotter countries? All the great invasions and displacements of races have been from North to South, rather than the reverse; and we have no record of there ever having existed, any more than there exists to-day, a solitary instance of an indigenous inter-tropical civilisation. The Mexican civilisation and government came from the North, and, as well as the Peruvian, was established, not in the rich tropical plains, but on the lofty and sterile plateaux of the Andes. The religion and civilisation of Ceylon were introduced from North India; the successive conquerors of the Indian peninsula came from the North-west; the northern Mongols conquered the more Southern Chinese; and it was the bold and adventurous tribes of the North that overran and infused new life into Southern Europe.

Extinction of Lower Races

It is the same great law of "the preservation of favoured races in the struggle for life," which leads to the inevitable extinction of all those low and mentally undeveloped populations with which Europeans come in contact. The red Indian in North America and in Brazil; the Tasmanian, Australian, and New Zealander in the southern hemisphere, die out, not from any one special cause, but from the inevitable effects of an unequal mental and physical struggle. The intellectual and moral, as well as the physical, qualities of the European are superior; the same powers and capacities which have made him rise in a few centuries from the condition of the wandering savage, with a scanty and stationary population, to his present state of culture and advancement, with a greater average longevity, a greater average strength, and a capacity of more rapid increase,—enable him when in contact with the savage man to conquer in the struggle for existence, and to increase at his expense, just as the better adapted increase at the expense of the less adapted varieties in the animal and vegetable kingdoms—just as the weeds of Europe overrun North America and Australia, extinguishing native productions by the inherent vigour of their organisation, and by their greater capacity for existence and multiplication.

The Origin of the Races of Man

If these view are correct,—if in proportion as man's social, moral, and intellectual faculties became developed, his physical structure would cease to be affected by the operation of natural selection,—we have a most important clue to the origin of races. For it will follow that those great modifications of structure and of external form, which resulted in the development of man out of some lower type of animal, must have occurred before his intellect had raised him above the condition of the brutes, at a period when he was gregarious, but scarcely social, with a mind perceptive but not reflective, era any sense of *right* or feelings of *sympathy* had been developed in him. He would be still subject, like the rest of the organic world, to the action of natural selection, which would retain his physical form and constitution in harmony with the surrounding universe. He was probably at a very early period a dominant race, spreading widely over the warmer regions of the earth as it then existed, and in agreement with what we see in the case of other dominant species, gradually becoming modified in accordance with local conditions. As he ranged farther from his original home, and became exposed to greater extremes of climate, to greater changes of food, and had to contend with new enemies, organic and inorganic, slight useful variations in his constitution would be selected and rendered permanent, and would, on the principle of "correlation of growth," be accompanied by corresponding external physical changes. Thus might have arisen those striking characteristics and special modifications which still distinguish the chief races of mankind. The red, black, yellow, or blushing white skin; the straight, the curly, the woolly hair; the scanty or abundant beard; the straight or oblique eyes; the various forms of the pelvis, the cranium, and other parts of the skeleton.

But while these changed had been going on, his mental development had, from some unknown cause, greatly advanced, and had now reached that condition in which it began powerfully to influence his whole existence, and would therefore become subject to the irresistible action of natural selection. This action would quickly give the ascendency to mind: speech would probably now be first developed, leading to a still further advance of the mental faculties; and from that moment man, as regards the form and structure of most parts of his body, would remain almost stationary. The are of making weapons, division of labour, anticipation of the future, restraint of the appetites, moral, social, and sympathetic feelings, would now have a preponderating influence on his well-being, and would therefore be that part of his nature on which natural selection would most powerfully act; and we should thus have explained that wonderful persistence of mere physical characteristics which is the stumbling-block of those who advocate the unity of mankind.

We are now, therefore, enabled to harmonise the conflicting views of anthropologists on this subject. Man may have been—indeed I believe must have been—once a homogeneous race; but it was at a period of which we have as yet discovered no remains—at a period so remote in his history that he had not yet acquired that wonderfully developed brain, the organ of the mind, which now, even in his lowest examples, raises him far above the highest brutes—at a period when he had the form but hardly the nature of man, when he neither possessed human speech, nor those sympathetic and moral feelings which in a greater or less degree everywhere now distinguish the race. Just in proportion as these truly human faculties became developed in him would his physical features become fixed and permanent, because the latter would be of less importance to his well-being; he would be kept in harmony with the slowly changing universe around him, by an advance in mind rather than by a change in body. If, therefore, we are of opinion that he was not really man till these higher faculties were fully developed, we may fairly assert that there were many originally distinct races of men; while, if we think that a being closely resembling us in form and structure, but with mental faculties scarcely raised above the brute, must still be considered to have been human, we are fully entitled to maintain the common origin of all mankind.

The Bearing of these Views on the Antiquity of Man

These considerations, it will be seen, enable us to place the origin of man at a much more remote geological epoch than has yet been thought possible. He may even have lived in the Miocene of Eocene period, when not a single other mammal was identical in form with any existing species. For, in the long series of ages during which these primeval animals were being slowly changed into the species which now inhabit the earth, the power which acted to modify them would only affect the mental organisation of man. His brain alone would have increased in size and complexity, and his cranium have undergone corresponding changes of form, while the whole structure of lower animals was being changed. This will enable us to understand how the fossil crania of Denise and Engis agree so closely with existing forms, although they undoubtedly existed in company with large mammalia now extinct. The Neanderthal skull may be a specimen of one of the lowest races then existing, just as the Australians are the lowest of our modern epoch. We have no reason to suppose that mind and brain and skull modification could go on quicker than that of the other parts of the organisation; and we must therefore look back very far in the past to find man in that early condition in which his mind was not sufficiently developed, to remove his body from the modifying influence of external conditions and the cumulative action of natural selection. I believe, therefore, that there is no *à priori* reason against our finding the remains of man or his works in the tertiary deposits. The absence of all such remains in the European beds of this age has little weight, because, as we go farther back in time, it is natural to suppose that man's distribution over the surface of the earth was less universal than at present.

Besides, Europe was in a great measure submerged during the tertiary epoch; and though its scattered islands may have been uninhabited by man, it by no means follows that he did not at the same time exist in warm or tropical continents. If geologists can point out to us the most extensive land in the warmer regions of the earth, which has not been submerged since Eocene or Miocene times, it is there that we may expect to find some traces of the very early progenitors of man. It is there that we may trace back the gradually decreasing brain of former races, till we come to a time when the body also begins materially to differ. Then we shall have reached the starting-point of the human family. Before that period he had not mind enough to preserve his body from change, and would, therefore, have been subject to the same comparatively rapid modifications of form as the other mammalia.

Their Bearing on the Dignity and Supremacy of Man

If the view I have here endeavoured to sustain have any foundation, they give us a new argument for placing man apart, as not only the head and culminating point of the grand series of organic nature, but as in some degree a new and distinct order of being. From those infinitely remote ages, when the first rudiments of organic life appeared upon the earth, every plant and every animal has been subject to one great law of physical change. As the earth has gone through its grand cycles of geological, climatal, and organic progress, every from of life has been subject to its irresistible action, and has been continually but imperceptibly moulded into such new shapes as would preserve their harmony with the ever-changing universe. No living thing could escape this law of its being; none (except, perhaps, the simplest and most rudimentary organisms) could remain unchanged and live, amid the universal change around it.

At length, however, there came into existence a being in whom that subtle force we term *mind,* became of greater importance than his mere bodily structure. Though with a naked and unprotected body, *this* gave him clothing against the varying inclemencies of the seasons. Though unable to compete with the deer in swiftness, or with the wild bull in strength, *this* gave him weapons with which to capture or overcome both. Though less capable than most other animals of living on the herbs and the fruits that unaided nature supplies, this wonderful faculty taught him to govern and direct nature to his own benefit, and make her produce food for him, when and where he pleased. From the moment when the first skin was used as a covering, when the first rude spear was formed to assist in the chase, when fire was first used to cook his food, when the first seed was sown or shoot planted, a grand revolution was effected in nature—a revolution which in all the previous ages of the earth's history had had no parallel, for a being had arisen who was no longer necessarily subject to change with the changing universe—a being who was in some degree superior to nature, inasmuch as he knew how to control and regulate her action, and could keep himself in harmony with her, not by a change in body, but by an advance of mind.

Here, then, we see the true grandeur and dignity of man. On this view of his special attributes, we may admit that even those who claim for him a position as an order, a class, or a sub-kingdom by himself, have some show of reason on their side. He is, indeed, a being part, since he is not influenced by the great laws which irresistibly modify all other organic beings. Nay more: this victory which he has gained for himself, gives him a directing influence over other existences. Man has not only escaped natural selection himself, but he is actually able to take away some of that power from nature which before his appearance she universally exercised. We can anticipate the time when the earth will produce only cultivated plants and domestic animals; when man's selection shall have supplanted natural selection; and when the ocean will be the only domain in which that power can be exerted, which for countless cycles of ages has ruled supreme over all the earth.

Their Bearing on the future Development of Man

We now find ourselves enabled to answer those who maintain that if Mr. Darwin's theory of the Origin of Species is true, man too must change in form, and become developed into some other animal as different from his present self as he is from the gorilla or the chimpanzee; and who speculate on what this form is likely to be. But it is evident that such will not be the case; for no change of conditions is conceivable which will render any important alteration of his form and organisation so universally useful and necessary to him, as to give those possessing it always the best chance of surviving, and thus lead to the development of a new species, genus, or higher group of man. On the other hand, we know that far greater changes of conditions and of his entire environment have been undergone by man than any other highly organised animal could survive unchanged, and have been met by mental, not corporeal adaptation. The difference of habits, of food, clothing, weapons, and enemies between savage and civilised man is enormous. Difference in bodily form and structure there is practically none, except a slightly increased size of brain, corresponding to his higher mental development.

We have every reason to believe, then, that man may have existed, and may continue to exist, through a series of geological periods which shall see all other forms of animal life again and again changed; while he himself remains unchanged, except in the two particulars already specified—the head and face, as immediately connected with the organ of the mind and as being the medium of expressing the most refined emotions of his nature,—and to a slight extent in colour, hair, and proportions, so far as they are correlated with constitutional resistance to disease.

Summary

Briefly to recapitulate the argument;—in two distinct ways has man escaped the influence of those laws which have produced unceasing change in the animal world. 1. By his superior intellect he is enabled to provide himself with clothing the weapons, and by cultivating the soil to obtain a constant supply of congenial food. This renders it unnecessary for his body to be modified in accordance with changing conditions—to gain a warmer natural covering, to acquire more powerful teeth or claws, or to become adapted to obtain and digest new kinds of food, as circumstances may require. 2.

By his superior sympathetic and moral feelings he becomes fitted for the social state; he ceases to plunder the weak and helpless of his tribe; he shares the game which he has caught with less active or less fortunate hunters, or exchanges it for weapons which even the weak or the deformed can fashion; he saves the sick and wounded from death; and thus the power which leads to the rigid destruction of all animals who cannot in every respect help themselves, is prevented from acting on him.

This power is natural selection; and, as by no other means can it be shown that individual variations can ever become accumulated and rendered permanent, so as to form well-marked races, it follows that the differences which now separate mankind from other animals must have been produced before he became possessed of a human intellect or human sympathies. This view also render possible, or even requires, the existence of man at a comparatively remote geological epoch. For, during the long periods in which other animals have been undergoing modification in their whole structure, to such an amount as to constitute distinct genera and families, man's *body* will have remained generically, or even specifically, the same, while his *head* and *brain* alone will have undergone modification equal to theirs. We can thus understand how it is that, judging from the head and brain, Professor Owen places man in a distinct sub-class of mammalia, while as regards the bony structure of his body, there is the closest anatomical resemblance to the anthropoid apes, "every tooth, every bone, strictly homologous—which makes the determination of the different between *Homo* and *Pithecus* the anatomist's difficulty." The present theory fully recognises and accounts for these facts; and we may perhaps claim as corroborative of its truth that it neither requires us to depreciate the intellectual chasm which separates man from the apes, nor refuses full recognition of the striking resemblances to them, which exist in other parts of his structure.

Conclusion

In concluding this brief sketch of a great subject, I would point out its bearing upon the future of the human race. If my conclusions are just, it must inevitably follow that the higher—the more intellectual and moral—must displace the lower and more degraded races; and the power of "natural selection," still acting on his mental organisation, must ever lead to the more perfect adaptation of man's higher faculties to the conditions of surrounding nature, and to the exigencies of the social state. While his external form will probably ever remain unchanged, except in the development of that perfect beauty which results from a healthy and well organised body, refined an ennobled by the highest intellectual faculties and sympathetic emotions, his mental constitution may continue to advance and improve, till the world is again inhabited by a single nearly homogeneous race, no individual of which will be inferior to the noblest specimens of existing humanity.

Our progress towards such a result is very slow, but it still seems to be a progress. We are just now living at an abnormal period of the world's history, owing to the marvellous developments and vast practical results of science having been given to societies too low morally and intellectually to know how to make the best use of them, and to whom they have consequently been curses as well as blessings. Among civilised nations at the present day it does not seem possible for natural selection to act in any way, so as to secure the permanent advancement of morality and intelligence; for it is indisputably the mediocre, if not the low, both a regards morality and intelligence, who succeed best in life and multiply fastest. Yet these is undoubtedly an advance—on the whole a steady and a permanent one—both in the influence on public opinion of a high morality, and in the general desire for intellectual elevation; and as I cannot impute this in any way to "survival of the fittest," I am forced to conclude that it is due to the inherent progressive power of those glorious qualities which raise us so immeasurably above our fellow animals, and at the same time afford us the surest proof that there are other and higher existences than ourselves, from whom these qualities may have been derived, and towards whom we may be ever tending.

IX

THE LIMITS OF NATURAL SELECTION AS APPLIED TO MAN

THROUGHOUT this volume I have endeavoured to show that the known laws of variation, multiplication, and heredity, resulting in a "struggle for existence" and the "survival of the fittest," have probably sufficed to produce all the varieties of structure, all the wonderful adaptations, all the beauty of form and of colour, that we see in the animal and vegetable kingdoms. To the best of my ability I have answered the most obvious and the most often repeated objections to this theory, and have, I hope, added to its general strength, by showing how colour—one of the strongholds of the advocates of special creation—may be, in almost all its modifications, accounted for by the combined influence of sexual selection and the need of protection.[1] I have also endeavoured to show how the same power which has modified animals has acted on man; and have, I believe, proved that, as soon as the human intellect became developed above a certain low stage, man's body would cease to be materially affected by natural selection, because the development of his mental faculties would render important modifications of its form and structure unnecessary. It will, therefore, probably excite some surprise among my readers to find that I do not consider that all nature can be explained on the principles of which I am so ardent an advocate; and that I am now myself going to state objections, and to place limits, to the power of natural selection. I believe, however, that there are such limits; and that just as surely as we can trace the action of natural laws in the development of organic forms, and can clearly conceive that fuller knowledge would enable us to follow step by step the whole process of that development, so surely can we trace the action of some unknown higher law, beyond and independent of all those laws of which we have any knowledge. We can trace this action more or less distinctly in many phenomena, the two most important of which are—the origin of sensation or consciousness, and the development of man from the lower animals. I shall first consider the latter difficulty as more immediately connected with the subject discussed in the volume.

[1] Since writing this in 1870 I have come to the conclusion that sexual selection has had little, if any, influence on colour. See chap. v. of "Tropical Nature" in this volume, and *Darwinism,* chap. x.

What Natural Selection can Not do

In considering the question of the development of man by known natural laws, we must ever bear in mind the first principle of natural selection, no less than of the general theory of evolution, that all changes of form or structure, all increase in the size of an organ or in its complexity, all greater specialisation or physiological division of labour, can only be brought about in as much as it is for the good of the being so modified. Mr. Darwin himself has taken care to impress upon us that natural selection has no power to produce absolute perfection, but only relative perfection,—no power to advance any being much beyond his fellow beings, but only just so much beyond them as to enable it to survive them in the struggle for existence. Still less has it any power to produce modifications which are in any degree injurious to its possessor, and Mr. Darwin frequently uses the strong expression, that a single case of this kind would be fatal to his theory. If, therefore, we find in man any characters, which all the evidence we can obtain goes to show would have been actually injurious to him on their first appearance, they could not possibly have been produced by natural selection. Neither could any specially developed organ have been so produced if it had been merely useless to him, or if its use were not proportionate to its degree of development. Such cases as these would prove that some other law, or some other power, than natural selection had been at work. But if, further, we could see that these very modifications, though hurtful or useless at the time when they first appeared, became in the highest degree useful at a much later period, and are now essential to the full moral and intellectual development of human nature, we should then infer the action of mind, foreseeing the future and preparing for it, just as surely as we do, when we see the breeder set himself to work with the determination to produce a definite improvement in some cultivated plant or domestic animal. I would further remark that this inquiry is as thoroughly scientific and legitimate as that into the origin of species itself. It is an attempt to solve the inverse problem, to deduce the existence of a new power of a definite character, in order to account for fact which, according to the theory of natural selection, ought not to happen. Such problems are well known to science, and the search after their solution has often led to the most brilliant results. In the case of man, there are facts of the nature above alluded to, and in calling attention to them, and in inferring a cause for them, I believe that I am as strictly within the bounds of scientific investigation as I have been in any other portion o my work.

The Brain of the Savage shown to be Larger then he Needs it to be

Size of Brain an important Element of Mental Power.—The brain is universally admitted to be the organ of the mind; and it is almost as universally admitted that size of brain is one of the most important of the elements which determine mental power or capacity. There seems to be no doubt that brains differ considerably in quality, as indicated by greater or less complexity of the convolutions, quantity of gray matter, and perhaps unknown peculiarities of organisation; but this difference of quality seems merely to increase or diminish the influence of quantity, not to neutralise it. Thus, all the most eminent modern writers see an intimate connection between the diminished size of the brain in the lower races of mankind, and their intellectual inferiority. The collections of Dr. J. B. Davis and Dr. Morton give the following as the average internal capacity of the cranium in the chief races: Teutonic family, 94 cubic inches; Esquimaux, 91 cubic inches; Negroes, 85 cubic inches; Australians and Tasmanians, 82 cubic inches; Bushmen, 77 cubic inches. These last numbers, however, are deduced from comparatively few specimens, and may be below the average, just as a small number of Finns and Cossacks give 98 cubic inches, or considerably more than that of the German races. It is evident, therefore, that the absolute bulk of the brain is not necessarily much less in savage than in civilised man, for Esquimaux skulls are known with a capacity of 113 inches, or hardly less than the largest among Europeans. But what is still more extraordinary, the few remains yet known of prehistoric man do not indicate any material diminution in the size of the brain case. A Swiss skull of the stone age, found in the lake dwelling of Meilen, corresponded exactly to that of a Swiss youth of the present day. The celebrated Neanderthal skull had a larger circumference than the average, and its capacity, indicating actual mass of brain, is estimated to have been not less than 75 cubic inches, or nearly the average of existing Australian crania. The Engis skull, perhaps the oldest known, and which, according to Sir John Lubbock, "there seems no doubt was really contemporary with the mammoth and the cave bear," is yet, according to Professor Huxley, "a fair average skull, which might have belonged to a philosopher, or might have contained the thoughtless brains of a savage." Of the cave men of Les Eyzies, who were undoubtedly contemporary with the reindeer in the south of France, Professor Paul Broca says (in a paper read before the Congress of Prehistoric Archæology in 1868): "The great capacity of the brain, the development of the frontal region, the fine elliptical form of the anterior part of the profile of the skull, are incontestible characteristics of superiority, such as we are accustomed to meet with in civilised races;" yet the great breadth of the face, the enormous development of the ascending ramus of the lower jaw, the extent and roughness of the surfaces for the attachment of the muscles, especially of the masticators, and the extraordinary development of the ridge of the femur, indicate great muscular power, and the habits of a savage and brutal race.

These facts might almost make us doubt whether the size of the brain is in any direct way an index of mental power, had we not the most conclusive evidence that it is so, in the fact that, whenever an adult male European has a skull less than 19 inches in circumference, or has less than 65 cubic inches of brain, he is invariably idiotic. When we join with this the equally undisputed fact that great men,—those who combine acute perception with great reflective power, strong passions, and general energy of character, such as Napolenon, Cuvier, and O'Connell,—have always heads far above the average size, we must feel satisfied that volume of brain is one, and perhaps the most important, measure of intellect; and this being the case, we cannot fail to be struck with the apparent anomaly that many of the lowest savages should have as much brains as average Europeans. The idea is suggested of a surplusage of power—of an instrument beyond the needs of its possessor.

Comparison of the Brains of Man and of Anthropoid Apes.—In order to discover if there is any foundation for this notion, let us compare the brain of man with that of animals. The adult male orang-utan is quite as bulky as a small sized man, while the gorilla is considerably above the average size of man, as estimated by bulk and weight; yet the former has a brain of only 28 cubic inches, the latter, one of 30, or, in the largest specimen yet known, of 34½ cubic inches. We have seen that the average cranial capacity of the lowest savages is probably not less than *five-sixths* of that of the highest civilised races, while the brain of the anthropoid apes scarcely amounts to *one-third* of that of man, in both cases taking the average; or the proportions may be more clearly represented by the following figures: Anthropoid apes, 10; savages, 26; civilised man, 32. But do these figures at all approximately represent the relative intellect of the three groups? Is the savage really no further removed from the philosopher, and so much removed from the ape, as these figures would indicate? In considering this question, we must not forget that the heads of savages vary in size almost as much as those of civilised Europeans. Thus, while the largest Teutonic skull in Dr. Davis's collection is 112.4 cubic inches, there is an Araucanian of 115.5, an Esquimaux of 113.1, a Marquesan of 110.6, a Negro of 105.8, and even an Australian of 104.5 cubic inches. We may, therefore, fairly compare the savage with the highest European on the one side, and with the orang, chimpanzee, or gorilla, on the other, and see whether there is any relative proportion between brain and intellect *Range of Intellectual Power in Man.*—First, let us consider what this wonderful instrument, the brain, is capable of in its higher developments. In Mr. Galton's interesting work on *Hereditary Genius,* he remarks on the enormous difference between the intellectual power and grasp of the well-trained mathematician or man of science, and the average Englishman. The number of marks obtained by high wranglers is often more than thirty times as great as that of the men at the bottom of the honour list, who are still of fair mathematical ability; and it is the opinion of skilled examiners that even this does not represent the full difference of intellectual power. If, now, we descend to those savage tribes who only count to three or five, and who find it impossible to comprehend the addition of two and three without having the objects actually before them, we feel that the chasm between them and the good mathematician is so vast that a thousand to one will probably not fully express it. Yet we know that the mass of brain might be nearly the same in both, or might not differ in a greater proportion than as 5 to 6; whence we may fairly infer that the savage possesses a brain capable, if cultivated and developed, of performing work of a kind and degree far beyond what he ever requires it to do.

Again, let us consider the power of the higher or even the average civilised man, of forming abstract ideas, and carrying on more or less complex trains of reasoning. Our languages are full or terms to express abstract conceptions. Our business and our pleasures involve the continual foresight of many contingencies. Our law, our government, and our science continually require us to reason through a variety of complicated phenomena to the expected result. Even our games, such as chess, compel us to exercise all these faculties in a remarkable degree. Compare this with the savage languages, which contain no words for abstract conceptions; the utter want of foresight of the savage man beyond his simplest necessities; his inability to combine, or to compare, or to reason on any general subject that does not immediately appeal to his senses. So, in his moral and æsthetic facutlies, the savage has none of these wide sympathies with all nature, those conceptions of the infinite, of the good, of the sublime and beautiful, which are so largely developed in civilised man. Any considerable development of these would, in fact, be useless or even hurtful to him, since they would to some extent interfere with the supremacy of those perceptive and animal faculties on which his very existence often depends, in the severe struggle he has to carry on against nature and his fellow-man. Yet the rudiments of all these powers and feelings undoubtedly exist in him, since one or other of them frequently manifest themselves in exceptional cases, or when some special circumstances call them forth. Some tribes, such as the Santals, are remarkable for as pure a love of truth as the most moral among civilised man. The Hindoo and the Polynesian have a high artistic feeling, the first traces of which are clearly visible in the rude drawings of the palæolithic men who were the contemporaries in France of the reindeer and the mammoth. Instances of unselfish love, of true gratitude, and of deep religious feeling, sometimes occur among most savage races.

On the whole, then, we may conclude that the general, moral, and intellectual development of the savage is not less removed from that of civilised man than has been shown to be the case in the one department of mathematics; and from the fact that all the moral and intellectual faculties do occasionally manifest themselves, we may fairly conclude that they are always latent, and that the large brain of the savage man is much beyond his actual requirements in the savage state.

Intellect of Savages and of Animals compared.—Let us now compare the intellectual wants of the savage, and the actual amount of intellect he exhibits, with those of the higher animals. Such races as the Andaman Islanders, the Australians, and the Tasmanians, the Digger Indians of North America, or the natives of Fuegia, pass their lives so as to require the exercise of few faculties not possessed in an equal degree by many animals. In the mode of capture of game or fish they by no means surpass the ingenuity or forethought of the jaguar, who drops saliva into the water, and seizes the fish as they come to eat it; or of wolves and jackals, who hunt in packs; or of the fox, who buries his surplus food till be requires it. The sentinels placed by antelopes and by monkeys, and the various modes of building adopted by field mice and beavers, as well as the sleeping-place of the orang-utan, and the tree-shelter of some of the African anthropoid apes, may well be compared with the amount of care and forethought bestowed by many savages in similar circumstances. His possession of free and perfect hands, not required for locomotion, enables man to form and use weapons and implements which are beyond the physical powers of brutes; but having done this, he certainly does not exhibit more mind in using them than do many lower animals. What is there in the life of the savage but the satisfying of the cravings of appetite in the simplest and easiest way? What thoughts, ideas, or actions are there that raise him many grades above the elephant or the ape? Yet he possesses, as we have seen, a brain vastly superior to theirs in size and complexity; and this brain gives him, in an undeveloped state, faculties which he never requires to use. And if this is true of existing savages, how much more true must it have been of the men whose sole weapons were rudely chipped flints, and some of whom, we may fairly conclude, were lower than any existing race; while the only evidence yet in our possession shows them to have had brains fully as capacious as those of the average of the lower savage races.

We see, then, that whether we compare the savage with the higher developments of man, or with the brutes around him, we are alike driven to the conclusion that in his large and well-developed brain he possesses an organ quite disproportionate to his actual requirements—an organ that seems prepared in advance, only to be fully utilised as he progresses in civilisation. A brain one-half larger than that of the gorilla would, according to the evidence before us, fully have sufficed for the limited mental development of the savage; and we must therefore admit that the large brain he actually possesses could never have been solely developed by any of those laws of evolution, whose essence is, that they lead to a degree of organisation exactly proportionate to the wants of each species, never beyond those wants—that no preparation can be made for the future development of the race—that one part of the body can never increase in size or complexity, except in strict co-ordination to the pressing wants of the whole.

The brain of prehistoric and of savage man seems to me to prove the existence of some power distinct from that which has guided the development of the lower animals through their ever-varying forms of being.

The Use of the Hairy Covering of Mammalia

Let us now consider another point in man's organisation, the bearing of which has been almost entirely overlooked by writers on both sides of this question. One of the most general external characters of the terrestrial mammalia is the hairy covering of the body, which, whenever the skin is flexible, soft, and sensitive, forms a natural protection against the severities of climate, and particularly against rain. That this is its most important function is well shown by the manner in which the hairs are disposed so as to carry off the water, by being invariably directed downwards from the most elevated parts of the body. Thus, on the under surface the hair is always less plentiful, and, in many cases, the belly is almost bare. The hair lies downwards, on the limbs of all walking mammals, from the shoulder to the toes; but in the orang-utan it is directed from the shoulder to the elbow, and again from the wrist to the elbow, in a reverse direction. This corresponds to the habits of the animal, which, when resting, holds its long arms upwards over its head, or clasping a branch above it, so that the rain would flow down both the arm and forearm to the long hair which meets at the elbow. In accordance with this principle, the hair is always longer or more dense along the spine or middle of the back from the nape to the tail, often rising into a crest of hair or bristles on the ridge of the back. This character prevails through the entire series of the mammalia, from the marsupials to the quadrumana, and by this long persistence it must have acquired such a powerful hereditary tendency that we should expect it to reappear continually even after it had been abolished by ages of the most rigid selection; and we may feel sure that it never could have been completely abolished under the law of nature selection, unless it had become so positively injurious as to lead to the almost invariable extinction of individuals possessing it.

The constant Absence of Hair from certain parts of Man's Body a remarkable Phenomenon

In man the hairy covering of the body has almost totally disappeared, and, what very remarkable, it has disappeared more completely from the back than from any other part of the body. Bearded and beardless races alike have the back smooth, and even when a considerable quantity of hair appears on the limbs and breast, the back, and especially the spinal region, is absolutely free, thus completely reversing the characteristics of all other mammalia. The Ainos of the Kurile Islands and Japan are said to be a hairy race; but Mr. Bickmore, who saw some of them, and described them in a paper read before the Ethnological Society, gives no details as to where the hair was most abundant, merely stating generally that "their chief peculiarity is their great abundance of hair, not only on the head and face, but over the whole body." This might very well be said of any man who had hairy limbs and breast, unless it was specially stated that his back was hairy, which is not done in this case. The hairy family in Birmah have, indeed, hair on the back rather longer than on the breast, thus reproducing the true mammalian character, but they have still longer hair on the face, forehead, and inside the ears, which is quite abnormal; and the fact that their teeth are all very imperfect shows that this is a case of monstrosity rather than one of true reversion to the ancestral type of man before he lost his hairy covering.

Savage Man feels the Want of this Hairy Covering

We must now inquire if we have any evidence to show, or any reason to believe, that a hairy covering to the back would be in any degree hurtful to savage man, or to man in any stage of his progress from his lower animal form; and if it were merely useless, could it have been so entirely and completely removed as not to be continually reappearing in mixed races? Let us look to savage man for some light on these points. One of the most common habits of savages is to use some covering for the back and shoulders, even when they have none on any other part of the body. The early voyagers observed with surprise that the Tasmanians, both men and women, wore the kangaroo-skin, which was their only covering, not from any feeling of modesty, but over the shoulders to keep the back dry and warm. A cloth over the shoulders was also the national dress of the Maories. The Patagonians wear a cloak or mantle over the shoulders, and the Fuegians often wear a small piece of skin on the back, laced on, and shifted from side to side as the wind blows. The Hottentots also were a somewhat similar skin over the back, which they never removed, and in which they were buried. Even in the tropics most savages take precautions to keep their backs dry. The natives of Timor use the leaf of a fan palm, carefully stitched up and folded, which they always carry with them, and which, held over the back, forms an admirable protection from the rain. Almost all the Malay races, as well as the Indians of South America, make great palm-leaf hats, four feet or more across, which they use during their canoe voyages to protect their bodies from heavy showers of rain; and they use smaller hats of the same kind when travelling by land.

We find, then, that so far from there being any reason to believe that a hairy covering to the back could have been hurtful or even useless to prehistoric man, the habits of modern savages indicate exactly the opposite view, as they evidently feel the want of it, and are obliged to provide substitutes of various kinds. The perfectly erect posture of man may be supposed to have something to do with the disappearance of the hair from his body while it remains on his head; but when walking, exposed to rain and wind, a man naturally stoops forwards and thus exposes his back; and the undoubted fact that most savages feel the effects of cold and wet most severely in that part of the body, sufficiently demonstrates that the hair could not have ceased to grow there merely because it was useless, even if it were likely that a character so long persistent in the entire order of mammalia could have so completely disappeared under the influence of so weak a selective power as a diminished usefulness.

Man's Naked Skin could not have been produced by Natural Selection

It seems to me, then, to be absolutely certain that natural selection could not have produced man's hairless body by the accumulation of variations from a hairy ancestor. The evidence all goes to show that such variations could not have been useful, but must, on the contrary, have been to some extent hurtful. If even, owing to an unknown correlation with other hurtful qualities, it had been abolished in the ancestral tropical man, we cannot conceive that, as man spread into colder climates, it should not have returned under the powerful influence of reversion to such a long persistent ancestral type. But the very foundation of such a supposition as this is untenable, for we cannot suppose that a character which, like hairiness, exists throughout the whole of the mammalia, can have become, in one form only, so constantly correlated with an injurious character as to lead to its permanent suppression—a suppression so complete and effectual that it never, or scarcely ever, reappears in mongrels of the most widely different races of man.

Two characters could hardly be wider apart than the size and development of man's brain and the distribution of hair upon the surface of his body, yet they both lead us to the same conclusion—that some other power than natural selection has been engaged in his production.

Feet and Hands of Man, considered as Difficulties on the Theory of Natural Selection

There are a few other physical characteristics of man that may just be mentioned as offering similar difficulties, though I do not attach the same importance to them as to those I have already dwelt on. The specialisation and perfection of the hands and feet of man seems difficult to account for. Throughout the whole of the quadrumana the foot is prehensile, and a very rigid selection must therefore have been needed to bring about that arrangement of the bones and muscles which has converted the thumb into a great toe, so completely, that the power of opposability is totally lost in every race, whatever some travellers may vaguely assert to the contrary. It is difficult to see why the prehensile power should have been taken away. It must certainly have been useful in climbing, and the case of the baboons shows that it is quite compatible with terrestrial locomotion. It may not be compatible with perfectly easy erect locomotion; but, then, how can we conceive that early man, *as an animal,* gained anything by purely erect locomotion? Again, the hand of man contains latent capacities and powers which are unused by savages, and must have been even less used by palæolithic man and his still ruder predecessors. It has all the appearance of an organ prepared for the use of civilised man, and on which was required to render civilisation possible. Apes make little use of their separate fingers and opposable thumbs. They grasp objects rudely and clumsily, and look as if a much less specialised extremity would have served their purpose as well. I do not lay much stress on this, but, if it be proved that some intelligent power has guided or determined the development of man, then we may see indications of that power in facts which, by themselves, would not serve to prove its existence.

The Voice of Man.—The same remark will apply to another peculiarly human character, the wonderful power, range, flexibility, and sweetness to the musical sounds producible by the human larynx, especially in the female sex. The habits of savages give no indication of how this faculty could have been developed by natural selection, because it is never required or used by them. The singing of savages is a more or less monotonous howling, and the females seldom sing at all. Savages certainly never choose their wives for fine voices, but for rude health, and strength, and physical beauty. Sexual selection could not therefore have developed this wonderful power, which only comes into play among civilised people. It seems as if the organ had been prepared in anticipation of the future progress of man, since it contains latent capacities which are useless to him in his earlier condition. The delicate correlations of structure that give it such marvelous powers could not therefore have been acquired by means of natural selection.

The Origin of some of Man's Mental Faculties, by the pre-servation of Useful Variations, not possible

Turning to the mind of man, we meet with many difficulties in attempting to understand how those mental faculties, which are especially human, could have been acquired by the preservation of useful variation. At first sight, it would seem that such feelings as those of abstract justice and benevolence could never have been so acquired, because they are incompatible with the law of the strongest, which is the essence of natural selection. But this is, I think, an erroneous view, because we must look, not to individuals, but to societies; and justice and benevolence exercised towards members of the same tribe would certainly tend to strengthen that tribe and give it a superiority over another in which the right of the strongest prevailed, and where, consequently, the weak and the sickly were left to perish, and the few strong ruthlessly destroyed the many who were weaker.

But there is another class of human faculties that do not regard our fellow-men, and which cannot, therefore, be thus accounted for. Such are the capacity to form ideal conceptions of space and time, of eternity and infinity—the capacity for intense artistic feelings of pleasure, in form, colour, and composition, and for those abstract notions of form and number which render geometry and arithmetic possible. How were all or any of these faculties first developed, when they could have been of no possible use to man in his early stages of barbarism? How could natural selection, or survival of the fittest in the struggle for existence, at all favour the development of mental powers so entirely removed from the material necessities of savage men, and which even now, with our comparatively high civilisation, are, in their farthest developments, in advance of the age, and appear to have relation rather to the future of the race than to its actual status?[1]

[1] This argument is extended and some new illustrations given in *Darwinism,* pp. 461–471.

Difficulty as to the Origin of the Moral Sense

Exactly the same difficulty arises when we endeavour to account for the development of the moral sense or conscience in savage man; for although the *practice* of benevolence, honesty, or truth may have been useful to the tribe possessing these virtues, that does not at all account for the peculiar *sanctity* attached to actions which each tribe considers right and moral, as contrasted with the very different feelings with which they regard what is merely *useful.* the utilitarian hypothesis (which is the theory of natural selection applied to the mind) seems inadequate to account for the development of the moral sense. This subject has been recently much discussed, and I will here only give one example to illustrate my argument. The utilitarian sanction for truthfulness is by no means very powerful or universal. Few laws enforce it. No very severe reprobation follows untruthfulness. In all ages and countries falsehood has been thought allowable in love, and laudable in war; while, at the present day, it is held to be venial by the majority of mankind in trade, commerce, and speculation. A certain amount of untruthfulness is a necessary part of politeness in the East and West alike, while even severe moralists have held a lie justifiable to elude an enemy or prevent a crime. Such being the difficulties with which this virtue has had to struggle, with so many exceptions to its practice, with so many instances in which it brought ruin or death to its too ardent devotee, how can we believe that considerations of utility could ever invest it with the mysterious sanctity of the highest virtue,—could ever induce men to value truth for its own sake, and practise it regardless of consequences?

Yet it is a fact that such a mystical sense of wrong does attach to untruthfulness, not only among the higher classes of civilised people, but among whole tribes of utter savages. Sir Walter Elliott tells us (in his paper "On the Characteristics of the Population of Central and Southern India," published in the *Journal of the Ethnological Society of London,* vol. i. p. 107) that the Kurubars and Santals, barbarous hill-tribes of Central India, are noted for veracity. It is a common saying that "a Kurubar *always* speaks the truth;" and Major Jervis says, "the Santals are the most truthful men I ever met with." As a remarkable instance of this quality the following fact is given. A number of prisoners, taken during the Santal insurrection, were allowed to go free on parole, to work at a certain spot for wages. After some time cholera attacked them and they were obliged to leave, but every man of them returned and gave up his earnings to the guard. Two hundred savages, with money in their girdles, walked thirty miles back to prison rather than break their word! My own experience among savages has furnished me with similar, although lest severely tested, instances; and we cannot avoid asking, How is it that in these few cases "experiences of utility" have left such an overwhelming impression, while in so many others they have left none? The experiences of savage men as regards the utility of truth must, in the long run, be pretty nearly equal. How is it, then, that in some cases the result is a sanctity which overrides all considerations of personal advantage, while in others there is hardly a rudiment of such a feeling?

The intuitional theory, which I am now advocating, explains this by the supposition that there is a feeling—a sense of right and wrong—in our nature, antecedent to and independent of experiences of utility. Where free play is allowed to the relations between man and man, this feeling attaches itself to those acts of universal utility of self-sacrifice which are the products of our affections and sympathies, and which we term moral; while it may be, and often is, perverted, to give the same sanction to acts of narrow and conventional utility which are really immoral,—as when the Hindoo will tell a lie, but will sooner starve than eat unclean food, and looks upon the marriage of adult females as gross immorality.

The strength of the moral feeling will depend upon individual or racial constitution, and on education and habit;—the acts to which its sanctions are applied will depend upon how far the simple feelings and affections of our nature have been modified by custom, by law, or by religion.

It is difficult to conceive that such an intense and mystical feeling of right and wrong (so intense as to overcome all ideas of personal advantage or utility), could have been developed out of accumulated ancestral experiences of utility; and still more difficult to understand how feelings developed by one set of utilities could be transferred to acts of which the utility was partial, imaginary, or altogether absent. But if a moral sense is an essential part of our nature, it is easy to see that its sanction may often be given to acts which are useless or immoral; just as the natural appetite for drink is perverted by the drunkard into the means of his destruction.

Summary of the Argument as to the Insufficiency of Natural Selection to account for the Development of Man

Briefly to resume my argument—I have shown that the brain of the lowest savages, and, as far as we yet know, of the prehistoric races, is little inferior in size to that of the highest types of man, and immensely superior to that of the higher animals; while it is universally admitted that quantity of brain is one of the most important, and probably the most essential, of the elements which determine mental power. Yet the mental requirements of savages, and the faculties actually exercised by them, are very little above those of animals. The higher feelings of pure morality and refined emotion, and the power of abstract reasoning and ideal conception, are useless to them, are rarely if ever manifested, and have no important relations to their habits, wants, desires, or well-being. They possess a mental organ beyond their needs. Natural selection could only have endowed savage man with a brain a few degrees superior to that of an ape, whereas he actually possesses one very little inferior to that of a philosopher.

The soft, naked, sensitive skin of man, entirely free from that hairy covering which is so universal among other mammalia, cannot be explained on the theory of natural selection. The habits of savages show that they feel the want of this covering which is most completely absent in man exactly where it is thickest in other animals. We have no reason whatever to believe that it could have been hurtful or even useless to primitive man; and, under these circumstances, its complete abolition, shown by its never reverting in mixed breeds, is a demonstration of the agency of some other power than the law of the survival of the fittest, in the development of man from the lower animals.

Other characters show difficulties of a similar kind, though not perhaps in an equal degree. The structure of the human foot and hand seem unnecessarily perfect for the needs of savage man, in whom they are as completely and as humanly developed as in the highest races. The structure of the human larynx, giving the power of speech and of producing musical sounds, and especially its extreme development in the female sex, are shown to be beyond the needs of savages, and, from their known habits, impossible to have been required either by sexual selection or by survival of the fittest.

The mind of man offers arguments in the same direction, hardly less strong than those derived from his bodily structure. A number of his mental faculties have no relation to his fellow-men, or to his material progress. The power of conceiving eternity and infinity, and all those purely abstract notions of form, number, and harmony, which play so large a part in the life of civilised races, are entirely outside of the world of thought of the savage, and have no influence on his individual existence or on that of his tribe. They could not, therefore, have been developed by any preservation of useful forms of thought; yet we find occasional traces of them amidst a low civilisation, and at a time when they could have had no practical effect on the success of the individual, the family, or the race; and the development of a moral sense or conscience by similar means is equally inconceivable.

But, on the other hand, we find that every one of these characteristics is necessary for the full development of human nature. The rapid progress of civilisation under favourable conditions would not be possible, were not the organ of the mind of man prepared in advance, fully developed as regards size, structure, and proportions, and only needing a few generations of use and habit to co-ordinate its complex functions. The naked and sensitive skin, by necessitating clothing and houses would lead to the more rapid development of man's inventive and constructive faculties; and, by leading to a more refined feeling of personal modesty, may have influenced, to a considerable extent, his moral nature. The erect form of man, by freeing the hands from all locomotive uses, has been necessary for his intellectual advancement; and the extreme perfection of his hands has alone rendered possible that excellence in all the arts of civilisation which raises him so far above the savage, and is perhaps but the forerunner of a higher intellectual and moral advancement. The perfection of his vocal organs has first led to the formation of articulate speech, and then to the development of those exquisitely toned sounds, which are only appreciated by the higher races, and which are probably destined for more elevated uses and more refined enjoyment in a higher condition than we have yet attained to. So, those faculties which enable us to transcend time and space, and to realise the wonderful conceptions of mathematics and philosophy, or which give us an intense yearning for abstract truth (all of which were occasionally manifested at such an early period of human history as to be far in advance of any of the few practical applications which have since grown out of them), are evidently essential to the perfect development of man as a spiritual being, but are utterly inconceivable as having been produced through the action of a law which looks only, and can look only, to the immediate material welfare of the individual or the race.

The inference I would draw from this class of phenomena is, that a superior intelligence has guided the development of man in a definite direction, and for a special purpose, just as man guides the development of many animal and vegetable forms. The laws of evolution alone would, perhaps, never have produced a grain so well adapted to man's use as wheat and maize; such fruits as the seedless banana and breadfruit; or such animals as the Guernsey milch cow, or the London dray-horse. Yet these so closely resemble the unaided productions of nature, that we may well imagine a being who had mastered the laws of development of organic forms through past ages, refusing to believe that any new power had been concerned in their production, and scornfully rejecting the theory (as my theory will be rejected by many who agree with me on other points) that in these few cases a controlling intelligence had directed the action of the laws of variation, multiplication, and survival, for his own purposes. We Know, however, that this has been done; and we must therefore admit the possibility that, if we are not the highest intelligences in the universe, some higher intelligence may have directed the process by which the human race was developed, by means of more subtle agencies than we are acquainted with. At the same time I must confess that this theory has the disadvantage of requiring the intervention of some distinct individual intelligence, to aid in the production of what we can hardly avoid considering as the ultimate aim and outcome of all organised existence—intellectual, ever-advancing, spiritual man. It therefore implies that the great laws which govern the material universe were insufficient for his production, unless we consider (as we may fairly do) that the controlling action of such higher intelligences is a necessary part of those law, just as the action of all surrounding organisms is one of the agencies in organic development. But even if my particular view should not be the true one, the difficulties I have put forward remain, and, I think, prove that some more general and more fundamental law underlies that of natural selection. The law of "unconscious intelligence" pervading all organic nature, put fourth by Dr. Laycock and adopted by Mr. Murphy, is such a law; but to my mind it has the double disadvantage of being both unintelligible and incapable of any kind of proof. It is more probable that the true law lies too deep for us to discover it; but there seems to me to be ample indications that such a law does exist, and is probably connected with the absolute origin of life and organisation.[1]

[1] Some of my critics seem quite to have misunderstood my meaning in this part of the argument. They have accused

me of unnecessarily and unphilosophically appealing to "first causes" in order to get over a difficulty—of believing that "out brains are made by God and our lungs by natural selection;" and that, in point of fact, "man in God's domestic animal." An eminent French critic, M. Claparède, makes me continually call in the aid of—"*une Force supérieure,*" the capital F meaning, I imagine, that this "higher Force" is Deity. I can only explain this misconception by the incapacity of the modern cultivated mind to realise the existence of any higher intelligence between itself and Deity. Angels and archangels, spirits and demons, have been so long banished from out belief as to have become actually unthinkable as actual existences, and nothing in modern philosophy takes their place. Yet the grand law of "continuity," the last outcome of modern science, which seems absolute throughout the realms of matter, force, and mind, so far as we can explore them, cannot surely fail to be true beyond the narrow sphere of our vision, and leave an infinite chasm between man and the Great Mind of the universe. Such a supposition seems to me in the highest degree improbable. Now, in referring to the origin of man, and its possible determining causes, I have used the words "some other power"—"some intelligent power"—"a superior intelligence"—"a controlling intelligence," and only in reference to the origin of universal forces and laws have I spoken of the will or power of "one Supreme Intelligence."

These are the only expressions I have used in alluding to the power which I believe has acted in the case of man, and they were purposely chosen to show that I reject the hypothesis of "first causes" for any and every *special* effect in the universe, except in the same sense that the action of man or of any other intelligent being is a first cause. In using such terms I wished to show plainly that I contemplated the possibility that the development of the essentially human portions of man's structure and intellect may have been determined by the directing influence of some higher intelligent beings, acting through natural and universal laws. A belief of this nature may or may not have a foundation, but it is an intelligible theory, and is not, *in its nature,* incapable of proof; and it rests on facts and arguments of an exactly similar kind to those which would enable a sufficiently powerful intellect to deduce, from the existence on the earth of cultivated plants and domestic animals, the presence of some intelligent being of a higher nature than themselves.

The Origin of Consciousness

The question of the origin of sensation and of thought can be but briefly discussed in this place, since it is a subject wide enough to require a separate volume for its proper treatment. No physiologist or philosopher has yet ventured to propound an intelligible theory of how sensation may possibly be a product of organisation; while many have declared the passage from matter to mind to be inconceivable. In his presidental address to the Physical Section of the British Association at Norwich, in 1868, Professor Tyndall expressed himself as follows:—

"The passage from the physics of the brain to the corresponding facts of consciousness is unthinkable. Granted that a definite thought and a definite molecular action in the brain occur simultaneously, we do not possess the intellectual organ, nor apparently any rudiment of the organ, which would enable us to pass by a process of reasoning from the one phenomenon to the other. They appear together, but we do not know why. Were our minds and senses so expanded, strengthened, and illuminated as to enable us to see and feel the very molecules of the brain,—were we capable of following all their motions, all their groupings, all their electric discharges, if such there be, and were we intimately acquainted with the corresponding states of thought and feeling,—we should be as far as ever from the solution of the problem, 'How are these physical processes connected with the facts of consciousness?' The chasm between the two classes of phenomena would still remain intellectually impassable."

In his latest work *(An Introduction to the Classification of Animals)*, published in 1869, Professor Huxley unhesitatingly adopts the "well founded doctrine that life is the cause and not the consequence of organisation." In his celebrated article "On the Physical Basis of Life," however, he maintains that life is a property of protoplasm, and that protoplasm owes its properties to the nature and disposition of its molecules. Hence he terms it "the matter of life," and believes that all the physical properties of organised beings are due to the physical properties of protoplasm. So far we might, perhaps, follow him, but he does not stop here. He proceeds to bridge over that chasm which Professor Tyndall has declared to be "intellectually impassable," and, by means which he states to be logical, arrives a the conclusion that our "*thoughts are the expression of molecular changes in that matter of life which is the source of our other vital phenomena.*" Not having been able to find any clue in Professor Huxley's writings to steps by which he passes from those vital phenomena, which consists only, in their last analysis, of movements of particles of matter, to those other phenomena which we term thought, sensation, or consciousness, but knowing that so positive an expression of opinion from him will have great weight with many persons, I shall endeavour to show, with as much brevity as is compatible with clearness, that this theory is not only incapable of proof, but is also, as it appears to me, inconsistent with accurate conceptions of molecular physics. To do this, and in order further to develop my views, I shall have to give a brief sketch of the most recent speculations and discoveries as to the ultimate nature and constitution of matter.

The Nature of Matter

It has been long seen by the deepest thinkers on the subject, that atoms,—considered as minute solid bodies from which emanate the attractive and repulsive forces which give what we term matter its properties,—could serve no purpose whatever; since it is universally admitted that the supposed atoms never touch each other, and it cannot be conceived that these homogeneous, indivisible, solid units are themselves the ultimate *cause* of the forces that emanate from their centers. As, therefore, none of the properties of matter can be due to the atoms themselves, but only to the forces which emanate from the points in space indicated by the atomic centers, it is logical continually to diminish their size till they vanish, leaving only localised centers of force to represent them. Of the various attempts that have been made to show how the properties of matter may be due to such modified atoms (considered as mere centers of force), the most successful, because the simplest and the most logical, is that of Mr. Bayma, who, in his *Molecular Mechanics,* has demonstrated how, from the simple assumption of such centres having attractive and repulsive forces (both varying according to the same law of the inverse squares as gravitation), and by grouping them in symmetrical figures, consisting of a repulsive center, an attractive nucleus, and one or more repulsive envelopes, we may explain all the general properties of matter; and, by more and more complex arrangements, even the special chemical, electrical, and magnetic properties of special forms of matter.[1] Each chemical element will thus consist of a molecule formed of simple atoms (or as Mr. Bayma terms them, to avoid confusion, "material elements") in greater or less number and of more or less complex arrangement; which molecule is in stable equilibrium, but liable to be changed in form by the attractive or repulsive influences of differently constituted molecules, constituting the phenomena of chemical combination, and resulting in new forms of molecule of greater complexity and more or less stability.

Those organic compounds of which organised beings are built up consist, as is well known, of matter of an extreme complexity and great instability; whence result the changes of form to which it is continually subject. This view enables us to comprehend the *possibility* of the phenomena of vegetative life being due to an almost infinite complexity of

[1] Mr. Bayma's work, entitled *The Elements of Molecular Mechanics,* was published in 1866, and has received less attention than it deserves. It is characterised by great lucidity, by logical arrangement, and by comparatively

simple geometrical and algebraical demonstrations, so that it may be understood and appreciated with a very moderate knowledge of mathematics. It consists of a series of Propositions, deduced from the known properties of matter; from these are derived a number of Theorems, by whose help the more complicated Problems are solved. Nothing is taken for granted throughout the work, and the only valid mode of escaping from its conclusions is, by either disproving the fundamental Propositions, or by detecting fallacies in the subsequent reasoning. molecular combinations, subject to definite changes under the stimuli of heat, moisture, light, electricity, and probably some unknown forces. But this greater and greater complexity, even if carried to an infinite extent, cannot, of itself, have the slightest tendency to originate consciousness in such molecules or groups of molecules. If a material element, or a combination of a thousand material elements in a molecule, are alike unconscious, it is impossible for us to believe that the mere addition of one, two, or a thousand other material elements to form a more complex molecule, could in any way tend to produce a self-conscious existence. The things are radically unlike, exclusive, and incommensurable. To say that mind is a product or function of protoplasm, or of its molecular changes, is to use words to which we can attach no clear conception; and those who argue thus should put forth a precise definition of matter with clearly enunciated properties, and show that the necessary result of a certain complex arrangement of the elements or atoms of that matter will be the production of self-consciousness. There is no escape from this dilemma,—either all matter is conscious, or consciousness is, or pertains to, something distinct from matter, and in the latter case its presence in material forms is a proof of the existence of conscious beings, outside of, and independent of, what we term matter.[1]

[1] A friend has suggested that I have not here explained myself sufficiently, and objects that *life* does not exist in matter any more than *consciousness,* and if the one can be produced by the laws of matter, why may not the other? I reply that there is a radical difference between the two. Organic or vegetative life consists essentially in chemical transformations and molecular motions, occurring under certain conditions and in a certain order. The matter and the forces which act upon it are for the most part known; and if there are any forced engaged in the manifestation of vegetative life yet undiscovered (which is a moot question), we can conceive them as analogous to such forces at heat, electricity, or chemical affinity, with which we are already acquainted. We can thus clearly *conceive* of the transition from dead matter to living matter. A complex mass which suffers decomposition or decay is dead, but if this mass has the power of attracting to itself from the surrounding medium, matter like that of which it is composed, we have the first rudiment

of vegetative life. If the mass can do this for a considerable time, and if its absorption of new matter more than replaces that lost by decomposition, and if it is of such a nature as to resist the mechanical or chemical forces to which it is usually exposed, and to retain a tolerably constant form, we term it a living organism. We can *conceive* an organism to be so constituted, and we can further conceive that any fragments, which may be accidentally broken from it, or which may fall away when its bulk has become too great for the cohesion of all its parts, may begin to increase anew and run the same course as the parent mass. This is growth and reproduction in their simplest forms; and from such a simple beginning it is possible to conceive a series of slight modifications of composition, and of internal and external forces, which should ultimately lead to the development of more complex organisms. The LIFE of such an organism may, perhaps, be nothing added to it, but merely the name we give to the result of a balance of internal and external forces in maintaining the permanence of the form and structure of the individual. The simplest conceivable form of such life would be the dew-drop, which owes its existence to the balance between the condensation of aqueous vapour in the atmosphere and the evaporation of its substance. If either is in excess, it soon ceases to maintain an individual existence. I do not maintain that vegetative life *is* wholly due to such a complex balance of forces, but only that it is *conceivable* as such.

With CONSCIOUSNESS the case is very different. Its phenomena are not comparable with those of any kind of *matter* subjected to any of the known or conceivable *forces* of nature; and we cannot *conceive* a gradual transition from absolute unconsciousness to consciousness, from an unsentient organism to a sentient being. The merest rudiment of sensation or self-consciousness is infinitely removed from absolutely non-sentient or unconscious matter. We can conceive of no physical addition to, or modification of, an unconscious mass which should create consciousness; no step in the series of changes organised matter may undergo, which should bring in sensation where there was no sensation or power of sensation at the preceding step. It is because the things are utterly incomparable and incommensurable that we can only conceive of *sensation* coming to matter from without, while *life* may be conceived as merely a specific combination and co-ordination of the matter and the forces that compose the universe, and with which we are separately acquainted. We may admit with Professor Huxley that *protoplasm* is the "matter of life" and the cause of organisation, but we cannot admit or conceive that *protoplasm* is the primary source of sensation and consciousness, or that it can ever of itself become *conscious* in the same way as we may perhaps conceive that it may become *alive.*

Matter is Force

The foregoing considerations lead us to the very important conclusion that matter is essentially force, and nothing but force; that matter, as popularly understood, does not exist, and is, in fact, philosophically inconceivable. When we touch matter, we only really experience sensations of resistance, implying repulsive force; and no other sense can give us such apparently solid proofs of the reality of matter as touch does. This conclusion, if kept constantly present in the mind, will be found to have a most important bearing on almost every high scientific and philosophical problem, and especially on such as relate to our own conscious existence.

All Force is probably Will-Force

If we are satisfied that force or forces are all that exist in the material universe, we are next led to inquire what is force? We are acquainted with two radically distinct or apparently distinct kinds of force—the first consists of the primary forces of nature, such as gravitation, cohesion, repulsion, heat, electricity, etc.; the second is our own will-force. Many persons will at once deny that the latter exists. It will be said that it is a mere transformation of the primary forces before alluded to; that the correlation of forces includes those of animal life, and that *will* itself is but the result of molecular change in the brain. I think, however, that it can be shown that this latter assertion has neither been proved, nor even been proved to be possible; and that in making it, a great leap in the dark has been taken from the known to the unknown. It may be at once admitted that the *muscular force* of animals and men is merely the transformed energy derived from the primary forces of nature. So much has been, if not rigidly proved, yet rendered highly probable, and it is in perfect accordance with all our knowledge of natural forces and natural laws. But it cannot be contended that the physiological balance-sheet has ever been so accurately struck, that we are entitled to say, not one-thousandth part of a grain more of force has been exerted by any organised body, or in any part of it, than has been derived from the known primary forces of the material world. If that were so, it would absolutely negative the existence of will; for if will is anything, it is a power that *directs* the action of the forces stored up in the body, and it is not conceivable that this *direction* can taken place, without the exercise of some force in some part of the organism. However delicately a machine may be constructed, with the most exquisitely contrived detents to release a weight or spring by the exertion of the smallest possible amount of force, *some* external force will always be required; so, in the animal machine, however minute may be the changes required in the cells or fibres of the brain, to set in motion the nerve currents which loosen or excite the pent-up forces of certain muscles, *some force* must be required to effect those changes.

If it is said, "those changes are automatic, and are set in motion by external causes," then one essential part of our consciousness, a certain amount of freedom in willing, is annihilated; and it is inconceivable how or why there should have arisen any consciousness or any apparent will, in such purely automatic organisms. If this were so, our apparent WILL would be a delusion, and Professor Huxley's belief "that our volition counts for something as a condition of events," would be fallacious, since our volition would then be but one link in the chain of events, counting for neither more nor less than any other link whatever.

If, therefore, we have traced one force, however minute, to an origin in our own WILL, while we have no knowledge of any other primary cause of force, it does not seem an improbable conclusion that all force may be will-force; and thus, that the whole universe is not merely dependent on, but actually is, the WILL of higher intelligences or of one Supreme Intelligence. It has been often said that the true poet is a seer; and in the noble verse of an American poetess we find expressed what may prove to be highest fact of science, the noblest truth of philosophy:

God of the Granite and the Rose!
Soul of the Sparrow and the Bee!
The mighty tide of Being flows
Through countless channels, Lord, from Thee.
It leaps to life in grass and flowers,
Through every grade of being runs,
While from Creation's radiant towers
Its glory flames in Stars and Suns.

Conclusion

These speculations are usually held to be far beyond the bounds of science; but they appear to me to be more legitimate deductions from the facts of science than those which consist in reducing the whole universe, not merely to matter, but to matter conceived and defined so as to be philosophically inconceivable. It is surely a great step in advance, to get rid of the notion that *matter* is a thing of itself, which can exist *per se,* and must have been eternal, since it is supposed to be indestructible and uncreated,—that force, or the forces of nature, are another thing, given or added to matter, or else its necessary properties,—and that mind is yet another thing, either a product of this matter and its supposed inherent forces, or distinct from and co-existent with it;—and to be able to substitute for this complicated theory, which leads to endless dilemmas and contradictions, the far simpler and more consistent belief, that matter, as an entity distinct from force, does not exist; and that FORCE is a product of MIND. Philosophy had long demonstrated our incapacity to prove the existence of matter, as usually conceived; while it admitted the demonstration to each of us of our own self-conscious, spiritual existence. Science has now worked its way up to the same result, and this agreement between them should give us some confidence in their combined teaching.

The view we have now arrived at seems to me more grand and sublime, as well as far simpler, than any other. It exhibits the universe as a universe of intelligence and will-power; and by enabling us to rid ourselves of the impossibility of thinking of mind, but as connected with our old notions of matter, opens up infinite possibilities of existence, connected with infinitely varied manifestations of force, totally distinct from, yet as real as, what we term matter.

The grand law of continuity which we see pervading our universe would lead us to infer infinite gradations of existence, and to people all space with intelligence and will-power; and, if so, we shall have no difficulty in believing that for so noble a purpose as the progressive development of higher and higher intelligences, those primal and general will-forces, which have sufficed for the production of the lower animals, should have been guided into new channels and made to converge in definite directions. And if, as seems to me probable, this has been done, I cannot admit that it in any degree affects the truth or generality of Mr. Darwin's great discovery. It merely shows that the laws of organic development have been occasionally used for a special end, just as man uses them for his special ends; and I do not see that the law of natural selection can be said to be disproved, if it can be shown that man does not owe his entire physical and mental development to its unaided action, any more than it is disproved by the existence of the poodle or the pouter pigeon, the production of which are equally beyond its undirected power.

The objections which in this essay I have taken to the view that the same law which appears to have sufficed for the development of animals has been alone the cause of man's superior physical and mental nature, will, I have no doubt, be overruled and explained away. But I venture to think they will nevertheless maintain their ground, and that they can only be met by the discovery of new facts or new laws, of a nature very different from any yet known to us. I can only hope that my treatment of the subject, though necessarily very meager, has been clear and intelligible; and that it may prove suggestive both to the opponents and to the upholders of the theory of natural selection.

TROPICAL NATURE AND OTHER ESSAYS

THE CLIMATE AND PHYSICAL ASPECTS OF THE EQUATORIAL ZONE

The three Climatal Zones of the Earht—Temperature of the Equatorial Zone—Causes of the Uniform High Temperature near the Equator—Influence of the Heat of the Soil—Influence of the Aqueour Vapour of the Atmosphere—Influence of Winds on the Temperature of the Equator—Heat due to the Condensation of Atmospheric Vapour—General features of the Equatorial Climate—Uniformity of the Equatorial Climate in all parts of the Globe—Effects of Vegetation on Climate—Short Twilight of the Equatorial Zone—The aspect of the Equatorial Heavens—Intensity of Meteorological Phenomena at the Equator—Concluding Remarks.

IT is difficult for an inhabitant of our temperate land to realise either the sudden and violent contrasts of the arctic seasons or the wonderful uniformity of the equatorial climate. The lengthening or the shortening days, the ever-changing tints of spring, summer, and autumn, succeeded by the leafless boughs of winter, are constantly recurring phenomena which represent to us the established course of nature. At the equator none of these changes occur; there is a perpetual equinox and a perpetual summer, and were it not for variations in the quantity of rain, in the direction and strength of the winds, and in the amount of sunshine, accompanied by corresponding slight changes in the development of vegetable and animal life, the monotony of nature would be extreme.

In the present chapter it is proposed to describe the chief peculiarities which distinguish the equatorial from the temperate climate, and to explain the causes of the difference between them,—causes which are by no means of so simple a nature as are usually imagined.

The Three Climatal Zones of the Earth

The three great divisions of the earth—the tropical, the temperate, and the frigid zones—may be briefly defined as the regions of uniform, of variable, and of extreme physical conditions respectively. They are primarily determined by the circumstance of the earth's axis not being perpendicular to the plane in which it moves round the sun; whence it follows that during one half of its revolution the north pole, and during the other half the south pole, is turned at a considerable angle towards the source of light and heat. This inclination of the axis on which the earth rotates is usually defined by the inclination of the equator to the plane of the orbit, termed the obliquity of the ecliptic. The amount of the is obliquity is 23½ degrees, and this measures the extent on each side of the equator of what are called the tropics, because within these limits the sun becomes vertical at noon twice a year, and at the extreme limit once a year, while beyond this distance it is never vertical. It will be evident however, from the nature of the case, that the two lines which mark the limits of the geographical "tropics" will not define any abrupt change of climate or physical conditions, such as characterise the tropical and temperate zones in their full development. There will be a gradual transition from one to the other, and in order to study them separately and contrast their special features we must only take into account the portion of each in which these are most fully exhibited. For the temperate zone we may take all countries situated between 35° and 60° of latitude, which in Europe will include every place between Christiana and Algiers, the districts farther south forming a transitional belt in which temperate and tropical features are combined. In order to study the special features of tropical nature, on the other hand, it will be advisable to confine out attention mainly to that portion of the globe which extends for about twelve degrees on each side of the equator, in which all the chief tropical phenomena dependent on astronomical causes are most fully manifested, and which we may distinguish as the "equatorial zone." In the debatable ground between these two well-contrasted belts local causes have a preponderating influence; and it would not be difficult to point out localities within the temperate zone of our maps, which exhibits all the chief characteristics of tropical nature to a great degree than other localities which are, as regards geographical position, tropical.

Temperature of the Equatorial Zone

The most characteristic, as it is the most important feature in the physical conditions of the great equatorial zone, is the wonderful uniformity of its temperature, alike throughout the changes of day and night, and from one part of the year to another. As a general rule, the greatest heat of the day does not exceed 90° or 91° Fahr., while it seldom falls during the night below 74deg; Fahr. It has been found by hourly observations carried on for three years at the meteorological observatory established by the Dutch government at Batavia, that the extreme range of temperature in that period was only 27deg; Fahr., the maximum being 95deg; and the minimum 68deg;. But this is, of course, very much beyond the usual daily range of the thermometer, which is, on the average, only a little more than 11deg; Fahr.; being 12.6°

in September, when it is greatest, and only 8.1° in January, when it is least.

Batavia, being situated between six and seven degrees south of the equator, may be taken as affording a fair example of the climate of the equatorial zone; though, being in an island, it is somewhat less extreme than many continental localities. Observations made at Para, which is on the South American Continent, and close to the equator, agree, however, very closely with those at Batavia; but at the latter place all the observations were made with extreme care and with the best instruments, and are therefore preferred as being thoroughly trustworthy.[1] The accompanying diagram, showing by curves the monthly means of the highest and lowest daily temperatures at Batavia and London, is very instructive; more especially when we consider that the maximum of temperature is by no means remarkably different in the two places, 90° Fahr. being sometimes reached with us and not being often much exceeded at Batavia.

1 "Observations made at the Magnetical and Meteorological Observatory at Batavia. Published by order of the Government of Netherlands, India. Vol. L. Meteorological, from Jan. 1866 to Dec. 1868; and Magnetical, from July 1867 to June 1870. By Dr. P.A. Bergsma, Batavia, 1871." This fine work is entirely in English.

Causes of the uniform High Temperature near the Equator

It is popularly supposed that the uniform high temperature of the tropics is sufficiently explained by the greater altitude, and therefore greater heating-power of the midday sun; but a little consideration will show that this alone by no means accounts for the phenomenon. The island of Java is situated in from six and a half to eight and a half degrees of south latitude, and in the month of June the Sun's altitude at noon will not be more than from 58° to 60°. In the same month at London, which is fifty-one and a half degrees of north latitude, the sun's noonday altitude is 62°. But besides this difference of altitude in favour of London there is a still more important difference, for in Java the day is only about eleven and a half hours long in the month of June, while at London it is sixteen hours long, so that the total amount of sun-heat received by the earth must be then very much greater at London than at Batavia. Yet at the former place the mean temperature of the day and night is under 60° Fahr., while in the latter place it is 80° Fahr., the daily maximum being on the average in the one case about 68° and in the other about 89°

Neither does the temperature at the same place depend upon the height of the sun at noon; for at Batavia it is nearly vertical during October and February, but these are far from being the hottest months, which are May, June, and September, while December, January, and February are the coldest months, although then the sun attains nearby its greatest altitude. It is evident, therefore, that a difference of 30° in the altitude of the sun at noon, at different times of the year, has no apparent influence in raising the temperature of a place near the equator, and hence we conclude that other agencies are at work which often completely neutralize the effect which increased altitude must undoubtedly exert.

There is another important difference between the temperate and tropical zones, in the direct heating effect of the sun's rays independently of altitude. In England the noonday sun in the month of June rarely inconveniences us or produces any burning of the skin, while in the tropics at almost any hour of the day, and when the sun has an elevation of only 40° or 50°, exposure to it for a few minutes will scorch a European so that the skin turns red, becomes painful, and often blisters or peels off. Almost every visitor to the tropics suffers from incautious exposure of the neck, the leg, or some other part of the body to the sun's rays, which there possess a power as new as it is at first sight inexplicable, for it is not accompanied by any extraordinary increase in the temperature of the air.

These very different effects, produced by the same amount of sun-heat poured upon the earth in different latitudes, is due to a combination of causes. The most important of these are, probably,—the constant high temperature of the soil and of the surface-waters of the ocean,—the great amount of aqueous vapour in the atmosphere,—the great extent of the intertropical regions which cause the winds that reach the equatorial zone to be always warm,—and the latent heat given out during the formation of rain and dew. We will briefly consider the manner in which each of these causes contributes to the high degree and great uniformity of the equatorial temperature.

Influence of the Heat of the Soil

It is well known that at a very moderate depth the soil maintains a uniform temperature during the twenty-four hours, while at a greater depth even the annual inequalities disappear, and a uniform temperature, which is almost exactly the mean temperature of the air in the same locality, is constantly maintained throughout the year. The depth at which this uniform temperature is reached is greater as the annual range of temperature is greater, so that it is least near the equator, and greatest in localities near the arctic circle, where the greatest difference between summer and winter temperature prevails. In the vicinity of the equator, where the annual range of the thermometer is so small as we have see it to be at Batavia, the mean temperature of about 80° Fahr. is reached at a depth of four or five feet. The surplus heat received during the day is therefore conducted downwards very slowly, the surface soil becomes greatly super-heated, and a large portion of this heat is given out at night and thus keeps up the high temperature of the air when the sun has ceased to warm the earth. In the temperate zones, on the other hand, the stratum of uniform earth-temperature lies very deep. At Geneva it is not less than from thirty to forty feet, and with us it is probably fifty or sixty feet, and the temperature found there is nearly forty degrees lower than at the equator. This great body of cool earth absorbs a large portion of the surface heat during the summer, and conducts it downwards with comparative rapidity, and it is only late in the year (in July and August), when the upper layers of the soil have accumulated a surplus store of solar heat, that a sufficient quantity is radiated at night to keep up a rather high temperature in the absence of the sun. At the equator, on the other hand, this radiation is always going on, and earth-heat is one of the most important of the agencies which tend to equalise the equatorial climate.

Influence of the Aqueous Vapour of the Atmosphere

The aqueous vapour which is always present in considerable quantities in the atmosphere, exhibits a singular and very important relation to solar and terrestrial heat. The rays of the sun pass through it unobstructed to the earth; but the warmth given off by the heated earth is very largely absorbed by it, thus raising the temperature of the air; and as it is the lower strata of air which contain most vapour, these act as a blanket to the earth, preventing it from losing heat at night by radiation into space. During a large part of the year the air in the equatorial zone is nearly saturated with vapour, so that, notwithstanding the heat, salt and sugar become liquid, and all articles of iron get thickly coated with rust. Complete saturation being represented by 100, the daily average of greatest humidity at Batavia reaches 96 in January and 92 in December. In January, which is the dampest month, the range of humidity is small (77 to 96), and at this time the range of temperature is also least; while in September, with a greater daily range of humidity (62 to 92) the range of temperature is the greatest, and the lowest temperatures are recorded in this and the preceding month. It is a curious fact that in many parts of England the degree of humidity, as measured by the comparative saturation of the air, is as great as that of Batavia or even greater. A register kept at Clifton during the years 1853-1862 shows a mean humidity in January of 90, while the highest monthly mean for the four years at Batavia was 88; and while the lowest of the monthly means at Clifton was 79.1, the lowest at Batavia was 78.9. These figures, however, represent an immense difference in the *quantity* of vapour in every cubic foot of air. In January at Clifton, with a temperature of 35° to 40° Fahr., there would be only about 4 to 4frac12; grains of vapour per cubic foot of air, while at Batavia, with a temperature from 80° to 90° Fahr., there would be about 20 grains in the same quantity of air. The most important fact, however, is, that the capacity of air for holding vapour in suspension increases more rapidly than temperature increases, so that a fall of ten degrees at 50° Fahr. will lead to the condensation of about 1frac12; grain of vapour per cubic foot, while a similar fall of 90° Fahr. will set free 6frac12; grains. We can thus understand how it is that the very moderate fall of the thermometer during a tropical night causes heavier dews and a greater amount of sensible moisture than are ever experienced during much greater variations of temperature in the temperate zone. It is this large quantity of vapour in the equatorial atmosphere that keeps up a genial warmth throughout the night by preventing the radiation into space of the heat absorbed by the surface soil during the day. That this is really the case is strikingly proved by what occurs in the plains of Northern India, where the daily maximum of heat is far beyond anything experienced near the equator, yet, owing to the extreme dryness of the atmosphere, the clear nights are very cold, radiation being sometimes so rapid that water placed in shallow pans becomes frozen over.

As the heated earth, and everything upon its surface, does not cool so fast when surrounded by moist as by dry air, it follows that even if the quantity and intensity of the solar rays falling upon two given portions of the earth's surface are exactly equal, yet the sensible and effective heat produced in the two localities may be very different according as the atmosphere contains much or little vapour. In the one case the heat is absorbed more rapidly than it can escape by radiation; in the other case it radiates away into space, and is lost, more rapidly than it is being absorbed. In both cases an equilibrium will be arrived at, but in the one case the resulting mean temperature will be much higher than in the other. Thus we can understand the burning effects of the sun's rays in the tropics, since it results from the inability of the skin to part with the heat, either by radiation, evaporation, or absorption, as fast as it is received, and thus a temperature is quickly reached which disorganizes the delicate structures of the epidermis.

Influence of Winds on the Temperature of the Equator

The distance from the northern to the southern tropics being considerably more than three thousand miles, and the area of the intertropical zone more than one-third the whole area of the globe, it becomes hardly possible for any currents of air too reach the equatorial belt without being previously warmed by contact with the earth or ocean, or by mixture with the heated surface-air which is found in all intertropical and sub-tropical lands. This warming of the air is rendered more certain and more effective by the circumstance that all currents of air coming from the north or south have their direction changed owing to the increasing rapidity of the earth's rotational velocity, so that they reach the equator as easterly winds, and thus pass obliquely over a great extent of the heated surface of the globe. The causes that produce the westerly monsoons act in a similar manner, so that on the equator direct north or south winds, except as local land and sea-breezes, are almost unknown. The Batavia observations show that for ten months in the year the average direction of the wind varies only between 5° and 30° from due east or west, and these are also the strongest winds. In the two months—March and October—when the winds are northerly, they are very light, and are probably in great part local sea-breezes, which, from the position of Batavia, must come from the north over about two thousand miles of warm land and sea. As a rule, therefore, every current of air at or near the equator has passed obliquely over an immense extent of tropical surface and is thus necessarily a warm wind.

In the north temperate zone, on the other hand, the winds are always cool, and often of very low temperature even in the height of summer, due probably to their coming from colder northern regions as easterly winds, or from the upper parts of the atmosphere as westerly winds; and this constant supply of cool air, combined with quick radiation through a dryer atmosphere, carries off the solar hear so rapidly that an equilibrium is only reached at a comparatively low temperature. In the equatorial zone, on the contrary, the heat accumulates, on account of the absence of any medium of sufficiently low temperature to carry it off rapidly, and it thus soon reaches a point high enough to produce those scorching effects which are so puzzling when the altitude of the sun or the indications of the thermometer are alone considered. Whenever, as is sometimes the case, exceptional cold occurs near the equator, it can almost always be traced to the influence of current of air of unusually low temperature. Thus in July near the Aru islands, the writer experienced a strong south-east wind which almost neutralized the usual effects of tropical heat, although the weather was bright and sunny. But the wind, coming direct from the southern ocean during its winter without acquiring heat by passing over land, was necessarily of a low temperature. Again, Mr. Bates informs us that in the Upper Amazon in the month of May there is a regularly recurring south wind which produces a remarkable lowering of the usual equatorial temperature. But owing to the increased velocity of the earth's surface at the equator a south wind there must have been a south-west wind at its origin, and this would bring it directly from the high chain of the Peruvian Andes during the winter of the southern hemisphere. It is therefore probably a cold mountain wind, and blowing as it does over a continuous forest, it has been unable to acquire the usual tropical warmth.

The cause of the striking contrast between the climates of equatorial and temperate lands at times when both are receiving an approximately equal amount of solar heat may perhaps be made clearer by an illustration. Let us suppose there to be two reservoirs of water, each supplied by a pipe which pours into it a thousand gallons a day, but with runs only during the daytime, being cut off at night. The reservoirs are both leaky, but while the one loses at the rate of nine hundred gallons in the twenty-four hours, the other loses at the rate of eleven hundred gallons in the same time, supposing that both are kept exactly half full and thus subjected to the same uniform water-pressure. If now both are left to be supplied by the above-mentioned pipes the result will be, that in the one which loses by leakage less than it receives the water will rise day by day till the increased pressure causes the leakage to increase to such an extent as exactly to balance the supply; while in the other the water will sink till the decreasing pressure causes the leakage to decrease till it also just balances the supply, when both will remain stationary, the one at a high the other at a low average level, each rising during the day and sinking again at night. Just the same thing occurs with that great heat-reservoir the earth, whose actual temperature at any spot will depend, not alone upon the quantity of heat it receives, but on the balance between its constantly varying waste and supply. We can thus understand how it is that, although in the months of June and July, Scotland in latitude 57° north receives as much sun-heat as Angola or Timor in latitude 10° south, and for a much greater number of hours daily, yet in the latter countries the mean temperature will be about 80° Fahr., with a daily maximum of 90° to 95°, while in the former the mean will be about 60° Fahr., with a daily maximum of 70° or 75°; and, while in Scotland exposure to the full noon-day sun produces no unpleasant heat-sensations, a similar exposure in Timor at any time between 9 A.M. and 3 P.M. would blister the skin in a few minutes almost as effectually as the application of scalding water.

Heat due to the Condensation of Atmospheric Vapour

Another cause which tends to keep up a uniform high temperature in the equatorial, as compared with the variable temperatures of the extra-tropical zones, is the large amount of heat liberated during the condensation of the aqueous vapour of the atmosphere in the form of rain and dew. Owing to the frequent near approach of the equatorial atmosphere to the saturation point, and the great amount of vapour its high temperature enables it to hold in suspension, a very slight fall of the thermometer is accompanied by the condensation of a large absolute quantity of atmospheric vapour, so that copious dews and heavy showers of rain are produced at comparatively high temperatures, and even at the sea level.

The drops of rain rapidly increase in size while falling through the saturated atmosphere; and during this process as well as by the formation of dew, the heat which retained the water in the gaseous form, and was insensible while doing so, is liberated, and thus helps to keep up the high temperature of the air. This production of heat is almost always going on. In fine weather the nights are always dewy, and the diagram on the preceding page, showing the mean monthly rainfall at Batavia and Greenwich, proves that this source of increased temperature is present during every month in the year, since the lowest monthly fall at the former place is almost equal to the highest monthly fall at the latter.

It may perhaps be objected that evaporation must absorb as much heat as is afterwards liberated by condensation, and this is true; but as evaporation and condensation occur usually at different times and in different places, the equalising effect is still very important. Evaporation occurs chiefly during the hottest sunshine, when it tends to moderate the extreme heat, while condensation takes place chiefly at night in the form of dew and rain, when the liberated heat helps to make up for the loss of the direct rays of the sun. Again the most copious condensation both of dew and rain is greatly influenced by vegetation and especially by forests, and also by the presence of hills and mountains, and is therefore greater on land than on the ocean, while evaporation is much greater on the ocean, both on account of the less amount of cloudy weather and because the air is more constantly in motion. This is particularly the case throughout that large portion of the tropical and subtropical zones where the trade-winds constantly blow, as the evaporation must there be enormous while the quantity of rain is very small. It follows, than, that on the equatorial land-surface there will be a considerable balance of condensation over evaporation, which must tend to the general raising of the temperature, and, owing to the condensation being principally at night, not less powerfully to its equalization.

General Features of the Equatorial Climate

The various causes now enumerated are sufficient to enable us to understand how the great characteristic features of the climate of the equatorial zone are brought about, how it is that so high a temperature is maintained during the absence of the sun at night, and why so little effect is produced by the sun's varying altitude during its passage from the northern to the southern tropic. In this favoured zone the heat is never oppressive, as it so often becomes on the borders of the tropics; and the large absolute amount of moisture always present in the air is almost as congenial to the health of man as it is favourable to the growth and development of vegetation.[1] Again, the lowering of the temperature at night is so regular and yet so strictly limited in amount, that, although never cold enough to be unpleasant, the nights are never so oppressively hot as to prevent sleep. During the wettest months of the year, it is rare to have many days in succession without some hours of sunshine, while even in the driest months there are occasional showers to cool and refresh the overheated earth. As a result of this condition of the earth and atmosphere, there is no check to vegetation, and little if any demarcation of the season. Plants are all evergreen; flowers and fruits, although more abundant at certain seasons, are never altogether absent; while many annual food-plants as well as some fruit-trees produce two crops a year. In other cases, more than one complete year is required to mature the large and massive fruits, so that it is not uncommon for fruit to be ripe at the same time that the tree is covered with flowers in preparation for the succeeding crop. This is the case with the Brazil nut tree in the forests of the Amazon, and with many other tropical as with a few temperate fruits.

[1] Where the inhabitants adapt their mode of life to the peculiarities of the climate, as is the case with the Dutch in the Malay Archipelago, they enjoy as robust health as in Europe both in the case of persons born in Europe and of those who for generations have lived under a vertical sun.

Uniformity of the Equatorial Climate in all Parts of the Globe

The description of the climatal phenomena of the equatorial zone here given has been in great part drawn from long personal experience in South America and in the Malay Archipelago. Over a large portion of these countries the same general features prevail, only modified by varying local conditions. Whether we are at Singapore or Batavia, in the Moluccas or New Guinea, at Para, at the sources of the

Rio Negro, or on the Upper Amazon, the equatorial climate is essentially the same, and we have no reason to believe that it materially differs in Guinea or the Congo. In certain localities, however, a more contrasted wet and dry season prevails, with a somewhat greater range of the thermometer. This is generally associated with a sandy soil, and a less dense forest, or with an open and more cultivated country. The open sandy country with scattered trees and shrubs or occasional thickets, which is found at Santarem and Monte-Alegre on the lower Amazon, are examples, as well as the open cultivated plains of Southern Celebes; but in both cases the forest country in adjacent districts has a moister and more uniform climate, so that it seem probable that the nature of the soil or the artificial clearing away of the forests, are important agents in producing the departure from the typical equatorial climate observed in such districts.

Effects of Vegetation on Climate

The almost rainless district of Ceara on the north-east coast of Brazil, and only a few degrees south of the equator, is a striking example of the need of vegetation to react on the rainfall. We have here no apparent cause but the sandy soil and bare hills, which, when heated by the equatorial sun, produce ascending currents of warm air and thus prevent the condensation of the atmospheric vapour, to account for such an anomaly; and there is probably no district where judicious planting would produce such striking and beneficial effects. In Central India the scanty and intermittent rainfall, with its fearful accompaniment of famine, is perhaps in great part due to the absence of a sufficient proportion of forest-covering to the earth's surface; and it is by a systematic planting of all the hill-tops, elevated ridges, and higher slopes that we shall probably cure the evil. This would almost certainly induce an increased rainfall; but even more important and more certain is the action of forests in checking evaporation from the soil and causing perennial springs to flow, which may be collected in vast storage tanks and serve to fertilise a great extent of country; whereas tanks without regular rainfall or permanent springs to supply them are worthless. In the colder parts of the temperate zones the absence of forests is not so much felt, because the hills and uplands are naturally clothed with a thick coating of turf or peat which absorbs moisture and does not become overheated by the sun's rays, and the rains are seldom violent enough to strip this protective covering from the surface. In tropical and even in warm-temperate countries, on the other hand, the rains are periodical and often of excessive violence for a short period; and when the forests are cleared away the torrents of rain soon strip off the vegetable soil, and thus destroy in a few years the fertility which has been the growth of many centuries. The bare subsoil becoming heated by the sun, every particle of moisture which does not flow off is evaporated, and this again react on the climate, producing long-continued droughts only relieved by sudden and violent storms, which add to the destruction and render all attempts at cultivation unavailing. Wide tracts of fertile land in the south of Europe have been devastated in this manner, and have become absolutely uninhabitable. Knowingly to produce such disastrous results would be a far more serious offence than any destruction of property which human labour has produced and can replace; yet we have ignorantly allowed such extensive clearings for coffee cultivation in India and Ceylon as to cause the destruction of much fertile soil, which human labour cannot replace, and which will surely, if not checked in time, lead to the deterioration of the climate and the permanent impoverishment of the country.[1]

[1] For a terrible picture of the irreparable devastation caused by the reckless clearing of forests, see the third chapter of Mr. Marsh's work, *The Earth as Modified by Human Action.*

Short Twilight of the Equatorial Zone

One of the phenomena which markedly distinguish the equatorial from the temperate and polar zones is the shortness of the twilight and consequent rapid transition from day to night and from night to day. As this depends only on the fact of the sun descending vertically instead of obliquely below the horizon, the difference is most marked when we compare our midsummer twilight with that of the tropics. Even with us the duration of twilight is very much shorter at the time of the equinoxes, and it is probably not much more than a third shorter than this at the equator. Travellers usually exaggerate the shortness of the tropical twilight, it being sometimes said that if we turn a page of the book we are reading when the sun disappears, by the time we turn over the next page it will be too dark to see to read. With an average book and an average reader this is certainly not true, and it will be well to describe as correctly as we can what really happens.

In fine weather the air appears to be somewhat more transparent near the equator than with us, and intensity of sunlight is usually very great up to the moment when the solar orb touches the horizon. As soon as it has disappeared the apparent gloom is proportionally great, but this hardly increases perceptibly during the first ten minutes. During the next ten minutes, however, it becomes rapidly darker, and at the end of about half and hour from sunset the complete darkness of night is almost reached. In the morning the changes are perhaps even more striking. Up to about a quarter past five o'clock the darkness is complete; but about that time a few cries of birds begin to break the silence of night, perhaps indicating that sings of dawn are perceptible in the eastern horizon. A little later the melancholy voices of the goat-suckers are heard, varied croakings of frogs, the plaintive whistle of mountain thrushes, and strange cries of birds or mammals peculiar to each locality. About half-past five the first glimmer of light becomes perceptible; if slowly becomes lighter, and then increases so rapidly that at about a quarter to six it seems full daylight. For the next quarter of an hour this changes very little in character; when, suddenly, the sun's rim appears above the horizon decking the dew-laden foliage with glittering gems, sending gleams of golden light far into the woods, and waking up all nature to life and activity. Birds chirp and flutter about, parrots scream, monkeys chatter, bees hum among the flowers, and gorgeous butterflies flutter lazily along or sit with fully expanded wings exposed to the warm and invigorating rays. The first hour of morning in the equatorial regions possesses a charm and a beauty that can never be forgotten. All nature seems refreshed and strengthened by the coolness and moisture of the past night; new leaves and buds unfold almost before the eye, and fresh shoots may often be observed to have grown many inches since the preceding day. The temperature is the most delicious conceivable. The slight chill of early dawn, which was itself agreeable, is succeeded by an invigorating warmth; and the intense sunshine lights up the glorious vegetation of the tropics, and realises all that the magic art of the painter or the growing words of the poet have pictured as their ideals of terrestrial beauty.

The Aspect of the Equatorial Heavens

Within the limits of the equatorial zone the noonday sun is truly vertical twice every year, and for several months it passes so near the zenith that the difference can hardly be detected without careful observation of the very short shadows of vertical objects. The absence of distinct horizontal shadows at noon, which thus characterises a considerable part of the year, is itself a striking phenomenon to an inhabitant of the temperate zones; and equally striking is the changed aspect of the starry heavens. The grand constellation Orion passes vertically overhead, while the Great Bear is only to be seen low down in the northern heavens, and the Pole star either appears close to the horizon or has altogether disappeared, according as we are north or south of the equator. Towards the south the Southern Cross, the Magellanic clouds, and the jet-black "coal sacks" are the most conspicuous objects invisible in our northern latitudes. The same cause that brings the sun overhead in its daily march equally affects the planets, which appear high up towards the zenith far more frequently than with us, thus affording splendid opportunities for telescopic observation.

Intensity of Meteorological Phenomena at the Equator

The excessive violence of meteorological phenomena generally supposed to be characteristic of the tropics is not by any means remarkable in the equatorial zone. Electrical disturbances are much more frequent, but not generally more violent then in the temperate regions. The wind-storms are rarely of excessive violence, as might in fact be inferred from the extreme steadiness of the barometer, whose daily range at Batavia rarely exceeds one-eighth, of an inch, while the extreme range during three years was less than one-third of an inch! The amount of the rainfall is very great, seventy or eighty inches in a year being a probable average; and as the larger part of this occurs during three or four months, individual rainfalls are often exceedingly heavy. The greatest fall recorded at Batavia during three years was three inches and eight-tenths in one hour,[1] but this was quite exceptional, and even half this quantity is very unusual. The greatest rainfall recorded in twenty-four hours is seven inches and a quarter; but more than four inches in one day occurs only on two or three occasions in a year. The blue colour of the sky if probably not so intense as in many parts of the temperate zone, while the brilliancy of the moon and stars is not perceptibly greater than on our clearest frosty nights, and is undoubtedly much inferior to what is witnesses in many desert regions, and even in Southern Europe.

On the whole, then, we must decide that uniformity and abundance, rather than any excessive manifestations, are the prevailing characteristic of all the climatal phenomena of the equatorial zone.

[1] On 10th January 1867, from 1 to 2 A.M.

Concluding Remarks

We cannot better conclude our account of the equatorial climate than by quoting the following vivid description of the physical phenomena which occur during the early part of the dry season at Para. It is taken from Mr. Bates' *Naturalist on the Amazons,* and clearly exhibits some of the more characteristic features of a typical equatorial day.

"At that early period of the day (the first two hours after

sunrise) the sky was invariably cloudless, the thermometer marking 72° or 73° Fahr.; the heavy dew or the previous night' rain, which lay on the moist foliage, becoming quickly dissipated by the glowing sun, which, rising straight out of the east, mounted rapidly towards the zenith. All nature was fresh, new leaf and flower-buds expanding rapidly. … The heat increased hourly, and towards two o'clock reached 92° to 93° Fahr., by which time every voice of bird and mammal was hushed. The leaves, which were so moist and fresh in early morning, now became lax and drooping, and flowers shed their petals. On most days in June and July a heavy shower would fall some time in the afternoon, producing a most welcome coolness. The approach of a rain-clouds was after a uniform fashion very interesting to observe. First, the cool sea-breeze which had commenced to blow about ten o'clock, and which had increased in force with the increasing power of the sun, would flag, and finally die away. The heat and electric tension of the atmosphere would then become almost insupportable. Languor and uneasiness would seize on every one, even the denizens of the forest betraying it by their motions. White clouds would appear in the east and gather into cumuli, with an increasing blackness along their lower portions. The whole eastern horizon would become almost suddenly black, and this would spread upwards, the sun at length becoming obscured. Then the rush of a mighty wind is heard through the forest, swaying the tree-tops; a vivid flash of lighting bursts forth, then a crash of thunder, and down streams the deluging rain. Such storms soon cease, leaving bluish-black motionless clouds in the sky until night. Meantime all nature is refreshed; but heaps of flower-petals and fallen leaves are seen under the trees. Towards evening life revives again, and the ringing uproar is resumed from bush and tree. The following morning the sun again rises in a cloudless sky; and so the cycle is completed; spring, summer, and autumn, as it were in one tropical day. The days are more or less like this throughout the year. A little difference exists between the dry and wet seasons; but generally the dry season, which lasts from July to December, is varied with showers, and the wet, from January to June, with sunny days. It results from this, that the periodical phenomena of plants and animals do not take place at about the same time in all species, or in the individuals of any given species, as they do in temperate countries. In Europe a woodland scene has its spring, its summer, its autumnal, and its winter aspects. In the equatorial forests the aspect is the same or nearly so every day in the year; budding, flowering, fruiting, and leaf-shedding are always going on in the one species or other. It is never either spring, summer, or autumn, but each day is a combination of all three. With the day and night always of equal length, the atmospheric disturbances of each day neutralising themselves before each succeeding morn; with the sun in its course proceeding midway across the sky, and the daily temperature almost the same throughout the year—how grand in its perfect equilibrium and simplicity is the march of Nature under the equator!"

II

EQUATORIAL VEGETATION

The Equatorial Forest-Belt and its causes—General features of the Equatorial Forests—Characteristics of the Larger Forest-trees—Flowering Trunks and their probable cause—Uses of Equatorial Forest-trees—The Climbing Plants of the Equatorial Forests—Palms—Uses of Palm-trees and their Products—Ferns—Ginger-worts and wild Bananas—Arums—Screw-pines—Orchids—Bamboos—Uses of the Bamboo—Mangroves—Sensitive-plantgs—Comparative Scarcity of Flowers—Concluding Remarks on Tropical Vegetation.

IN the following sketch of the characteristics of vegetable life in the equatorial zone, it is not intended to enter into any scientific details or to treat the subject in the slightest degree from a botanical point of view; but merely to describe those general features of vegetation which are almost or quite peculiar to this region of the globe, and which are so general as to be characteristic of the greater part of it rather than of any particular country or continent within its limits.

The Equatorial Forest-Belt and its Causes

With but few and unimportant exceptions a great forest band from a thousand to fifteen hundred miles in width girdles the earth at the equator, clothing hill, plain, and mountain with an evergreen mantle. Lofty peaks and precipitous ridges are sometimes bare, but often the woody covering continues to a height of eight or ten thousand feet, as in some of the volcanic mountains of Java and on portions of the Eastern Andes. Beyond the forest both to the north and south, we meet first with woody and then open country, soon changing into arid plains or even deserts which form an almost continuous band in the vicinity of the two tropics. On the line of the tropic of Cancer we have, in America, the deserts and dry plains of New Mexico; in Africa the Sahara; and in Asia, the Arabian deserts, those of Beloochistan and Western India, and farther east the dry plains of North China and Mongolia. On the tropic of Capricorn we have, in America, the Grand Chaco desert and the Pampas; in Africa, the Kalahari desert and the dry plains north of the Limpopo; while the deserts and waterless plains of Central Australia complete the arid zone. These great contrasts of verdure and barrenness occurring in parallel bands all round the globe, must evidently depend on the general laws which determine the distribution of moisture over the earth, more or less modified by local causes. Without going into meteorological details, some of which have been given in the preceding chapter, the main facts may be explained by the mode in which the great aerial currents are distributed. The trade winds passing over the ocean from north-east to south-west, and from south-east to north-west, with an oblique tendency towards the equator, become saturated with vapour, and are ready to give out moisture whenever they are forced upwards or in any other way have their temperature lowered. The entire equatorial zone becomes thus charged with vapour-laden air, which is the primary necessity of a luxuriant vegetation. The surplus air (produced by the meeting of the two trade winds) which is ever rising in the equatorial belt and giving up its store of vapour, flows off north and south as dry, cool air, and descends to the earth in the vicinity of the tropics. Here it sucks up whenever moisture it meets with and thus tends to keep this zone in an arid condition. The trades themselves are believed to be supplied by descending currents from the temperate zones, and these are at first equally dry and only become vapour-laden when they have passed over some extent of moist surface. At the solstices the sun passes vertically over the vicinity of the tropics for several weeks, and this further aggravates the aridity; and wherever the soil is sandy and there are no lofty mountain chains to supply ample irrigation, the result is a more or less perfect desert. Analogous causes, which a study of aerial currents will render intelligible, have produced other great forest-belts in the northern and southern parts or the temperate zones; but owing to the paucity of land in the southern hemisphere these are best seen in North America and Northern Euro-Asia, where they form the great northern forests of deciduous trees and of Coniferæ. These being comparatively well known to us, will form the standard by a reference to which we shall endeavour to point out and render intelligible the distinctive characteristic of the equatorial forest vegetation.

General Features of the Equatorial Forests

It is not easy to fix upon the most distinctive features of these virgin forests, which nevertheless impress themselves upon the beholder as something quite unlike those of temperate lands, and as possessing a grandeur and sublimity altogether their own. Amid the countless modifications in detail which these forests presents, we shall endeavour to point out the chief peculiarities as well as the more interesting phenomena which generally characterise them.

The observer new to the scene would perhaps be first struck by the varied yet symmetrical trunks, which rise up with perfect straightness to a great height without a branch, and which, being placed at a considerable average distance apart, give an impression similar to that produced by the columns of some enormous building. Overhead, at a height, perhaps, of a hundred and fifty feet, is an almost unbroken canopy of foliage formed by the meeting together of these great trees and their interlacing branches; and this canopy is usually so dense that but an distinct glimmer of the sky is to be seen, and even the intense tropical sunlight only penetrates to the ground subdued and broken up into scattered fragments. There is a weird gloom and a solemn silence, which combine to produce a sense of the vast—the primeval—almost of the infinite. It is a world in which man seems an intruder, and where he feels overwhelmed by the contemplation of the ever-acting forces which, form the simple elements of the atmosphere, build up the great mass of vegetation which overshadows and almost seems to oppress the earth.

Characteristics of the Larger Forest Trees

Passing from the general impression to the elements of which the scene is composed, the observer is struck by the great diversity of the details amid the general uniformity. Instead of endless repetitions of the same forms of trunk such as are to be seen in our pine, or oak, or beechwoods, the eye wanders from one tree to another and rarely detects two together of the same species. All are tall and upright columns, but they differ from each other more than do the columns of Gothic, Greek, and Egyptian temples. Some are almost cylindrical, rising up out of the ground as if their bases were concealed by accumulations of the soil; others get much thicker near the ground like our spreading oaks; others again, and these are very characteristic, send out towards the base flat and wing-like projections. These projections are thin slabs radiating from the main trunk, from which they stand out like the buttresses of Gothic cathedral. They rise to various heights on the tree, from five or six to twenty or thirty feet; they often divide as they approach the ground, and sometimes twist and curve along the surface for a considerable distance, forming elevated and greatly compressed roots. These buttresses are sometimes so large that the spaces between them if roofed over would form huts capable of containing several persons. Their use is evidently to give the tree an extended base, and so assist the subterranean roots in maintaining in an erect position so lofty a column crowned by a broad and massive head of branches and foliage. The buttressed trees belong to a variety of distinct groups. Thus, many of the Bombaceæ or silk-cotton trees, several of the Leguminosæ, and perhaps many trees belonging to other natural orders, possess these appendages.

There is another form of tree, hardly less curious, in which the trunk, though generally straight and cylindrical, is deeply furrowed and indented, appearing as if made up of a number of small trees grown together at the centre. Sometimes the junction of what seem to be the component parts is so imperfect that gaps or holes are left by which you can see through the trunk in various places. At first one is disposed to think this is caused by accident or decay, but repeated examination shows it to be due to the natural growth of the tree. The accompanying outline sections of one of these trees that was cut down exhibits its character. It was a noble forest tree, more than two hundred feet high, but rather slender in proportion, and it was by no means an extreme example of its class. This peculiar form is probably produced by the downward growth of aerial roots, like some New Zealand trees whose growth has been traced, and of whose different stages drawings may be seen at the Library of the Linnæan Soceity. These commence their existence as parasitical climbers, which take root in the fork of some forest tree and send down aerial roots which clasp round the steam that upholds them. As these roots increase in size and grow together laterally they cause the death of their foster-parent. The climber then grows rapidly, sending out large branches above and spreading roots below, and as the supporting tree decays away the aerial roots grow together and form a new trunk, more or less furrowed and buttressed, but exhibiting no other marks of its exceptional origin. Aerial-rooted forest tress—like that figured in my *Malay Archipelago* (vol

.i. p. 131)—and the equally remarkable fig-trees of various species, whose trunks are formed by a miniature forest of aerial roots, sometimes separate, sometimes matted together, are characteristic of the Eastern tropics, but appear to be rare or altogether unknown in America, and can therefore hardly be included among the general characteristics of the equatorial zone.

Besides the varieties of form, however, the tree-trunks of these forests present many peculiarities of colour and texture. The majority are rather smooth-barked, and many are of peculiar whitish, green, yellowish, or brown colours, or occasionally nearly black. Some are perfectly smooth, others deeply cracked and furrowed, while in a considerable number the bark splits off in flakes or hangs down in long fibrous ribands. Spined or prickly trunks (excepts of palms) are rare in the damp equatorial forests. Turning our gaze upwards from the stems to the foliage, we find two types of leaf not common in the temperate zone, although the great mass of the trees offer nothing very remarkable in this respect. First, we have many trees with large, thick, and glossy leaves, like those of the cherry-laurel or the magnolia, but even larger, smoother, and more symmetrical. The leaves of the Asiatic caoutchouc tree (Ficus elastica,) so often cultivated in houses, is a type or this class, which has a very fine effect among the more ordinary-looking foliage. Contrasted with this is the fine pinnate foliage of some of the largest forest trees, which, seen far aloft against the sky, looks as delicate as that of the sensitive mimosa.

Forest Trees of Low Growth

The great trees we have hitherto been describing form, however, but a portion of the forest. Beneath their lofty canopy there often exists a second forest of moderate-seized trees, whose crowns, perhaps forty or fifty feet high, do not touch the lowermost branches of those above them. These are of course shade-loving trees, and their presence effectually prevents the growth of any young trees of the larger kinds, until overcome by age and storms, some monarch of the forest falls down, and, carrying destruction in its fall, opens up a considerable space into which sun and air can penetrate. Then comes a race for existence among the seedlings of the surrounding trees, in which a few ultimately prevail and fill up the space vacated by their predecessor. Yet beneath this second set of medium-sized forest trees there is often a third undergrowth of small trees, from six to ten feet high, of dwarf palms, of tree-ferns, and of gigantic herbaceous ferns. Yet lower, on the surface of the ground itself, we find much variety. Sometimes the earth is completely bare, a mass of decaying leaves and twigs and fallen fruts. More frequently it is covered with a dense carpet of selaginella or other lycopodiaceæ,and these sometimes give place to a variety of herbaceous plants, sometimes with pretty, but rarely with very conspicuous flowers.

Flowering Trunks and their Probable Cause

Among the minor but not unimportant peculiarities that characteristic these lofty forests is the curious way in which many of the smaller trees have their flowers situated on the main trunk or larger branches instead of on the upper part of the tree. The cacao-tree is a well-known example of this peculiarity, which is not uncommon in tropical forests; and some of the smaller trunks are occasionally almost hidden by the quantity of fruit produced on them. One of the most beautiful examples of this mode of flowering is a small tree of the genus Polyalthea, belonging to the family of the custard-apples, not uncommon in the forests of north-western Borneo. Its slender trunk, about fifteen or twenty feet high, was completely covered with star-shaped flowers, three inches across and of a rich orange-red colour, making the trees look as if they had been artificially decorated with brilliant garlands. The recent discoveries as to the important part played by insects in the fertilisation of flowers offers a very probable explanation of this peculiarity. Bees and butterflies are the greatest flower-haunters. The former love the sun and frequent open grounds or the flowery tops of the lofty forest trees fully exposed to the sun and air. The forest shades are frequently by thousands of butterflies, but these mostly keep near the ground, where they have a free passage among the tree-trunks and visit the flowering shrubs and herbaccous plants. To attract these it is necessary that flowers should be low down and conspicuous. If they grew in the usual way on the tops of these smaller trees overshadowed by the dense canopy above them they would be out of sight of both groups of insects; but being placed openly on the stems, and in the greatest profusion, they cannot fail to attract the attention of the wandering butterflies.

Uses of Equatorial Forest Trees

Amid this immense variety of trees, the native have found out such as are best adapted to certain purposes. The wood of some is light and soft, and is used for floats or for carving rude images, stools, and ornaments for boats and houses. The flat slabs of the buttresses are often used to make paddles. Some of the trees with furrowed stems are exceedingly strong and durable, serving as posts for houses or as piles on which the water-villages are built. Canoes, formed from a trunk hollowed out and spread open under the action of heat require one kind of wood, those built up with planks another; and as the species of trees in these forests are so much more numerous than the wants of a semi-civilised population, there are probably a large number of kinds of timber which will some day be found to be well adapted to the special requirements of the arts and sciences. The products of the trees of the equatorial forests, notwithstanding our imperfect knowledge of them, are already more useful to civilised man than to the indigenous inhabitants. To mention only a few of those whose names are tolerably familiar to us, we have such valuable woods as mahogany, teak, ebony, lignum-vitæ, purple-heart, iron-wood, sandal-wood, and satin-wood; such useful gums as india-rubber, gutta-percha, tragacanth, copal, lac, and dammar; such dyes as are yielded by log-wood, brazil-wood, and sappan-wood; such drugs as the balsams of Capivi and Tolu, camphor, benzoin, catechu or terra-japonica, cajaput oil, gamboges, quinine, Angostura bark, quassia, and the urari and upas poisons; of species we have cloves, cinnamon, and nutmegs; and of fruits, brazil-nuts, tamarinds, guavas, and the valuable cacao; while residents in our tropical colonies enjoy the bread-fruit, avocado-pear, custard-apple, durian, mango, mangosteen, sour-sop, papaw, and many others. This list of useful products from the exogenous trees alone of the equatorial forests, excluding those from the palms, shrubs, herbs, and creepers, might have been multiplied many times over by the introduction of articles whose names would be known only to those interested in special arts of sciences; but imperfect as it is, it will serve to afford a notion of the value of this vast treasure-house, which is as yet but very partially explored.

The Climbing Plants of the Equatorial Forests

Next to the trees themselves the most conspicuous and remarkable feature of the tropical forests is the profusion of woody creepers and climbers that everywhere meet the eye. They twist around the slender stems, they drop down pendent from the branches, they stretch tightly from tree to tree, they hang looped in huge festoons from bough to bough, they twist in great serpentine coils or lie in entangled masses on the ground. Some are slender, smooth, and root-like; others are rugged or knotted; often they are twined together into veritable cables; some are flat like ribands, others are curiously waved and indented. Where they spring from or how they grow is at first a complete puzzle. They pass overhead from tree to tree, they stretch in tight cordage like the rigging of a ship from the top of one tree to the base of another, and the upper regions of the forest often seem full of them without our being able to detect any earth-growing stem from which they arise. The conclusion is at length forced upon us that these woody climbers must possess the two qualities of very long life and almost indefinite longitudinal growth, for by these suppositions alone can we explain their characteristic features. The growth of climbers, even more than all other plants, is upward towards the light. In the shade of the forest they rarely or never flower, and seldom even produce foliage, but when they have reached the summit of the tree that supports them, they expand under the genial influence of light and air, and often cover their foster-parent with blossoms not is own. Here, as a rule, the climber's growth would cease; but the time comes when the supporting tree rots and falls, and the creeper comes with it in torn and tangled masses to the ground. But through its foster-parent is dead it has itself received no permanent injury, but shoots out again till if finds a fresh support, mounts another tree, and again puts forth its leaves and flowers. In time the old tree rots entirely away and the creeper remains tangled on the ground. Sometimes branches only fall and carry a portion of the creeper tightly stretched to an adjoining tree; at other times the whole tree is arrested by a neighbour, to which the creeper soon transfer itself in order to reach the upper light. When by the fall of a branch the creepers are left hanging in the air, they may be blown about by the wind and catch hold of trees growing up beneath them, and thus become festooned from one tree to another. When these accidents and changes have been again and again repeated the climber may have travelled very far from its parents stem, and may have mounted to the tree tops and descended again to the earth several times over. Only in this way does it seem possible to explain the wonderfully complex manner in which these climbing plants wander up and down the forest as if guided by the strangest caprices, or how they become so crossed and tangled together in the wildest confusion.

The variety in the length, thickness, strength, and toughness of these climbers enables the natives of tropical countries to put them to various uses. Almost every kind of cordage is supplied by them. Some will stand in water without rotting, and are used for cables, for lines to which are attached fish-traps, and to bind and strengthen the wooden anchors used generally in the East. Boats and even large sailing vessels are built, whose planks are entirely fastened together by this kind of cordage skilfully applied to internal ribs. For the better kinds of houses, smooth and uniform varieties are chosen, so that the beams and rafters can be bound together with neatness, strength, and uniformity, as is especially observable among the indigenes of the Amazonian forests. When baskets of great strength are required special kinds of creepers are used; and to serve almost every purpose for which we should need rope or a chain, the tropical savage adopts some one of the numerous forest-ropes which long experience has shown to have qualities best adapted for it. Some are smooth and supple; some are tough and will bear twisting or tying; some will last longest in salt water, others in fresh; one is uninjured by the heat and smoke of fires, while another is bitter or otherwise prejudicial to insect enemies.

Besides these various kinds of trees and climbers, which form the great mass of the equatorial forests and determine their general aspect, there are a number of forms of plants which are always more or less present, though in some parts scarce and in others in great profusion, and which largely aid in giving a special character to tropical as distinguished from temperate vegetation. Such are the various groups of palms, ferns, ginger-worts, and wild plantains, arums, orchids, and bamboos; and under these heads we shall give a short account of the part they take in giving a distinctive aspect to the equatorial forests.

Palms

Although these are found throughout the tropics, and a few species even extend into the warmer parts of the temperate regions, they are yet so much more abundant and varied within the limits of the region we are discussing that they may be considered as among the most characteristic forms of vegetation of the equatorial zone. They are, however, by no means generally present, and we may pass through miles of forest without even seeing a palm. In other parts they abound; either forming a lower growth in the lofty forest, or in swamps and on hillsides sometimes rising up above the other trees. On river-banks they are especially conspicuous and elegant, bending gracefully over the stream, their fine foliage waving in the breeze, and their stems often draped with hanging creepers.

The chief feature of the palm tribe consists in the cylindrical trunk crowned by a mass of large and somewhat rigid leaves. They vary in height from a few feet to that of the loftiest forest trees. Some are stemless, consisting only of a spreading crown of large pinnate leaves; but the great majority have a trunk slender in proportion to its height. Some of the smaller species have stems no thicker than a lead pencil, and four or five feet high; while the great Mauritia of the Amazon has a trunk full two feet in diameter, and more than one hundred feet high. Some species probably reach a height of two hundred feet, for Humboldt states that in South America he measured a palm, which was one hundred and ninety-two English feet high. The leaves of palms are often of immense size. Those of the Manicaria saccifera of Para are thirty feet long and four or five feet wide, and are not pinnate but entire and very rigid. Some of the pinnate leaves are much larger, those of the Raphia tædigera and Maximiliana regia being both sometimes more than fifty feet long. The fan-shaped leaves of other species are ten or twelve feet in diameter. The trunks of palms are sometimes smooth and more or less regularly ringed, but they are frequently armed with dense prickles which are sometimes eight inches long. In some species the leave fall to the ground as they decay, leaving a clean scar, but in most cases they are persistent, rotting slowly away, and leaving a mass of fibrous stumps attached to the upper part of the stem. This rotting mass forms an excellent soil for ferns, orchids, and other semi-parasitical plants, which form an attractive feature on what would otherwise be an unsightly object. The sheathing margins of the leaves often break up into a fibrous material, sometimes resembling a coarse cloth, and in other cases more like horse-hair. The flowers are not individually large, but form large spikes or racemes, and the fruits are often beautifully scaled and hang in huge bunches, which are sometimes more than a load for a strong man. The climbing palms are very remarkable, their tough, slender, prickly stems mounting up by means of the hooked midribs of the leaves to the tops of the loftiest forest trees, above which they send up an elegant spike of foliage and flowers. The most important are the American Desmoncus and the Eastern Calamus, the latter being the well-known rattan or cane of which chair-seats are made, from the Malay name "rotang." The rattan-palms are the largest and most remarkable of the climbing group. They are very abundant in the drier equatorial forests, and more than sixty species are known from the Malay Archipelago. The stems (when cleaned from the sheathing leaves and prickles) vary in size, from the thickness of a quill to that of the wrist; and where abundant they render the forest almost impassable. They lie about the ground coiled and twisted and looped in the most fantastic manner. They hang in festoons from trees and branches, they rise suddenly through mid air up to the top of the forest, or coil loosely over shrubs and in thickness like endless serpents. They must attain an immense age, and apparently have almost unlimited powers of growth, for some are said to have been found which were six hundred or even one thousand feet long, and if so they are probably the longest of all vegetable growths. The mode in which such great lengths and tangled convolutions have been attained has already been explained in the general account of woody climbers. From the immense strength of these canes and the facility with which they can be split, they are universally used for cordage in the countries where they grow in preference to any other climbers, and immense quantities are annually exported to all parts of the world.

Uses of Palm-trees and their Products

To the natives of the equatorial zone the uses of palms are both great and various. The fruits of several species—

more especially the coca-nut of the East and the peachnut (Guilielma speciosa) of America—furnish abundance of wholesome food, and the whole interior of the trunk of the sago palm is converted into a edible starch—our sago. Many other palm-fruits yield a thin pulp, too small in quantity to be directly eaten, but which, when rubbed off and mixed with a proper quantity of water, forms an exceedingly nutritious and agreeable article of food. The most celebrated to these is the assai of the Amazon, made from the fruit of Euterpe oleracea, and which, as a refreshing, nourishing, and slightly stimulating beverage for a tropical country takes the place of our chocolate and coffee. A number of other palms yield a similar product, and many that are not eaten by man are greedily devoured by a variety of animals, so that the amount of food produced by this tribe of plants is much larger than is generally supposed.

The sap which pours out of the cut flower-stalk of several species of palm, when slightly fermented, forms palm-wine or toddy, a very agreeable drink; and when mixed with various bitter herbs or roots which check fermentation, a fair imitation of beer is produced. If the same fluid is at once boiled and evaporated it produces a quantity of excellent sugar. The Arenga saccharifera, or sugar-palm of the Malay countries, is perhaps the most productive of sugar. A single tree will continue to pour out several quarts of sap daily for weeks together, and where the trees are abundant this forms the chief drink and most esteemed luxury of the natives. A Dutch chemist, Mr. De Vry, who has studied the subject in Java, believes that great advantages would accrue from the cultivation of this tree in place of the sugar-cane. According to his experiments it would produce an equal quantity of sugar of good quality with far less labour and expense, because no manure and no cultivation would be required, and the land will never be impoverished, as it so rapidly becomes by the growth of sugar-cane. The reason of this difference is, that the whole produce of a cane-field is taken off the ground, the crushed canes being burnt; and the soil thus becomes exhausted of the various salts and minerals which form part of the woody fibre and foliage. These must be restored by the application of manure, and this, together with the plating, weeding, and necessary cultivation, is very expensive. With the sugar-palm, however, nothing whatever is taken away but the juice itself; the foliage falls on the ground and rots, giving back to it what it had taken; and the water an sugar in the juice being almost wholly derived from the carbonic acid and aqueous vapour of the atmosphere, there is no impoverishment; and a plantation of these palms may be kept up on the same ground for an indefinite period. Another most important consideration is, that these trees will grows on poor rocky soil and on the steep slopes of ravines and hillsides, where any ordinary cultivation is impossible, and a great extent of fertile land would thus be set free for other purposes. Yet further, the labour required for such sugar plantations as these would be of light and intermittent kind, exactly suited to a semi-civilised people, to whom severe and long-continued labour is never congenial. This combination of advantages appears to be so great that it seems possible that the sugar of the world may in the future be produced from what would otherwise be almost waste ground; and it is to be hoped that the experiment will soon be tried in some of our tropical colonies, more especially as an Indian palm, Phœnix sylvestris, also produces abundance of sugar, and might be tried in its native country.

Other articles of food produced from palms are, cooking-oil from the cocoa-nut and baccaba palm, salt from the fruit of a South American palm (Leopoldinia major), while the terminal bud or "cabbage" of many species is an excellent and nutritious vegetable; so that palms may be said to supply bread, oil, sugar, salt, fruit, and vegetables. Oils for various other purposes are made from several distinct palms, especially from the celebrated oil palm of West Africa, while wax is secreted from the leaves of some South American species; the resin called Dragon's blood is the product of one of the rattan palm. The fruit of the Areca palm is the "betel-nut" so universally chewed by the Malays as a gentle stimulant, and is their substitute for the opium of the Chinese, the tobacco of Europeans, and the coca-leaf of South America.

For thatching purposes the leaves of palms are invaluable, and are universally used wherever they are abundant; and the petioles or leaf stalks, often fifteen or twenty feet long, are used as rafters, or when fastened together with pegs form doors, shutters, partitions, or even the walls of entire houses. They are wonderfully light and strong, being formed of a dense pith covered with a hard rind or bark, and when split up and pegged together serve to make many kinds of boxes, which, when covered with the broad leaves of a species of screw-pine and painted or stained of various colours, are very strong and serviceable as well as very ornamental. Ropes and cables are woven from the black fibrous matter that fringes the leaves of the sugar-palm and some other species, while fine strings of excellent quality, used even for bow-strings, fishing-lines and hammocks, are made of fibres obtained from the unopened leaves of some American species. The fibrous sheath at the base of the leaves of the cocoa-nut palm is so compact and cloth-like that it is used for a variety of purposes, as for strainers, for wrappers, and to make very good hats. The great woody spathes of the larger palms serve as natural baskets, as cradles, or even as cooking-vessels in which water may be safely boiled. The trunks form excellent posts and fencing, and when split make good flooring. Some species are used for bows, others for blow-pipes; the smaller palm-spines are sometimes used as needles or to make fish-hooks, and the larger as arrows. To describe in detail all the uses to which palm-trees and their products are applied in various parts of the world might occupy a volume; but the preceding sketch will serve to give an idea of how important a part is filled by this noble family of plants, whether we regard them as a portion of the beautiful vegetation of the tropics, or in relation to the manners and customs, the lives and the well-being, of the indigenous inhabitants.

Ferns

The type of plants which, next to palms, most attracts attention in the equatorial zone is perhaps that of the ferns, which here display themselves in vast profusion and variety. They grow abundantly on rocks and on decaying trees; they clothe the sides of ravines and the margins of streams; they climb up the trees and over bushes; they form tufts and hanging festoons among the highest branches. Some are as small as mosses, others have huge fronds eight or ten feet long, while in mountainous districts the most elegant of the group, the tree-ferns, bear their graceful crowns on slender stems twenty to thirty, or even fifty feet high. It is this immense variety rather than any special features that characterises the fern-vegetation of the tropics. We have here almost every conceivable modification of size, form of fronds, position of spores, and habit of growth, in plants that still remain unmistakably ferns. Many climb over shrubs and bushes in a most elegant manner; others cling closely to the bark of trees like ivy. The great birds'-nest fern (Platycerium) attaches its shell-like fronds high up on the trunks of lofty trees. Many small terrestrial species have digitate, or ovate, or ivy-shaped, or even whorled fronds, resembling at first sight those of some herbaceous flowering plants. Their numbers may be judged from the fact that in the vicinity of Tarrapoto, in Peru, Dr. Spruce gathered two hundred and fifty species of ferns, while the single volcanic mountains of Pangerango in Java (ten thousand feet high) is said to have produced three hundred species.

Ginger-worts and wild Bananas

These plants, forming the families Zingiberaceæ and Musaceæ of botanists, are very conspicuous ornaments of the equatorial forests, on account of their large size, fine foliage, and handsome flowers. The bananas and plantains are well known as among the most luxuriant and beautiful productions of the tropics. Many species occur wild in the forests; all have majestic foliage and handsome flowers, while some produce edible fruit. Of the ginger-worts (Zingiberaceæ and Marantaceæ), the well-known cannas of out sub-tropical gardens may be taken as representatives, but the equatorial species are very numerous and varied, often forming dense thickets in damp places, and adorning the forest shades with their elegant and curious or showy flowers. The maranths produce "arrowroot," while the ginger-worts are highly aromatic, producing ginger, cardamums, grains of paradise, turmeric, and several medicinal drugs. The Musaceæ produce the most valuable of tropical fruits and foods. The banana is the variety which is always eaten as a fruit, having a delicate aromatic flavour; the plantain is a larger variety, which is best cooked. Roasted in the green state it is an excellent vegetable, resembling roasted chestnuts; when ripe it is sometimes pulped and boiled with water, making a very agreeable sweet soup; or it is roasted, or cut into slices and fried, in either form being a delicious tropical substitute for fruit pudding. These plants are annuals, producing one immense bunch of fruit. This bunch is sometimes four or five feet long, containing near two hundred plantains, and often weighs about a hundredweight. They grow very close together, and Humboldt calculated that an acre of plantains would supply more food than could be obtained from the same extent of ground by any other known plant. Well may it be said that the plantain is the glory of the tropics, and well was the species named by Linnæus—Musa Paradisiaca!

Arums

Another very characteristic and remarkable group of tropical plants are the epiphytal and climbing arums. These are known by their large, arrow-shaped, dark green and glossy leaves, often curiously lobed or incised, and sometimes reticulated with large open spaces, as if pieces had been regularly eaten out of them by some voracious insects. Sometimes they form clusters of foliage on living or dead trees, to which they cling by their aerial roots. Others climb up the smooth bark of large trees, sending out roots as they ascend which clasp around the trunk. Some mount straight up, others wind round the supporting trunks, and their large, handsome, and often highly remarkable leaves, which spread out profusely all along the stem, render them one of the most striking forms of vegetation which adorn the damper and more luxuriant parts of the tropical forests of both hemispheres.

Screw-pines

These singular plants, constituting the family Pandanaceæ of botanists, are very abundant in many parts of the Eastern tropics, while they are comparatively scarce in America. They somewhat resemble Yuccas, but have larger leaves, which grow in a close spiral screw on the stem. Some are large and palm-like, and it is a curious sight to stand under these and look up at the huge vegetable screw formed by the bases of the long, drooping leaves. Some have slender branched trunks, which send out aerial roots; others are stemless, consisting of an immense spiral cluster of stiff leaves ten or twelve feet long and only two or three inches wide. They abound most in sandy islands, while the larger species grow in swampy forests. Their large-clustered fruits, something like pine-apples, are often of a red colour; and their long, Stiff leaves are of great use for covering boxes and for many other domestic uses.

Orchids

These interesting plants, so well known from the ardour with which they are cultivated on account of their beautiful and singular flowers, are pre-eminently tropical, and are probably more abundant in the mountains of the equatorial zone than in any other region. Here they are almost omnipresent in some of their countless forms. They grow on the stems, in the forks or on the branches of trees; they abound on fallen trunks; they spread over rocks, or hang down the face of precipices; while some, like our northern species, grow on the ground among grass and herbage. Some trees whose bark is especially well adapted for their support are crowded with them, and these form natural orchid-gardens. Some orchids are particularly fond of the decaying leaf-stalks of palms or of tree-ferns. Some grow best over water, others must be elevated on lofty trees and well exposed to sun and air. The wonderful variety in the form, structure, and colour of the flowers of orchids is well known; but even our finest collections give an inadequate idea of the numbers of these plants that exist in the tropics, because a large proportion of them have quite inconspicuous flowers and are not worth cultivation. More than thirty years ago the number of known orchids was estimated by Dr. Lindley at three thousand species, in Bentham and Hooker's *Genera Plantarum*at five thousand, and it is not improbable that they may be now nearly six thousand. But whatever may be numbers of the collected and described orchids, those that still remain to be discovered must be enormous. Unlike ferns, the species have a very limited range, and it would require the systematic work of a good botanical collector during several years to exhaust any productive district—say such an island as Java—of its orchids. It is not therefore at all improbable that this remarkable group may ultimately prove to be the most numerous in species of all the families of flowering plants.

Although there is a peculiarity of habit that enables one soon to detect an orchidaceous plant even when not in flower, yet they vary greatly in size and aspect. Some of the small creeping species are hardly larger than mosses, while the larger Grammatophyllums of Borneo, which grow in the forks of trees, form a mass of leafy stems ten feet long, and some of the terrestrial species—as the American Sobralias—grow erect to an equal height. The fleshy aerial roots of most species give them a very peculiar aspect, as they often grow to a great length in the open air, spread over the surface of rocks, or attach themselves loosely to the bark of trees, extracting nourishment from the rain and from the aqueous vapour of the atmosphere. Yet notwithstanding the abundance and variety of orchids in the equatorial forests, they seldom produce much effect by their flowers. This is due partly to the very large proportion of the species having quite inconspicuous flowers; and partly to the fact that the flowering season for each kind lasts but a few weeks, while different species flower almost every month in the year. It is also due to the manner of growth of orchids, generally in single plants or clumps, which are seldom large or conspicuous as compared with the great mass of vegetation around them. It is only at long intervals that the traveller meets with anything which recalls the splendour of our orchid-houses and flower shows. The slender-stalked golden Oncidiums of the flooded forests of the Upper Amazon; the grand Cattleyas of the drier forests; the Cælogynes of the swamps, and the remarkable Vanda lowii of the hill forests of Borneo,—are the chief examples of orchid beauty that have impressed themselves on the memory of the present writer during twelve years' wandering in tropical forests. The last-named plant is unique among orchids, its comparatively small cluster of leaves sending out numerous flower-stems, which hang down like cords to a length of eight feet, and are covered with numbers of large star-like crimson-spotted flowers.

Bamboos

The gigantic grasses called bamboos can hardly be classed as typical plants of the tropical zone, because they appear to be rare in the entire African continent and are comparatively scarce in South America. They also extend beyond the geographical tropics in China and Japan as well as in Northern India. It is, however, within the tropics and towards the equator that they attain their full size and beauty, and it is here that the species are most numerous and offer that variety of form, size, and quality which renders them so admirable a boon to man. A fine clump of large bamboos is perhaps the most graceful of all vegetable forms, resembling the light and airy plumes of the bird of paradise copied on a gigantic scale in living foliage. Such clumps are often eighty or a hundred feet high, the glossy stems, perhaps six inches thick at the base, springing up at first straight as an arrow, tapering gradually to a slender point, and bending over in elegant curves with the weight of the slender branches and grassy leaves. The various species differ greatly in size and proportions, in the comparative length of the joints, in the thickness and strength of the stem-walls, in their straightness, smoothness, hardness, and durability. Some are spiny, others are unarmed; some have simple stems, others are thickly set with branches; while some species even grow in such an irregular, zigzag, branched manner as to form veritable climbing bamboos. They generally prefer dry and upland stations, though some grow near the banks of rivers, and a few in the thick forests, and, in South America, in flooded tracts. They often form dense thickets where the forests have been cleared away, and owing to their great utility they are cultivated or preserved near native houses and villages, and in such situations often give a finishing charm to the landscape.

Uses of the Bamboo

Perhaps more than any other single type of vegetation, the bamboo seems specially adapted for the use of half-civilised man in a wild tropical country; and the purposes to which it is applied are almost endless. It is a natural column or cylinder, very straight, uniform in thickness, of a compact and solid texture, and with a smooth, flinty, naturally-polished, external skin. It is divided into ringed joints at regular intervals which correspond to *septa* or partitions within, so that each joint forms a perfectly closed and water-tight vessel. Owing to its hollowness, the hardness of the external skin, and the existence of the joints and partitions, it is wonderfully strong in proportion to its weight. It can be found of many distinct sizes and proportions, light or heavy, long or short-jointed, and varying from the size of a reed to that of a tall and slender palm-tree. It can be split with great facility and accuracy, and, owing to its being hollow, it can be easily cut across or notched with a sharp knife or hatchet. It is excessively strong and highly elastic, and whether green or dry is almost entirely free from any peculiar taste or smell. The way in which these various qualities of the bamboo render it so valuable will be best shown by giving a brief account of some of the uses to which it is applies in the Malay Archipelago.

Several effective weapons are easily made from bamboo. By cutting off the end very obliquely just beyond a joint, a very sharp cutting point is produced suitable for a spear, dagger, or arrow-head, and capable of penetrating an animal's body as readily as iron. Such spears are constantly used by many of the Malay tribes. In the eastern half of the Archipelago, where bows and arrows are used, these weapons are often formed entirely of bamboo. The harder and thicker sorts, split and formed with tapering ends, make a very strong and elastic bow, while a narrow strip of the outer skin of the same is used for the string, and the slender reed-like kinds make excellent arrows. One of the few agricultural tools used by the Papuans—a spud or hoe for planting or weeding—is made of a stout bamboo cut somewhat like the spear.

For various domestic purpose the uses of bamboo are endless. Ladders are rapidly made from two bamboo poles of the required length, by cutting small notches just above each ring, forming holes to receive the rungs or steps formed of a slenderer bamboo. For climbing lofty trees to get beeswax, a temporary ladder reaching to any height is ingeniously formed of bamboo. One of the hardest and thickest sorts is chosen, and from this a number of pegs about foot long are made. These are sharpened at one end and then driven into the tree in a vertical line about three feet apart. A tall and slender bamboo is then placed upright on the ground and securely tied with rattan or other cords to the heads of these pegs, which thus with the tree itself form a ladder. A man mounts these steps and builds up the ladder as he goes, driving in fresh pegs and splicing on fresh bamboos till he reaches the lower branches of the tree, which is sometimes eighty or a hundred feet from the ground. As the weight of the climber is thrown on several of the pegs which are bound together and supported by the upright bamboo, this ladder is much safer than it looks at first sight, and it is made with wonderful rapidity. When a path goes up a steep hill over smooth ground, bamboo steps are often laid down to prevent slipping while carrying heavy loads. These are made with uniform lengths of stout bamboo in which opposite notches are cut at each end just within a joint. These notches allow strong bamboo pegs to be driven through into the ground, thus keeping the steps securely in place. The masts and yards of native vessels are almost always formed of bamboo, as it combines lightness, strength, and elasticity in an unequalled degree. Two or three large bamboos also form the best outriggers to canoes on account of their great buoyancy. They also serve to form rafts; and in the city of Palembang, in Sumatra, there is a complete street of floating houses supported on rafts formed of huge bundles of bamboos. Bridges across streams or to carry footpaths along the fact of precipices are constructed by the Dyaks of Borneo wholly of bamboos, and some of these are very ingeniously hung from overhanging trees by diagonal rods of bamboo, so as to form true suspension bridges. The flooring of Malay houses is almost always of bamboo, but is constructed in a variety of ways. Generally large bamboos are used, split lengthways twice and the pieces ties down with rattan. This forms a grated floor, slightly elastic, and very pleasant to the barefooted natives. A superior floor is sometimes formed of slabs, which are made from very stout bamboos cut into lengths of about three or four feet and split down one side. The joints are then deeply and closely notched all round with a sharp chopping-knife, so that the piece can be unrolled as it were and pressed flat, when it forms a hard board with a natural surface, which, with a little wear, becomes beautifully smooth and polished. Blinds, screens, and mats are formed pf bamboos in a variety of ways,—sometimes of thin kinds crushed flat and plaited, but more frequently of narrow strips connected together with cords of bamboo-bark or rattan. Strips of bamboo supported on cross-pieces form an excellent bed, which from its elasticity supplies the purpose of a mattress as well, and only required a mat laid over it to insure a comfortable night's repose. Every kind of basket, too, is made of bamboo, from the coarsest heavy kinds to such as are fine and ornamental. In such countries as Lombock and Macassar, where the land is much cultivated and timber scarce, entire houses are built of bamboo,—posts, walls, floors, and roofs all being constructed of this one material; and perhaps in no other way can so elegant and well-finished a house be built so quickly and so cheaply. Almost every kind of furniture is also made of the same material, excellent bamboo chairs, sofas, and bedsteads being made in the Moluccas, which, for appearance, combined with cheapness, are probably unsurpassed in the world. A chair costs sixpence, and a sofa two shillings.

Among simpler uses bamboos are admirably adapted for water-vessels. Some of the lighter sorts are cut into lengths of about five feet, a small hole being knocked through the septa of the joints. This prevents the water from running out too quickly, and facilitates its being poured out in a regulated stream to the last drop. Three or four of these water-vessels are tied together and carried on the back, and they stand very conveniently in a corner of the hut. Water-pipes and aqueducts are also readily made from bamboo tubes supported at intervals on two smaller pieces tied crosswise. In this way a stream of water is often conveyed from some distance to the middle of a village. Measures for rice or palm-wine, drinking-vessels, and water-dippers, are to be found almost ready-made in a joint of bamboo; and when fitted with a cap or lid they form tobacco or tinder-boxes. Perches for parrots, with food and water vessels, are easily made out of a single piece of bamboo, while with a little more labour elegant bird-cages are constructed. In Timor a musical instrument is formed from a single joint of a large bamboo by carefully raising seven strips of the hard skin to form strings, which remain attached at both ends and are elevated by small pegs wedged underneath, the strings being prevented from splitting off by a strongly-plaited ring of a similar material bound round each end. An opening cut on one side allows the bamboo to vibrate in musical notes when the harplike strings are sharply pulled with the fingers. In

Java strips of bamboo supported on stretched strings and struck with a small stick produce the higher notes in the "gamelung" or native band, which consists mainly of sets of gongs and metallic plates of various sizes. Almost all the common Chinese paper is made from the foliage and stems of some species of bamboo, while the young shoots, as they first spring out of the ground, are an excellent vegetable, quite equal to artichokes. Single joints of bamboo make excellent cooking-vessels while on a journey. Rice can be boiled in them to perfection, as well as fish and vegetables. They serve too for jars in which to preserve sugar, salt, fruit, molasses, and cooked provisions; and for the smoker, excellent pipes and hookahs can be formed in a few minutes out of properly chosen joints of bamboo.

These are only a sample of the endless purposes to which the bamboo is applied in the countries of which it is a native, its chief characteristic being that in a few minutes it can be put to uses which, if ordinary wood were used, would require hours or even days of labour. There is also a regularity and a finish about it which is found in hardly any other woody plant; and its smooth and symmetrically ringed surface gives an appearance of fitness and beauty to its varied applications. On the whole, we may perhaps consider it as the greatest boon which nature gives to the natives of the eastern tropics.

Mangroves

Among the forms of plants which are sure to attract attention in the tropics are the mangroves, which grow between tide-marks on coasts and estuaries. These are low trees with widely-spreading branches and a network of aerial roots a few feet above the ground; but their most remarkable peculiarity is, that their fruits germinate on the tree, sending out roots and branches before falling into the muddy soil—a completely formed plant. In some cases the root reaches the ground before the seed above falls off. These trees greatly aid the formation of new land, as the mass of aerial roots which arch out from the stem to a considerable distance collects mud and floating refuse, and so raises and consolidates the shore; while the young plants, often dropping from the farthest extremity of the branches, rapidly extend the domain of vegetation to the farthest possible limits. The branches, too, send down slender roots like those of the banyan, and become independent trees. Thus a complete woody labyrinth is formed; and the network of tough roots and stems resists the action of the tides, and enables the mud brought down by great tropical rivers to be converted into solid land far more rapidly than it could be without this aid.

Sensitive Plants

Among the more humble forms of vegetation that attract the traveller's notice none are more interesting than the sensitive species of Mimosa. These are almost all natives of South America, but one species, Mimosa pudica, has spread to Africa and Asia, so that sensitive plants now abound as wayside weeds in many parts both of the eastern and western tropics, sometimes completely carpeting the ground with their delicate foliage. Where a large surface of ground is thus covered the effect of walking over it is most peculiar. At each step the plants for some distance round suddenly droop, as if struck with paralysis, and a broad track of prostrate herbage, several feet wide, is distinctly marked out by the different colour of the closed leaflets. The explanation of this phenomenon given by botanists is not very satisfactory;[1] while the purpose or use of the peculiarity is still more mysterious, seeing that out of more than two hundred species belonging to this same genus Mimosa, only a small number are sensitive in any remarkable degree, and in the whole vegetable kingdom there are but few other plants which possess more than the rudiments of a similar property. The true sensitive plants are all low-growing herbs or shrubs with delicate foliage, which might possibly be liable to destruction by herbivorous animals, a fate which they may perhaps escape by their singular power of suddenly collapsing before the jaws opened to devour them. The fact that one species has been naturalised as a weed over so wide an area in the tropics, seems to show that it possesses some advantage over the generality of tropical weeds. It is, however, curious that, as the most sensitive species of Mimosa are somewhat prickly, so easy and common a mode of protection as the development of stronger spines should here have failed; and that its place should be supplied by so singular a power as that of simulating death in a manner which suggests the possession of both sensation and voluntary motion.

[1] See *Nature,* vol.xvi. p. 349, where the German botanist Pfeffer's theory is given.

Comparative Scarcity of Flowers

It is a very general opinion among inhabitants of our temperate climes that amid the luxuriant vegetation of the tropics there must be a grand display of floral beauty, and this idea is supported by the number of large and showy flowers cultivated in our hothouses. The fact is, however, that in proportion as the general vegetation becomes more luxuriant, flowers form a less and less prominent feature; and this rule applies not only to the tropics but to the temperate and frigid zones. It is amid the scanty vegetation of the higher mountains and towards the limits of perpetual snow that the alpine flowers are most brilliant and conspicuous. Our own meadows and pastures and hillsides produce more gay flowers than our woods and forests; and, in the tropics, it is in the parts where vegetation is less dense and luxuriant that flowers most abound. In the damp and uniform climate of the equatorial zone the mass of vegetation is greater and more varied than in any other part of the globe, but in the great virgin forests themselves flowers are rarely seen. After describing the forests of the Lower Amazon, Mr. Bates asks: "But where were the flowers? To our great disappointment we saw none, or only such as were insignificant in appearance. Orchids are rare in the dense forests of the lowlands, and I believe it is now tolerably well ascertained that the majority of the forest trees in equatorial Brazil have small and inconspicuous flowers."[1] My friend Dr. Richard Spruce assured me that by far the greater part of the plants gathered by him in equatorial America had inconspicuous green or white flowers. My own observations in the Aru Islands for six months, and in Borneo for more than a year, while living almost wholly in the forests, are quite in accordance with this view. Conspicuous masses of showy flowers are so rare that weeks and months may be passed without observing a single flowering plant worthy of special admiration.

[1] *The Naturalist on the River Amazons,* 2d ed., p. 38.

Occasionally some tree or shrub will be seen covered with magnificent yellow or crimson or purple flowers, but it is usually an oasis of colour in a desert of verdure, and therefore hardly affects the general aspect of the vegetation. The equatorial forest is too gloomy for flowers or generally even for much foliage, except of ferns and other shade-loving plants; and were it not that the forests are broken up by rivers and streams, by mountain ranges, by precipitous rocks and by deep ravines, there would be far fewer flowers visible than there are. Some of the great forest trees have showy blossoms, and when these are seen from an elevated point looking over an expanse of tree tops the effect is very grand; but nothing is more erroneous than the statement sometimes made that tropical forest trees *generally* have showy flowers, for it is doubtful whether the proportion is at all greater in tropical than in temperate zones. On such natural exposures as steep mountain sides, the banks of rivers, or ledges of precipices, and on the margins of such artificial openings as roads and forest clearings, whatever floral beauty is to be found in the more luxuriant parts of the tropics is exhibited. But even in such favourable situations it is not the abundance and beauty of the flowers but the luxuriance and the freshness of the foliage, and the grace and infinite variety of the forms of vegetation, that will most attract the attention and extort the admiration of the traveller. Occasionally indeed you will come upon shrubs gay with blossoms or trees festooned with flowering creepers; but, on the other hand, you may travel for a hundred miles and see nothing but the varies greens of the forest foliage and the deep gloom of its tangled recesses. In Mr. Belt's *Naturalist in Nicaragua,* he thus describes the great virgin forests of that country which, being in a mountainous region and on the margin of the equatorial zone, are among the most favourable examples. "On each side of the road great trees towered up, carrying their crowns out of sight amongst a canopy of foliage, and with lianas hanging from nearly every bough, and passing from tree to tree, entangling the giants in a great network of coiling cables. Sometimes a tree appears covered with beautiful flowers which do not belong to it but to one of the lianas that twines through its branches and sends down great rope-like stems to the ground. Climbing ferns and vanilla cling to the trunks, and a thousand epiphytes perch themselves on the branches. Amongst these are large arums that send down long aerial roots, tough and strong, and universally used instead of cordage by the natives. Amongst the undergrowth several small species of palms, varying in height from two to fifteen feet, are common; and now and then magnificent tree ferns sending off their feathery crowns twenty feet from the ground delight the sight by their graceful elegance. Great broadleaved heliconias, leathery melastomæ, and succulent-stemmed, lop-sided, leaved, and flesh-coloured begonias are abundant, and typical of tropical American forests; but not less so are the cecropia trees, with their white stems and large palmated leaves standing up like great candelabra. Sometimes the ground is carpeted with large flowers, yellow, pink, or white, that have fallen from some invisible tree-top above; or the air is filled with a delicious perfume, the source of which one seeks around in vain, for the flowers that cause it are far overhead out of sight, lost in the great overshadowing crown of verdure."

Although, as has been shown elsewhere, it may be doubted whether light directly produces floral colour, there can be no doubt that it is essential to the growth of vegetation and to the full development of foliage and of flowers. In the forests all trees, and shrubs, and creepers struggle upwards to the light, there to expand their blossoms and ripen their fruit. Hence, perhaps, the abundance of climbers which make use of their more sturdy companions to reach this necessary of vegetable life. Yet even on the upper surface of the forest, fully exposed to the light and heat of the tropical sun, there is no special development of coloured flowers. When from some elevated point you can gaze down upon an unbroken expanse of woody vegetation, it often happens that not a single patch of bright colour can be discerned. At other times, and especially at the beginning of the dry season, you may behold scattered at wide intervals over the mottled-green surface a few masses of yellow, white, pink, or more rarely of blue colour, indicating the position of handsome flowering trees.

The well-established relation between coloured flowers and the need of insects to fertilise them may perhaps be connected with the comparative scarcity of the former in the equatorial forests. The various forms of life are linked together in such mutual dependence that no one can inordinately increase without bringing about a corresponding increase or diminution of other forms. The insects which are best adapted to fertilise flowers cannot probably increase much beyond definite limits, because in doing so they would lead to a corresponding increase of insectivorous birds and other animals which would keep them down. The chief fertilisers—bees and butterflies—have enemies at every stage of their growth, from the egg to the perfect insect, and their numbers are, therefore, limited by causes quite independent of the supply of vegetable food. It may, therefore, be the case that the numbers of suitable insects are totally inadequate to the fertilisation of the countless millions of forest trees over such vast areas as the equatorial zone presents, and that, in consequence, a large proportion of the species have become adapted either for self-fertilisation, or for cross-fertilisation by the agency of the wind. Were there not some such limitation as this, we should expect that the continued struggle for existence among the plants of the tropical forests would have led to the acquisition, by a much larger proportion of them, of so valuable a character as bright-coloured flowers, this being almost a necessary preliminary to a participation in the benefits which have been proved to arise from cross-fertilisation by insect agency.

Concluding Remarks on Tropical Vegetation

In concluding this general sketch of the aspects of tropical vegetation, we will attempt briefly to summarise its main features. The primeval forests of the equatorial zone are grand and overwhelming by their vastness, and by the display of a force of development and vigour of growth rarely or never witnessed in temperate climates. Among their best distinguishing features are the variety of forms and species which everywhere meet and grow side by side, and the extent to which parasites, epiphytes, and creepers fill up every available station with peculiar modes of life. If the traveller notices a particular species and wishes to find more like it, he may often turn his eyes in vain in every direction. Trees of varied forms, dimensions, and colours are around him, but he rarely sees any one of them repeated. Time after time he goes towards a tree which looks like the one he seeks, but a closer examination proves it to be distinct. He may at length, perhaps, meet with a second specimen half a mile off, or may fail altogether, till on another occasion he stumbles on one by accident.

The absence of the gregarious or social habit, so general in the forests of extra-tropical countries, is probably dependent on the extreme equability and permanence of the climate. Atmospheric conditions are much more important to the growth of plants than any others. Their severest struggle for existence is against climate. As we approach towards regions of polar cold or desert aridity the variety of groups and species regularly diminishes; more and more are unable to sustain the extreme climatal conditions, till at last we find only a few specially organised forms which are able to maintain their existence. In the extreme north, pine or birch trees—in the desert, a few palms and prickly shrubs or aromatic herbs—alone survive. In the equable equatorial zone there is no such struggle against climate. Every form of vegetation has become alike adapted to its genial heat and ample moisture, which as probably changed little even throughout geological periods; and the never ceasing struggle for existence between the various species in the same area has resulted in a nice balance of organic forces, which gives the advantage, now to one, now to another species, and prevents any one type of vegetation from monopolising territory to the exclusion of the rest. The same general causes have led to the filling up of every place in nature with some specially adapted form. Thus we find a forest of smaller trees adapted to grow in the shade of greater trees. Thus we find every tree supporting numerous other forms of vegetation, and some so crowded with epiphytes of various kinds that their forks and horizontal branches are veritable gardens. Creeping ferns and arums run up the smoothest trunks; an immense variety of climbers hang in tangled masses from the branches and mount over the highest tree-tops. Orchids, bromelias, arums, and ferns grow from every boss and crevice, and cover the fallen and decaying trunks with a graceful drapery. Even these parasites have their own parasitical growth, their leaves often supporting an abundance of minute creeping mosses and hepaticæ. But the uniformity of climate which has led to this rich luxuriance and endless variety of vegetation is also the cause of a monotony that in time becomes oppressive. To quote the words of Mr. Belt: "Unknown are the autumn tints, the bright browns and yellows of English woods; much less the crimsons, purples, and yellows of Canada, where the dying foliage rivals, nay excels, the expiring dolphin in splendour. Unknown the cold sleep of winter; unknown the lovely awakening of vegetation at the first gentle touch of spring. A ceaseless round of ever-active life weaves the fairest scenery of the tropics into one monotonous whole, of which the component parts exhibit in detail untold variety and beauty."[1]

To the student of nature the vegetation of the tropics will ever be of surpassing interest, whether for the variety of forms and structures which it presents, for the boundless energy with which the life of plants is therein manifested, or for the help which it gives us in our search after the laws which have determined the production of such infinitely varied organisms. When, for the first time, the traveller wanders in these primeval forests, he can scarcely fail to experience sensations of awe, akin to those excited by the trackless ocean or the alpine snowfields. There is a vastness, a solemnity, a gloom, a sense of solitude and of human insignificance, which for a time overwhelm him; and it is only when the novelty of these feelings have passed away that he is able to turn his attention to the separate constituents that combine to produce these emotions, and examine the varied and beautiful forms of life which, in inexhaustible profusion, are spread around him.

[1] *The Naturalist in Nicaragua*, p. 58.

III

ANIMAL LIFE IN THE TROPICAL FORESTS

Difficulties of the Subject—General Aspect of the Animal Life of Equatorial Forests—Diurnal Lepidoptera or Butterflies—Peculiar Habits of Tropical Butterflies—Ants, Wasps, and Bees—Ants—Special Relations between Ants and Vegetation—Wasps and Bees—Orthoptera and other Insects—Beetles—Wingless Insects—General Observations on Tropical Insects—Birds: Parrots—Pigeons—Picariæ—Cuckoos—Trogons, Barbets, Toucans, and Hornbills—Passeres—Reptiles and Amphibia: Lizards—Snakes—Frogs and Toads—Mammalia: Monkeys—Bats—Summary of the Aspects of Animal Life in the Tropics.

THE attempt to give some account of the general aspects of animal life in the equatorial zone presents far greater difficulties than in the case of plants. On the one hand, animals rarely play any important part in scenery, and their entire absence may pass quite unnoticed; while the abundance, variety, and character of the vegetation are among those essential features that attract every eye. On the other hand, so many of the more important and characteristic types of animal life are restricted to one only out of the three great divisions of equatorial land, that they more extensive zoological groups which have a wide range in the tropics and do not equally abound in the temperate zones, are few in number, and often include such a diversity of forms, structures, and habits as to render any typical characterisation of them impossible. We must then, in the first place, suppose that our traveller is on the look-out for all signs of animal life; and that, possessing a general acquaintance as an out-door observer with the animals of our own country, he carefully notes those points in which the forests of the equatorial zone offer different phenomena. Here, as in the case of plants, we exclude all zoological science, classifications, and nomenclature, except in as far as it is necessary for a clear understanding of the several groups of animals referred to. We shall therefore follow no systematic order in our notes, except that which would naturally arise from the abundance or prominence of the objects themselves. We further suppose our traveller to have no prepossessions, and to have no favourite group, in the search after which he passes by other objects which, in view of their frequent occurrence in the landscape, are really more important.

General Aspect of the Animal Life of Equatorial Forests

Perhaps the most general impression produced by a first acquaintance with the equatorial forests is the comparative absence of animal life. Beast, bird, and insect alike require looking for, and it very often happens that we look for them in vain. On this subject Mr. Bates, describing one of his early excursions into the primeval forests of the Amazon valley, remarks as follows: "We were disappointed in not meeting with any of the larger animals of the forest. There was no tumultuous movement or sound of life. We did not see or hear monkeys, and no tapir jaguar crossed our path. Birds also appeared to be exceedingly scarce." Again: "I afterwards saw reason to modify my opinion, founded on first impressions, with regard to the amount and variety of animal life in this and other parts of the Amazonian forests. There is, in fact, a great variety of mammals, birds, and reptiles, but they are widely scattered and all excessively shy of man. The region is so extensive and uniform in the forest clothing of its surface, that it is only at long intervals that animals are seen in abundance, where some particular spot is found which is more attractive than others. Brazil, moreover, is throughout poor in terrestrial mammals, and the species are of small size; they do not, therefore, form a conspicuous feature in the forests. The huntsman would be disappointed who expected to find here flocks of animals similar to the buffalo-herds of North America, or the swarms of antelopes and herds of ponderous pachyderms of Southern Africa. We often read in books of travel of the silence and gloom of the Brazilian forests. They are realities, and the impression deepens on a longer acquaintance. The few sounds of birds are of that pensive and mysterious character which intensifies the feeling of solitude rather than imparts a sense of life and cheerfulness. Sometimes in the midst of the stillness a sudden yell or scream will startle one; this comes from some defenceless fruit-eating animal which is pounced upon by a tiger-cat or a boa-constrictor. Morning and evening the howling monkeys make a most fearful and harrowing noise, under which it is difficult to keep up one's buoyancy of spirit. The feeling of inhospitable wildness which the forest is calculated to inspire is increased tenfold under this fearful uproar. Often, even in the still mid-day hours, a sudden crash will be heard resounding after through the wilderness, as some great bough or entire tree falls to the ground." With a few verbal alterations these remarks will apply equally to the primeval forests of the Malay Archipelago; and it is probable that those of West Africa offer no important differences in this respect. There is, nevertheless, one form of life which is very rarely absent in the more luxuriant parts of the tropics, and which is more often so abundant as to form a decided feature in the scene. It is therefore the group which best characterises the equatorial zone, and should form the starting-point for our review. This group is that of the

Diurnal Lepidoptera or Butterflies

Wherever in the equatorial zone a considerable extent of the primeval forest remains, the observer can hardly fail to be struck by the abundance and the conspicuous beauty of the butterflies. Not only are they abundant in individuals, but their large size, their elegant forms, their rich and varied colours, and the number of distinct species almost everywhere to be met with, are equally remarkable. In many localities near the northern or southern tropics they are perhaps equally abundant, but these spots are more or less exceptional, whereas within the equatorial zone, and with the limitations above stated, butterflies form one of the most constant and most conspicuous displays of animal life. They abound most in old and tolerably open roads and pathways through the forest, but they are also very plentiful in old settlements in which fruit-trees and shrubbery offer suitable haunts. In the vicinity of such old towns as Malacca and Amboyna in the East, and of Para and Rio de Janeiro in the West, they are especially abundant, and comprise some of the handsomest and most remarkable species in the whole group. Their aspect is altogether different from that presented by the butterflies of Europe and of most temperate countries. A considerable proportion of the species are very large, six to eight inches across the wings being not uncommon among the Papilionidæ and Morphidæ, while several species are even larger. This great expanse of wings is accompanied by a slow flight; and, as they usually keep near the ground and often rest, sometimes with closed and sometimes with expanded wings, these noble insects really look larger and are much more conspicuous objects than the majority of our native birds. The first sight of the great blue Morphos flapping slowly along in the forest roads near Para, of the large white-and-black semitransparent Ideas floating airily about in the woods near Malacca, and of the golden-green Ornithopteras sailing on bird-like wing over the flowering shrubs which adorn the beach of the Ké and Aru islands, can never be forgotten by any one with a feeling of admiration for the new and beautiful in nature. Next to the size, the infinitely varied and dazzling hues of these insects most attract the observer. Instead of the sober browns, the plain yellows, and the occasional patches of red, or blue, or orange that adorn our European species, we meet with the most intense metallic blues, the purest satiny greens, the most gorgeous crimsons, not in small spots but in large masses, relieved by a black border or background. In others we have contrasted bands of blue and orange, or of crimson and green, or of silky yellow relieved by velvety black. In not a few the wings are powdered over with scales and spangles of metallic green, deepening occasionally into blue or golden or deep red spots. Others again have spots and markings as of molten silver or gold, while several have changeable hues, like shot-silk or richly-coloured opal. The form of the wings, again, often attracts attention. Tailed hind wings occur in almost all the families, but vary much in character. In some the tails are broadly spoon-shaped, in others long and pointed. Many have double or triple tails, and some of the smaller species have them immensely elongated and often elegantly curled. In some groups the wings are long and narrow, in others strongly falcate; and though many fly with immense rapidity, a large number flutter lazily along, as if they had no enemies to fear, and therefore no occasion to hurry.

The number of species of butterflies inhabiting any one locality is very variable, and is, as a rule, far larger in America than in the Eastern hemisphere; but it everywhere very much surpasses the numbers in the temperate zone. A few months' assiduous collecting in any of the Malay islands will produce from 150 to 250 species of butterflies, and thirty or forty species may be obtained any fine day in good localities. In the Amazon valley, however, much greater results may be achieved. A good day's collecting will produce from forty to seventy species, while in one year at Para about 600 species were obtained. More than 700 species of butterflies actually inhabit the district immediately around the city of Para, and this, as far as we yet know, is the richest spot on the globe for diurnal lepidoptera. At Ega, during four years' collecting, Mr. Bates obtained 550 species, and these, on the whole, surpassed those of Para in variety and beauty. Mr. Bates thus speaks of a favourite locality on the margin of the lake near Ega: "The number and variety of gaily-tinted butterflies, sporting about in this grove on sunny days, were so great that the bright moving flakes of colour gave quite a character to the physiognomy of the place. It was impossible to walk far without disturbing flocks of them from the damp sand at the edge of the water, where they congregated to imbibe the moisture. They were of almost all colours, sizes, and shapes; I noticed here altogether eighty species, belonging to twenty-two distinct genera. The most abundant, next to the very common sulphur-yellow and orange-coloured kinds, were about a dozen species of Eunica, which are of large size and conspicuous from their liveries of glossy dark blue and purple. A superbly adorned creature, the Callithea markii, having wings of a thick texture, coloured sapphire-blue and orange, was only an occasional visitor. On certain days, when the weather was very calm, two small gilded species (Symmachia trochilus and colubris) literally swarmed on the sands, their glittering wings lying wide open on the flat surface."[1]

when we consider that only sixty-four species of butterflies have been found in Britain and about 150 in

Germany, many of which are very rare and local, so that these numbers are the result of the work of hundreds of collectors for a long series of years, we see at once the immense wealth of the equatorial zone in this form of life.

[1] *The Naturalist on the River Amazons,* 2d ed., p. 331.

Peculiar Habits of Tropical Butterflies

The habits of the butterflies of the tropics offer many curious points rarely or never observed among those of the temperate zone. The majority, as with us, are truly diurnal, but there are some Eastern Morphidæ and the entire American family Brassolidæ, which are crecuspular, coming out after sunset and flitting about the roads till it is nearly dark. Others, though flying in the daytime, are only found in the gloomiest recesses of the forest, where a constant twilight may be said to prevail. The majority of the species fly at a moderate height (from five to ten feet above the ground), while a few usually keep higher up and are difficult to capture; but a large number, especially the Satyridæ, many Erycinidæ, and some few Nymphalidæ, keep always close to the ground and usually settle on or among the lowest herbage. As regards the mode of flight, the extensive and almost exclusively tropical families of Heliconidæ and Danaidæ fly very slowly, with a gentle undulating or floating motion which is almost peculiar to them. Many of the strong-bodied Nymphalidæ and Hesperidæ, on the other hand, have an excessively rapid flight, darting by so swiftly that the eye cannot follow them, and in some cases producing a deep sound louder than that of the humming-birds.

The places they frequent, and their mode of resting, are various and often remarkable. A considerable number frequent damp open places, especially river-sides and the margins of pools, assembling together in flocks of hundreds of individuals; but these are almost entirely composed of males, the females remaining in the forests, where, towards the afternoon, their partners join them. The majority of butterflies settle upon foliage and on flowers, holding their wings erect and folded together, though early in the morning, or when newly emerged from the chrysalis, they often expand them to the sun. Many, however, have special stations and attitudes. Some settle always on tree-trunks, usually with the wings erect, but the Ageronias expand them and always rest with the head downwards. Many Nymphalidæ prefer resting on the top of a stick; others choose bushes with dead leaves; others settle on rocks or sand or in dry forest paths. Pieces of decaying animal or vegetable matter are very attractive to certain species, and if disturbed they will sometimes return to the same spot day after day. Some Hesperidæ, as well as species of the genera Cyrestis and Symmachia, and a few others, rest on the ground with their wings fully expanded and pressed closely to the surface, as if exhibiting themselves to the greatest advantage. The beautiful little Erycinidæ of South America vary remarkably in their mode of resting. The majority always rest on the under surface of leaves with their wings expanded, so that when they settle they suddenly disappear from sight. Some, however, as the elegant gold-spotted Helicopis cupido, rest beneath leaves with closed wings. A few, as the genera Charis and Themone, for example, sit on the upper side of leaves with their wings expanded; while the gorgeously-coloured Erycinas rest with wings erect and exposed as in the majority of butterflies. The Hesperidæ vary in a somewhat similar manner. All rest on the upper side of leaves or on the ground, but some close their wings, others expand them, and a third group keep the upper pair of wings raised while the hind wings are expanded, a habit found in some of our European species. Many of the Lycænidæ especially the Theclas, have the curious habit, while sitting with their wings erect, of moving the lower pair over each other in opposite directions, giving them the strange appearance of eccentrically revolving discs.

The great majority of butterflies disappear at night, resting concealed amid foliage, or on sticks or trunks, or in such places as harmonise with their colours and markings; but the gaily-coloured Heliconidæ and Daniadæ seek no such concealment but rest at night hanging at the ends of slender twigs or upon fully exposed leaves. Being uneatable they have no enemies and need no concealment. Day-flying moths of brilliant or conspicuous colours are also comparatively abundant in the tropical forests. Most magnificent of all are the Uranias, whose long-tailed green-and-gold powdered wings resemble those of true swallow-tailed butterflies. Many Agaristidæ of the East are hardly inferior in splendour, while hosts of beautiful clearwings and ægeriidæ add greatly to the insect beauty of the equatorial zone.

The wonderful examples afforded by tropical butterflies of the phenomena of sexual and local variation, of protective modifications, and of mimicry, have been fully discussed elsewhere. For the study of the laws of variation in all its forms, these beautiful creatures are unsurpassed by any class of animals, both on account of their great abundance, and the assiduity with which they have been collected and studied. Perhaps no group exhibits the distinctions of species and genera with such precision and distinctness, due, as Mr. Bates has well observed, to the fact that all the superficial signs of change in the organisation are exaggerated, by their affecting the size, shape, and colour of the wings, and the distribution of the ribs or veins which form their framework. The minute scales or feathers with which the wings are clothed are coloured in regular patterns, which vary in accordance with the slightest change in the conditions to which the species are exposed. These scales are sometimes absent in spots or patches, and sometimes over the greater part of the wings, which then become transparent, relieved only by the dark veins and by delicate shades or small spots of vivid colour, producing a special form of delicate beauty characteristic of many South American butterflies. The following remark by Mr. Bates will fitly conclude our sketch of these lovely insects. "It may be said, therefore, that on these expanded membranes Nature writes, as on a tablet, the story of the modifications of species, so truly do all the changes of the organisation register themselves thereon. And as the laws of Nature must be the same for all beings, the conclusions furnished by this group of insects must be applicable to the whole organic world; therefore the study of butterflies—creatures selected as the types of airiness and frivolity—instead of being despised, will some day be valued as one of the most important branches of biological science."[1]

[1] Bates, *The Naturalist on the River Amazons,* 2d ed., p. 413.

Next after the butterflies in importance, as giving an air of life and interest to tropical nature, we must place the birds; but to avoid unnecessary passage, to and fro, among unrelated groups, it will be best to follow on with a sketch of such other groups of insects as from their numbers, variety, habits, or other important features, attract the attention of the traveller from colder climates. We begin then with a group which, owing to their small size and obscure colours, would attract little attention, but which nevertheless, by the universality of their presence, their curious habits, and the annoyance they often cause to man, are sure to force themselves upon the attention of every one who visits the tropics.

Ants, Wasps, and Bees

The hymenopterous insects of the tropics are, next to the butterflies, those which come most prominently before the traveller, as they love the sunshine, frequent gardens, houses, and roadways as well as the forest shades, never seek concealment, and are many of them remarkable for their size or form, or are adorned with beautiful colours and conspicuous markings. Although ants are, perhaps, on the whole, the smallest and the least attractive in appearance of all tropical insects, yet, owing to their being excessively abundant and almost omnipresent, as well as on account of their curious habits and the necessity of being ever on the watch against their destructive powers, they deserve our first notice.

Ants are found everywhere. They abound in houses,

some living underground, others in the thatched roofs, on the under surface of which they make their nests, while covered ways of earth are often constructed upon the posts and doors. In the forests they live on the ground, under leaves, on the branches of trees, or under rotten bark; while others actually dwell in living plants, which seem to be specially modified so as to accommodate them. Some sting severely, others only bite; some are quite harmless, others exceedingly destructive. The number of different kinds is very great. In India and the Malay Archipelago nearly 500 different species have been found, and other tropical countries are no doubt equally rich. I will first give some account of the various species observed in the Malay islands, and afterwards describe some of the more interesting South American groups, which have been so carefully observed by Mr. Bates on the Amazons and by Mr. Belt in Nicaragua.

Among the very commonest ants in all parts of the world are the species of the family Formicidæ, which do not sting, and are most of them quite harmless. Some make delicate papery nests, others live under stones or among grass. Several of them accompany Aphides to feed upon the sweet secretions from their bodies. They vary in size from the large Formica gigas, more than an inch long, to minute species so small as to be hardly visible. Those of the genus Polyrachis, which are plentiful in all Eastern forests, are remarkable for the extraordinary hooks and spines with which their bodies are armed, and they are also in many cases beautifully sculptured or furrowed. They are not numerous individually, and are almost all arboreal, crawling about bark and foliage. One species has processes on its back just like fish-hooks, others are armed with long, straight spines. They generally form papery nests on leaves, and when disturbed they rush out and strike their bodies against the nest so as to produce a loud rattling noise; but the nest of every species differs from those of all others either in size, shape, or position. As they all live in rather small communities in exposed situations, are not very active, and are rather large and conspicuous, they must be very much exposed to the attacks of insectivorous birds and other creatures, and having no sting or powerful jaws with which to defend themselves, they would be liable to extermination without some special protection. This protection they no doubt obtain by their hard smooth bodies, and by the curious hooks, spines, points, and bristles with which they are armed, which must render them unpalatable morsels, very liable to stick in the jaws or throats of their captors.

A curious and very common species in the Malay islands is the green ant (œcophylla Smaragdina), a rather large, longlegged, active, and intelligent-looking creature, which lives in large nests formed by gluing together the edges of leaves, especially of Zingiberaceous plants. When the nest is touched a number of the ants rush out, apparently in a great rage, stand erect, and make a loud rattling noise by tapping against the leaves. This no doubt frightens away many enemies, and is their only protection; for though they attempt to bite, their jaws are blunt and feeble, and they do not cause any pain.

Coming now to the stinging groups, we have first a number of solitary ants of the great genus Odontomachus, which are seen wandering about the forest and are conspicuous by their enormously long and slender hooked jaws. These are not powerful, but severe admirably to hold on by while they sting, which they do pretty severely. The Poneridæ are another group of large-sized ants which sting acutely. They are very varied in species but are not abundant individually. The Ponera clavata in Guiana is one of the worst stinging ants known. It is a large species, frequenting the forests on the ground, and is much dreaded by the natives, as its sting produces intense pain and illness. I was myself stung by this or an allied species when walking barefoot in the forest on the Upper Rio Negro. It caused such pain and swelling of the leg that I had some difficulty in reaching home, and was confined to my room for two days. Sir Robert Schomburgh suffered more; for he fainted with the pain, and had an attack of fever in consequence.

We now come to the Myremecidæ, which may be called the destroying ants, from their immense abundance and destructive propensities. Many of them sting most acutely, causing a pain like that of a sudden burn, whence they are often called "fire-ants." They often swarm in houses and devour everything eatable. Isolation by water is the only security, and even this does not always succeed, as a little dust on the surface will enable the smaller species to get across. Oil is, however, an effectual protection, and after many losses of valuable insect specimens, for which ants have a special affection, I always used it. One species of this group, a small black Crematogaster, took possession of my house in New Guinea, building nests in the roof and making covered ways down the posts and across the floor. They also occupied the setting boards I used for pinning out my butterflies, filling up the grooves with cells and storing them with small spiders. They were in constant motion, running over my table, in my bed, and all over my body. Luckily, they were diurnal, so that on sweeping out my bed at night I could get on pretty well; but during the day I could always feel some of them running over my body, and every now and then one would give me a sting so sharp as to make me jump and search instantly for the offender, who was usually found holding on tight with his jaws and thrusting in his sting with all his might. Another genus, Pheidole, consists of forest ants, living under rotten bark or in the ground, and very voracious. They are brown or blackish, and are remarkable for their great variety of size and form in the same species, the largest having enormous heads many times larger than their bodies, and being at least a hundred times as bulky as the smallest individuals. These great-headed ants are very sluggish and incapable of keeping up with the more active small workers, which often surround and drag them along as if they were wounded soldiers. It is difficult to see what use they can be in the colony, unless, as Mr. Bates suggests, they are mere baits to be attacked by insect-eating birds, and thus save their more useful companions. These ants devour grubs, white ants, and other soft and helpless insects, and seem to take the place of the foraging ants of America and driver ants of Africa, though they are far less numerous and less destructive. An allied genus, Solenopsis, consists of red ants, which, in the Moluccas, frequent houses, and are a most terrible pest. They form colonies underground, and work their way up through the floors, devouring everything eatable. Their sting is excessively painful, and some of the species are hence called fire-ants. When a house is infested by them, all the tables and boxes must be supported on blocks of wood or stone placed in dishes of water, as even clothes not newly washed are attractive to them; and woe to the poor fellow who puts on garments in the folds of which a dozen of these ants are lodged. It is very difficult to preserve bird skins or other specimens of natural history where these ants abound, as they gnaw away the skin round the eyes and the base of the bill, and if a specimen is laid down for even half an hour in an unprotected place it will be ruined. I remember once entering a native house to rest and eat my lunch; and having a large tin collecting-box full of rare butterflies and other insects, I laid it down on the bench by my side. On leaving the house I noticed some ants on it, and on opening the box found only a mass of detached wings and bodies, the latter in process of being devoured by hundreds of fire-ants.

The celebrated Sauba ant of America (œcodoma cephalotes) is allied to the preceding, but is even more destructive, though it seems to confine itself to vegetable products. It forms extensive underground galleries, and the earth brought up is deposited on the surface, forming huge mounds sometimes thirty or forty yards in circumference and from one to three feet high. On first seeing these vast deposits of red or yellow earth in the woods near Para, it was hardly possible to believe they were not the work of man, or at least of some large burrowing animal. In these underground caves the ants store up large quantities of leaves, which they obtain from living trees. They gnaw out circular pieces and carry them away along regular paths a few inches wide, forming a stream of apparently animated leaves. The great extent of the subterranean workings of these ants is no doubt due in part to their permanence in one spot, so that when portions of the galleries fall in or are otherwise rendered useless, they are extended in another direction. When in the island of Marajo, near Para, I noticed a path along which a stream of Saübas were carrying leaves from a neighbouring thicket; and a relation of the proprietor assured me that he had known that identical path to be in constant use by the ants for twenty years. Thus we can account for the fact mentioned by Mr. Bates, that the underground galleries were traced by smoke for a distance of seventy yards in the Botanic Gardens at Para; and for the still more extraordinary fact related by the Rev. Hamlet Clark, that an allied species in Rio De Janeiro has excavated a tunnel under the bed of the river Parahyba, where it is about a quarter of a mile wide! These ants seem to prefer introduced to native trees; and young plantations of orange, coffee, or mango trees are sometimes destroyed by them, so that where they abound cultivation of any kind becomes almost impossible. Mr. Belt ingeniously accounts for this preference by supposing that for ages there has been a kind of struggle going on between the trees and the ants; those varieties of trees which were in any way distasteful or unsuitable escaping destruction, while the ants were becoming slowly adapted to attack new trees. Thus in time the great majority of native trees have acquired some protection against the ants, while foreign trees, not having been so modified, are more likely to be suitable for their purposes. Mr. Belt carried on war against them for four years to protect his garden in Nicaragua, and found that carbolic acid and corrosive sublimate were most effectual in destroying or driving them away.

The use to which the ants put the immense quantities of leaves they carry away has been a great puzzle, and is, perhaps, not yet quite understood. Mr. Bates found that the Amazon species used them to thatch the domes of earth covering the entrances to their subterranean galleries, the pieces of leaf being carefully covered and kept in position by a thin layer of grains of earth. In Nicaragua Mr. Belt found the underground cells full of a brown flocculent matter, which he considers to be the gnawed leaves connected by a delicate fungus which ramifies through the mass and which serves as food for the larvæ; and he believes that the leaves are really gathered as manure-heaps to favour the growth of this fungus!

When they enter houses, which they often do at night, the Saübas are very destructive. Once, when travelling on the Rio Negro, I had bought about a peck of rice, which was tied up in a large cotton handkerchief and placed on a bench in a native house where we were spending the night. The next morning we found about half the rice on the floor, the remainder having been carried away by the ants; and the empty handkerchief was still on the bench, but with hundreds of neat cuts in it reducing it to a kind of sieve.[1]

The foraging ants of the genus Eciton are another remarkable group, especially abundant in the equatorial forests of America. They are true hunters, and seem to be continually roaming about the forests in great bands in search of insect prey. They especially devour maggots, caterpillars, white ants, cockroaches, and other soft insects; and their bands are always accompanied by flocks of insectivorous birds, who prey upon the winged insects that are continually trying to escape from the ants. They even attack wasps' nest, which they cut to pieces and then drag out the larvæ. They bite and sting severely, and the traveller who accidentally steps into a horde of them will soon be overrun, and must make his escape as quickly as possible. They do not confine themselves to the ground, but swarm up bushes and low trees, hunting every branch, and clearing them of all insect life. Sometimes a band will enter a house, like the driver ants in Africa, and clear it of cockroaches, spiders, centipedes, and other insects. They seem to have no permanent abode, and to be ever wandering about in search of prey, but they make temporary habitations in hollow trees or other suitable places.

[1] For a full and most interesting description of the habits and instincts of this ant, see Bates' *Naturalist on the River Amazons,* 2d, ed., pp. 11–18; and Belt's *Naturalist in Nicaragua,* pp. 71–84.

Perhaps the most extraordinary of all ants are the blind species of Eciton discovered by Mr. Bates, which construct a covered way or tunnel as they march along. On coming near a rotten log, or any other favourable hunting ground, they pour into all its crevices in search of booty, their covered way serving as a protection to retire to in case of danger. These creatures, of which two species are known, are absolutely without eyes; and it seems almost impossible to imagine that the loss of so important a sense-organ can be otherwise than injurious to them. Yet on the theory of natural selection the successive variations by which the eyes were reduced and ultimately lost must all have been useful. It is true they do manage to exist without eyes; but that is probably because, as sight became more and more imperfect, new instincts or new protective modifications were developed to supply its place, and this does not in any way account for so widespread and invaluable a sense having become permanently lost, in creatures which still roam about and hunt for prey very much as do their fellows who can see.

Special Relations between Ants and Vegetation

Attention has recently been called to the very remarkable relations existing between some trees and shrubs and the ants which dwell upon them. In the Malay islands are several curious shrubs belonging to the Cinchonaceæ, which grow parasitically on other trees, and whose swollen stems are veritable ants' nests. When very young the stems are like small, irregular, prickly tubers, in the hollows of which ants establish themselves; and these in time grow into irregular masses the size of large gourds, completely honeycombed with the cells of ants.[1] In America there are some analogous cases occurring in several families of plants, one of the most remarkable being that of certain Melastomas which have a kind of pouch formed by an enlargement of the petiole of the leaf, and which is inhabited by a colony of small ants. The hollow stems of the Cecropias (curious trees with pale bark and large palmate leaves which are white beneath) are always tenanted by ants, which make small entrance holes through the bark; but here there seems no *special* adaptation to the wants of the insect. In a species of Acacia observed by Mr. Belt, the thorns are immensely large and hollow, and are always tenanted by ants. When young these thorns are soft and full of a sweetish pulpy substance, so that when the ants first take possession they find a store of food in their house. Afterwards they find a special provision of honey-glands on the leaf-stalks, and also small yellow fruit-like bodies which are eaten by the ants; and this supply of food permanently attaches them to the plant. Mr. Belt believes, after much careful observation, that these ants protect the plant they live on from leaf-eating insects, especially from the destructive Saüba ants,—that they are in fact a standing army kept for the protection of the plant! This view is supported by the fact that other plants—Passion-flowers for example—have honey-secreting glands on the young leaves and on the sepals of the flower-buds which constantly attract a small black ant. If this view is correct, we see that the need of escaping from the destructive attacks of the leaf-cutting ants has led to strange modifications in many plants. Those in which the foliage was especially attractive to these enemies were soon weeded out unless variations occurred which tended to preserve them. Hence the curious phenomenon of insects specially attracted to certain plants to protect them from other insects; and the existence of the destructive leaf-cutting ant in America will thus explain why these specially modified plants are so much more abundant there than in the Old World, where no ants with equally destructive habits appear to exist.

[1] These form two genera, Myrmecodia and Hydnophytum. For description and figures see Mr. H. O. Forbes' *Naturalist's Wanderings in the Eastern Archipelago*, p. 79.

Wasps and Bees

These insects are excessively numerous in the tropics, and, from their large size, their brilliant colours, and their great activity, they are sure to attract attention. Handsomest of all, perhaps, are the Scoliadæ, whose large and rather broad hairy bodies, often two inches long, are richly banded with yellow or orange. The Pompilidæ comprise an immense number of large and handsome insects, with rich blue-black bodies and wings and exceedingly long legs. They may often be seen in the forests dragging along large spiders, beetles, or other insects they have captured. Some of the smaller species enter houses and build earthen cells, which they store with small green spiders rendered torpid by stinging, to feed the larvæ. The Eumenidæ are beautiful wasps with very long pedunculated bodies, which build papery cones covering a few cells in which the eggs are deposited. Among the bees the Xylocopas, or wood-boring bees, are remarkable. They resemble large humble-bees, but have broad, flat, shining bodies, either black or banded with blue; and they often bore large cylindrical holes in the posts of houses. True honeybees are chiefly remarkable in the East for their large semicircular combs suspended from the branches of the loftiest trees without any covering. From these exposed nests large quantities of wax and honey are obtained, while the larvæ afford a rich feast to the natives of Borneo, Timor, and other islands where bees abound. They are very pugnacious, and, when disturbed will follow the intruders for miles, stinging severely.

Orthoptera and other Insects

Next to the butterflies and ants, the insects that are most likely to attract the attention of the stranger in the topics are the various forms of Mantidæ and Phasmidæ, some of which are remarkable for their strange attitudes and bright colours; while others are among the most singular of known insects, owing to their resemblance to sticks and leaves. The Mantidæ—usually called "praying insects," from their habit of sitting with their long fore-feet held up as if in prayer—are really tigers among insects, lying in wait for their prey, which they seize with their powerful serrated fore-feet. They are usually so coloured as to resemble the foliage among which they live, and as they sit quite motionless, they are not easily perceived.

The Phasmidæ are perfectly inoffensive leaf-eating insects of very varied forms; some being broad and leaf-like, while others are long and cylindrical, so as to resemble sticks, whence they are often called walking-stick insects. The imitative resemblance of some of these insects to the plants on which they live is marvellous. The true leaf-insects of the East, forming the genus Phyllium, are the size of a moderate leaf, which their large wing-covers and the dilated margins of the head, thorax, and legs cause them exactly to resemble. The veining of the wings and their green tint exactly correspond to that of the leaves of their food-plant;

and as they rest motionless during the day, only feeding at night, they the more easily escape detection. In Java they are often kept alive on a branch of the guava tree; and it is common thing for a stranger, when asked to look at this curious insect, to inquire where it is, and on being told that it is close under his eyes, to maintain that there is no insect at all, but only a branch with green leaves.

The larger wingless stick-insects are often eight inches to a foot long. They are abundant in the Moluccas; hanging on the shrubs that line the forest-paths; and they resemble sticks so exactly, in colour, in the small rugosities of the bark, in the knots and small branches, imitated by the joints of the legs which are either pressed close to the body, or struck out at random, that it is absolutely impossible, by the eye alone, to distinguish the real dead twigs which fall down from the trees overhead, from the living insects. The writer has often looked at them in doubt, and has been obliged to use the sense of touch to determine the point. Some are small and slender like the most delicate twigs; others again have wings, and it is curious that these are often beautifully coloured, generally bright pink, sometimes yellow, and sometimes finely banded with black; but when at rest the wings fold up so as to be completely concealed under the narrow wing-covers, and the whole insect is then green or brown, and almost invisible among the twigs or foliage. To increase the resemblance to vegetation, some of these Phasmas have small green processes in various parts of their bodies looking exactly like moss. These inhabit damp forests both in the Malay islands and in America, and they are so marvellously like moss-grown twigs that the closest examination is needed to satisfy oneself that it is really a living insect we are looking at.

Many of the locusts are equally well-disguised, some resembling green leaves, others those that are brown and dead; and the latter often have small transparent spots on the wings, looking like holes eaten through them. That these disguises deceive their natural enemies is certain, for otherwise the Phasmidæ would soon be exterminated. They are large and sluggish, and very soft and succulent; they have no means of defence or of flight, and they are eagerly devoured by numbers of birds, especially by the numerous cuckoo tribe, whose stomachs are often full of them; yet numbers of them escape destruction, and this can only be due to their vegetable disguises. Mr. Belt records a curious instance of the actual operation of this kind of defence in a leaf-like locust, which remained perfectly quiescent in the midst of a host of insectivorous ants, which ran over it without finding out that it was an insect and not a leaf! It might have flown away from them, but it would then instantly have fallen a prey to the numerous birds which always accompany these roaming hordes of ants to feed upon the insects that endeavour to escape. Far more conspicuous than any of these imitative species are the large locusts, with rich crimson or blue-and-black spotted wings. Some of these are nearly a foot in expanse of wings; they fly by day, and their strong spiny legs probably serve as a protection against all the smaller birds. They cannot be said to be common; but when met with they fully satisfy our notions as to the large size and gorgeous colours of tropical insects.[1]

[1] It has now been ascertained that these conspicuously coloured locusts are protected by inedibility. See *Darwinism,* p. 267.

Beetles

Considering the enormous numbers and endless variety of the beetle tribe that are known to inhabit the tropics, they form by no means so prominent a feature in the animal life of the equatorial zone as we might expect. Almost every entomologist is at first disappointed with them. He finds that they have to be searched for almost as much as at home, while those of large size (except one or two very common species) are rarely met with. The groups which most attract attention, from their size and beauty, are the Buprestidæ and the Longicorns. The former are usually smooth insects of an elongate ovate form, with very short legs and antennæ, and adorned with the most glowing metallic tints. They abound on fallen tree-trunks and on foliage, in the hottest sunshine, and are among the most brilliant ornaments of the tropical forests. Some parts of the temperate zone, especially Australia and Chili, abound in Buprestidæ which are equally beautiful; but the largest species are only found within the tropics, those of the Malay islands being the largest of all.

The Longicorns are elegantly shaped beetles, usually with long antennæ and legs, varied in form and structure in an endless variety of ways, and adorned with equally varied colours, spots and markings. Some are large and massive insects three or four inches long, while others are no bigger than our smaller ants. The majority have sober colours, but often delicately marbled, veined, or spotted; while others are red, or blue, or yellow, or adorned with the richest metallic tints. Their antennæ are sometimes excessively long and graceful, often adorned with tufts of hair, and sometimes pectinated. They especially abound where timber trees have been recently felled in the primeval forests; and while extensive clearings are in progress their variety seems endless. In such a locality in the island of Borneo, nearly 300 different species were found during one dry season, while the number obtained during eight years' collecting in the whole Malay Archipelago was about a thousand species.

Among the beetles that always attract attention in the tropics are the large, horned Copridæ and Dynastidæ, corresponding to our dung-beetles. Some of these are of great size, and they are occasionally very abundant. The immense horn-like protuberances on the head and thorax of the males in some of the species are very extraordinary, and, combined with their polished or rugose metallice colours, render them perhaps the most conspicuous of all the beetle tribe. The weevils and their allies are also very interesting, from their immense numbers, endless variety, and the extreme beauty of many of the species. The Anthribidæ, which are especially abundant in the Malay Archipelago, rival the Longicorns in the immense length of their elegant antennæ; while the diamond beetles of Brazil, the Eupholi of the Papuan islands, and the Pachyrhynchi of the Philippines, are veritable living jewels.

Where a large extent of virgin forest is cut down in the early part of the dry season, and some hot sunny weather follows, the abundance and variety of beetles attracted by the bark and foliage in various stages of drying is amazing. The air is filled with the hum of their wings. Golden and green Buprestidæ are flying about in every direction, and settling on the bark in full sunshine. Green and spotted rose-chafers hum along near the ground; long-horned Anthribidæ are disturbed at every step; elegant little Longicorns circle about the drying foliage, while larger species fly slowly from branch to branch. Every fallen trunk is full of life. Strange mottled, and spotted, and rugose Longicorns, endless

Curculios, queer-shaped Brenthidæ, velvety brown or steel-blue Cleridæ, brown or yellow or whitish click beetles (Elaters), and brilliant metallic Carabidæ. Close by, in the adjacent forest, a whole host of new forms are found. Elegant tiger-beetles, leaf-hunting Carabidæ, musk-beetles of many sorts, scarlet Telephori, and countless Chrysomelas, Hispas, Coccinellas, with strange Heteromera, and many curious species which haunt fungi, rotten bark, or decaying leaves. With such variety and beauty the most ardent entomologist must be fully satisfied; and when, every now and then, some of the giants of the tropics fall in his way—grand Prionidæ or Lamiidæ several inches long, a massive golden Buprestis, or a monster horned Dynastes—he feels that his most exalted notions of the insect-life of the tropics are at length realised.

Wingless Insects

Passing on to other orders of insects, the hemiptera dragon-flies and true files hardly call for special remark. Among them are to be found a fair proportion of large and handsome species, but they require much searching after in their special haunts, and seldom attract so much attention as the groups already referred to. More prominent are the wingless tribes, such as spiders, scorpions, and centipedes. The wanderer in the forest often finds the path closed by large webs almost as strong as silk, inhabited by gorgeous spiders with bodies nearly two inches long and legs expanding six inches. Others are remarkable for their hard flat bodies, terminating in horned processes which are sometimes long, slender, and curved like a pair of miniature cow's horns. Hairy terrestial species of large size are often met with, the largest belonging to the South American genus Mygale, which sometimes actually kill birds, a fact which had been stated by Madame Merian and others, but was discredited till Mr. Bates succeeded in catching one in the act. The small jumping spiders are also noticeable from their immense numbers, variety, and beauty. They frequent foliage and flowers, running about actively in pursuit of small insects; and many of them are so exquisitely coloured as to resemble jewels rather than spiders. Scorpions and centipedes make their presence known to every traveller. In the damp forests of the Malay islands are huge scorpions of a greenish colour and eight or ten inches long; while in huts and houses smaller species lurk under boxes and boards, or secret themselves in almost every article not daily examined. Centipedes of immense size and deadly venom harbour in the thatch of houses and canoes, and will even ensconce themselves under pillows and in beds, rendering a thorough examination necessary before retiring to rest. Yet with moderate precautions there is little danger from these disgusting insects, as may be judged by the fact that during twelve years' wanderings in American and Malayan forests the author was never once bitten or stung by them.

General Observations on Tropical Insects

The characteristics of tropical insects that will most attract the ordinary traveller are, their great numbers, and the large size and brilliant colours often met with. But a more extended observation leads to the conclusion that the average of size is probably but little greater in tropical than in temperate zones, and that, to make up for a certain proportion of very large, there is a corresponding increase in the numbers of very small species. The much greater size reached by many tropical insects is no doubt due to the fact that the supply of food is always in excess of their demands in the larva state, while there is no check from the ever-recurring cold of winter; and they are thus able to acquire the dimensions that may be on the whole most advantageous to the race, unchecked by the annual or periodical scarcities which in less favoured climates would continually threaten their extinction. The colours of tropical insects are, probably, on the average more brilliant than those of temperate countries, and some of the causes which may have led to this have been discussed in another part of this volume.[1] It is in the tropics that we find, most largely developed, whole groups of insects which are unpalatable to almost all insectivorous creatures, and it is among these that some of the most gorgeous colours prevail. Others obtain protection in a variety of ways; and the amount of cover or concealment always afforded by the luxuriant tropical vegetation is probably a potent agent in permitting a full development of colour.

BIRDS

Although the number of brilliantly-coloured birds in almost every part of the tropics is very great, yet they are by no means conspicuous, and as a rule they can hardly be said to add much to the general effect of equatorial scenery. The traveller is almost always disappointed at first with the birds, as he is with the flowers and the beetles; and it is only when, gun in hand, he spends days in the forest, that he finds out how many beautiful living things are concealed by its dense foliage and gloomy thickets. A considerable number of the handsomest tropical birds belong to family groups which are confined to one continent with its adjacent islands, and we shall therefore be obliged to deal for the most part with such large divisions as tribes and orders, by means of which to define the characteristics of tropical bird-life. We find that there are three important orders of birds which, though by no means exclusively tropical, are yet so largely developed there in proportion to their scarcity in extra-tropical regions, that, more than any others, they serve to give a special character to equatorial ornithology. These are the Parrots, the Pigeons, and the Picariæ, to each of which groups we will devote some attention.

[1] Chapters v. and vi., *post. The colours of Animals and Plants.*

Parrots

The parrots, forming the order Psittaci of naturalists, are a remarkable group of fruit-eating birds, of such high and peculiar organisation that they are often considered to stand at the head of the entire class. They are pre-eminently characteristic of the intertropical zone, being nowhere absent within its limits (except from absolutely desert regions), and they are generally so abundant and so conspicuous as to occupy among birds the place assigned to butterflies among insects. A few species range far into the temperate zones. One reaches Carolina in North America, another the Magellan Straits in South America; in Africa they only extend a few degrees beyond the southern tropic; in North-Western India they reach 35° north latitude, but in the Australian region they range farthest towards the pole, being found not only in New Zealand, but as far as the Macquarie islands in 54° south, where the climate is very cold and boisterous, but sufficiently uniform to supply vegetable food throughout the year. There is hardly any part of the equatorial zone in which the traveller will not soon have his attention called to some members of the parrot tribe. In Brazil the great blue and yellow or crimson macaws may be seen every evening wending their way homeward in pairs, almost as commonly as rooks with us, while innumerable parrots and parraquets attract attention by their harsh cries when disturbed from some favourite fruit-tree. In the Moluccas and New Guinea White cockatoos and gorgeous lories in crimson and blue are the very commonest of birds.

No group of birds—perhaps no other group of animals—exhibits within the same limited number of genera and species so wide a range and such an endless variety of colour. As a rule, parrots may be termed green birds, the majority of the species having this colour as the basis of their plumage relieved by caps, gorgets, bands, and wing-spots of other and brighter hues. Yet this general green tint sometimes changes into light or deep blue, as in some macaws; into pure yellow or rich orange, as in some of the American macaw-parrots (Conurus); into purple, gray, or dove-colour, as in some American, African, and Indian species; into the purest crimson, as in some of the lories;

into rosy-white and pure white, as in the cockatoos; and into a deep purple, ashy, or black, as in several Papuan, Australian, and Mascarene species. There is, in fact, hardly a single distinct and definable colour that cannot be fairly matched among the 400 species of known parrots. Their habits, too, are such as to bring them prominently before the eye. They usually feed in flocks; they are noisy, and so attract attention; they love gardens, orchards, and open sunny places; they wander about far in search of food, and towards sunset return homewards in noisy flocks, or in constant pairs. Their forms and motions are often beautiful and attractive. The immensely long tails of the macaws, and the more slender tails of the Indian parraquets; the fine crest of the cockatoos; the swift flight of many of the smaller species, and the graceful motions of the little love-birds and allied forms, together with their affectionate natures, aptitude for domestication, and powers of mimicry—combine to render them at once the most conspicuous and the most attractive of all the specially tropical forms of bird-life.

The number of species of parrots found in the different divisions of the tropics is very unequal. Africa is by far the poorest; since along with Madagascar and the Mascarene islands, which have many peculiar forms, it scarcely numbers two dozen species. Asia, along with the Malay islands as far as Java and Borneo, is also very poor, with about thirty species. Tropical America is very much richer, possessing about 140 species, among which are many of the largest and most beautiful forms. But of all parts of the globe the tropical islands belonging to the Australian region (from Celebes eastward), together with the tropical parts of Australia, are richest in the parrot tribe, possessing more than 150 species, among which are many of the most remarkable and beautiful of the entire group. The whole Australian region, whose extreme limits may be defined by Celebes, the Marquesas, and the New Zealand group, possesses over 200 species of parrots.

Pigeons

These are such common birds in all temperate countries that it may surprise many readers to learn that they are nevertheless a characteristic tropical group. That such is the case, however, will be evident from the fact that only sixteen species are known from the whole of the temperate parts of Europe, Asia, and North America, while about 330 species inhabit the tropics. Again, the great majority of the species are found congregated in the equatorial zone, whence they diminish gradually toward the limits of the tropics, and then suddenly fall off in the temperate zones. Yet although they are pre-eminently tropical or even equatorial as a group, they are not, from our present point of view, of much importance, because they are so shy and so generally inconspicuous that in most parts of the tropics an ordinary observer might hardly be aware of their existence. The remark applies especially to America and Africa, where they are neither very abundant nor peculiar; but in the Eastern hemisphere, and especially in the Malay Archipelago and Pacific islands, they occur in such profusion and present such singular forms and brilliant colours, that they are sure to attract attention. Here we find the extensive group of fruit-pigeons, which, in their general green colours adorned with patches and bands of purple, white, blue, or orange, almost rival the parrot tribe; while the golden-green Nicobar pigeon, the great crowned pigeons of New Guinea as large as turkeys, and the golden-yellow fruit-dove of the Fijis, can hardly be surpassed for beauty.

Pigeons are especially abundant and varied in tropical archipelagoes, so that if we take the Malay and Pacific islands, the Madagascar group, and the Antilles or West Indian islands, we find that they possess between them more different kinds of pigeons than all the continental tropics combined. Yet further, that portion of the Malay Archipelago east of Borneo, together with the Pacific islands, is exceptionally rich in pigeons; and the reason seems to be that monkeys and all other arboreal mammals that devour eggs and young birds are entirely absent from this region. Even in South America pigeons are scarce where monkeys are abundant, and *vice versâ,* so that here we seem to get a glimpse of one of the curious interactions of animals on each other, by which their distribution, their habits, and even their colours, may have been influenced, for the most conspicuous pigeons, whether by colour or by their crests, are all found in countries where they have the fewest enemies.

Picariæ

The extensive and heterogeneous series of bird till recently comprised under this term includes most of the fissirostral and scansorial groups of the older naturalists. They may be described as, for the most part, arboreal birds, of a low grade of organisation, with weak or abnormally developed feet, and usually less active than the true Passeres or perching birds of which our warblers, finches, and crows may be taken as the types. The order Picariæ comprises twenty-five families, some of which are very extensive. All are either wholly or mainly tropical, only two of the families—the woodpeckers and the kingfishers—having a few representatives which are permanent residents in the temperate regions, while our summer visitor, the cuckoo, is the sole example in Northern Europe of one of the most abundant and widespread tropical families of birds. Only four of the families have a general distribution over all the warmer countries of the globe—the cuckoos, the kingfishers, the swifts, and the goatsuckers;

while two others—the trogons and the woodpeckers—are only wanting in the Australian region, ceasing suddenly at Borneo and Celebes respectively.

Cuckoos

Whether we consider their wide range, their abundance in genera and species, or the peculiarities of their organisation, the cuckoos may be taken as the most typical examples of this extensive order of birds; and there is perhaps no part of the tropics where they do not form a prominent feature in the ornithology of the country. Their chief food consists of soft insects, such as caterpillars, grasshoppers, and the defenceless stick- and leaf-insects; and in search after these they frequent the bushes and lower parts of the forest, and the more open tree-clad plains. They vary greatly in size and appearance, from the small and beautifully metallic golden-cuckoos of Africa, Asia, and Australia, no larger than sparrows, to the pheasant-like ground cuckoo of Borneo, the Scythrops of the Moluccas, which almost resembles a hornbill, the Rhamphococcyx of Celebes with its richly-coloured bill, and the Goliath cuckoo of Gilolo with its enormously long and ample tail.

Cuckoos, being invariably weak and defenceless birds, conceal themselves as much as possible among foliage or herbage; and as a further protection, many of them have acquired the coloration of rapacious or combative birds. In several parts of the world cuckoos are coloured exactly like hawks, while some of the small Malayan cuckoos closely resemble the pugnacious drongo-shrikes.

Trogons, Barbets, Toucans, and Hornbills

Many of the families of Picariæ are confined to the tropical forests, and are remarkable for their varied and beautiful colouring. Such are the trogons of America, Africa, and Malaya, whose dense puffy plumage exhibits the purest tints of rosy-pink, yellow, and white, set off by black heads and a golden-green or rich brown upper surface. Of more slender forms, but hardly less brilliant in colour, are the jacamars and motmots of America, with the bee-eaters and rollers of the East, the latter exhibiting tints of pale-blue or verditer-green, which are very unusual. The barbets are rather clumsy fruit-eating birds, found in all the great tropical regions except that of the Austro-Malay islands, and they exhibit a wonderful variety as well as strange combinations of colours. Those of Asia and Malaya are mostly green, but adorned about the head and neck with patches of the most vivid reds, blues, and yellows in endless combinations. The African species are usually black or greenish-black, with masses of intense crimson, yellow, or white, mixed in various proportions and patterns; while the American species combine both styles of colouring, but the tints are usually more delicate, and are often more varied and more harmoniously interblended. In the Messrs. Marshall's fine work[1] all the species are described and figures, and few more instructive examples can be found than are exhibited in their beautifully-coloured plates, of the endless ways in which the most glaring and inharmonious colours are often combined in natural objects with a generally pleasing result.

[1] *A Monograph of the Capitonidæ or Scansorial Barbets,* by C. F. T. Marshall and G. F. L. Marshall. 1871.

We will next group together three families which, although quite distinct, may be said to represent each other in their respective countries,—the toucans of America, and plaintaineaters of Africa, and the hornbills of the East,—all being large and remarkable birds, and certain to attract the traveller's attention. The toucans are the most beautiful on account of their large and richly-coloured bills, their delicate breast-plumage, and the varied bands of colour with which they are often adorned. Though feeding chiefly on fruits, they also devour birds' eggs and young birds; and they are remarkable for the strange habit of sleeping with the tail laid flat upon their backs, in what seems a most unnatural and inconvenient position. What can be the use of their enormous bills has been a great puzzle to naturalists, the only tolerably satisfactory solution yet arrived at being that suggested by Mr. Bates,—that it simply enables them to reach fruit at the ends of slender twigs which, owing to their weight and clumsiness, they would otherwise be unable to obtain. At first sight it appears very improbable that so large and remarkable an organ should have been developed for such a purpose; but we have only to suppose that the original toucans had rather large and thick bills, not unlike those of the barbets (to which group they are undoubtedly allied), and that as they increased in size and required more food, only those could obtain a sufficiency whose unusually large beaks enabled them to reach farthest. So large and broad a bill as they now possess would not be required; but the development of the bill naturally went on as it had begun, and, so that it was light and handy, the large size was no disadvantage if length was obtained. The plantain-eaters of Africa are less remarkable birds, though adorned with rich colours and elegant crests. The hornbills, though less beautiful than the toucans, are more curious, from the strange forms of their huge bills, which are often adorned with ridges, knobs, or recurved horns. They are bulky and heavy birds, and during flight beat the air with prodigious force, producing a rushing sound very like the puff of a locomotive, and which can sometimes be heard a mile off. They mostly feed on fruits; and as their very short legs render them even less active than the toucans, the same explanation may be given of the large size of their bills, although it will not account for the curious horns and processes from which they derive their distinctive name. The largest hornbills are more than four feet long, and their laboured noisy flight and huge bills, as well as their habit of perching on the top of bare or isolated trees, render them very conspicuous objects.

The Picariæ comprise many other interesting families—as, for example, the puff-birds, the todies, and the humming-birds; but as these are all confined to America we can hardly claim them as characteristic of the tropics generally. Others, though very abundant in the tropics, like the kingfishers and the goatsuckers, are too well known in temperate lands to allow of their being considered as specially characteristic of the equatorial zone. We will therefore pass on to consider what are the more general characteristics of the tropical as compared with the temperate bird-fauna, especially as exemplified among the true perchers or Passeres, which constitute about three-fourths of all terrestrial birds.

Passeres

This great order comprises all our most familiar birds, such as the thrushes, warblers, tits, shrikes, flycatchers, starlings, crows, wagtails, larks, and finches. These families are all more or less abundant in the tropics; but there are a number of other families which are almost or quite peculiar to tropical lands and give a special character to their bird-life. All the peculiarly tropical families are, however, confined to some definite portion of the tropics, a number of them being American only, others Australian, while others again are common to all the warm countries of the Old World; and it is a curious fact that there is no single family of this great order of birds that is common to all tropical regions and confined to them, or that is even especially characteristic of the tropical zone, like the cuckoos among the Picariæ. The tropical families of passerine birds being very numerous, and their peculiarities not easily understood by any but ornithologists, it will be better to consider the series of fifty families of Passeres as one compact group, and endeavour to point out what external peculiarities are most distinctive of those which inhabit tropical countries.

Owing to the prevalence of forests and the abundance of flowers, fruits, and insects, tropical and especially equatorial birds have become largely adapted to these kinds of food; while the seed-eaters, which abound in temperate lands where grasses cover much of the surface, are proportionately scarce. Many of the peculiarly tropical families are therefore either true insect-eaters or true fruit-eaters, whereas in the temperate zones a mixed diet is more general.

One of the features of tropical birds that will first strike the observer is the prevalence of crests and of ornamental plumage in various parts of the body, and especially of extremely long or curiously shaped feathers in the tails, tail-coverts, or wings of a variety of species. As examples we may refer to the red paradise-bird, whose middle tail-feathers are like long ribands of whalebone; to the wire-like tail feathers of the king bird-of-paradise of New Guinea, and of the wire-tailed manakin of the Amazons; and to the long waving tail plumes of the whydah finch of West Africa and paradise flycatcher of India; to the varied and elegant crests of the cock-of-the-rock, the king-tyrant, the umbrella-bird, and the six-plumed bird-of-paradise; and to the wonderful side plumes of most of the true paradise-birds. In other orders of birds we have such remarkable examples as the racquet-tailed kingfishers of the Moluccas, and the racquet-tailed parrots of Celebes; the enormously developed tail-coverts of the peacock and the Mexican trogon; and the excessive wing-plumes of the argus-pheasant of Malacca and the long-shafted goatsucker of West Africa.

Still more remarkable are the varied styles of coloration in the birds of tropical forests, which rarely or never appear in those of temperate lands. We have intensely lustrous metallic plumage in the jacamars, trogons, humming-birds, sun-birds, and paradise-birds; as well as in some starlings, pittas or ground thrushes, and drongo-shrikes. Pure green tints occur in parrots, pigeons, green bulbuls, greenlets, and in some tanagers, finches, chatterers, and pittas. These undoubtedly tend to concealment; but we have also the strange phenomenon of white forest birds in the tropics, a colour only found elsewhere among the aquatic tribes and in the arctic regions. Thus, we have the bell-bird of South America, the white pigeons and cockatoos of the East, with a few starlings, woodpeckers, kingfishers, and goatsuckers, which are either very light-coloured or in great part pure white.

But besides these strange and new and beautiful forms of bird life, which we have attempted to indicate as characterising the tropical regions, the traveller will soon find that there are hosts of dull and dingy birds, not one whit different, so far as colour is concerned, from the sparrows, warblers, and thrushes of our northern climes. He will, however, if observant, soon note that most of these dull colours are protective; the groups to which they belong frequenting low thickets, or the ground, or the trunks of trees. He will find groups of birds specially adapted to certain modes of tropical life. Some live on ants upon the ground, others pick minute insects from the bark of trees; one group will devour bees and wasps, others prefer caterpillars; while a host of small birds seek for insects in the corollas of flowers. The air, the earth, and undergrowth, the tree-trunks, the flowers, and the fruits, all support their specially adapted tribes of birds. Each species fills a place in nature, and can only continue to exist so long as that place is open to it; and each has become what it is in every detail of form, size, structure, and even of colour, because it has inherited through countless ancestral forms all those variations which have best adapted it among its fellows to fill that place, and to leave behind it equally well adapted successors.

REPTILES AND AMPHIBIA

Next to the birds, or perhaps to the less observant eye even before them, the abundance and variety of reptiles form the chief characteristic of tropical nature; and the three groups—lizards, snakes, and frogs—comprise all that, from our present point of view, need to noticed.

Lizards

Lizards are by far the most abundant in individuals and the most conspicuous; and they constitute one of the first attractions to the visitor from colder lands. They literally swarm everywhere. In cities they may be seen running along walls and up palings; sunning themselves on logs of wood, or creeping up to the eaves of cottages. In every garden, road, or dry sandy path, they scamper aside as you walk along. They crawl up trees, keeping at the farther side of the trunk and watching the passer-by with the caution of a squirrel. Some will walk up smooth walls with the greatest ease; while in houses the various kinds of Geckos cling to the ceilings, along which they run back downwards in pursuit of flies, holding on by means of their dilated toes with suctorial discs, though sometimes, losing hold, they fall upon the table or on the upturned face of the visitor. In the forests large, flat, and marbled Geckos cling to the smooth trunks; small and active lizards rest on the foliage; while occasionally the larger kinds, three or four feet long, rustle heavily as they move among the fallen leaves.

Their colours vary much, but are usually in harmony with their surroundings and habits. Those that climb about walls and rocks are stone-coloured, and sometimes nearly black; the house lizards are gray or pale-ashy, and are hardly visible on a palm-leaf thatch, or even on white-washed ceiling. In the forest they are often mottled with ashy-green, like lichen-grown bark. Most of the ground-lizards are yellowish or brown; but some are of beautiful green colours, with very long and slender tails. These are among the most active and lively; and instead of crawling on their bellies like many lizards, they stand well upon their feet and scamper about with the agility and vivacity of kittens. Their tails are very brittle; a slight blow causing them to snap off, when a new one grows, which is, however, not so perfectly formed and completely scaled as the original member. It is not uncommon, when a tail is half broken, for a new one to grow out of the wound producing the curious phenomenon of a forked tail. There are about 1300 different kinds of lizards known, the great majority of which inhabit the tropics, and they probably increase in numbers towards the equator. A rich vegetation and a due proportion of moisture and sunshine seem favourable to them, as shown by their great abundance and their varied kinds at Para and in the Aru islands—places which are nearly the antipodes of each other, but which both enjoy the fine equatorial climate in perfection, and are alike pre-eminent in the variety and beauty of their insect life.

Three peculiar forms of lizard may be mentioned as specially characteristic of the American, African, and Asiatic tropical zones respectively. The iguanas of South America are large, arboreal, herbivorous lizards of a beautiful green colour, which renders them almost invisible when resting quietly among foliage. They are distinguished by the serrated back, deep dew-lap, and enormously long tail, and are one of the few kinds of lizards whose flesh is considered a delicacy. The chameleons of Africa are also arboreal lizards, and they have the prehensile tail, which is more usually found among American animals. They are excessively slow in their motions, and are protected by the wonderful power of changing their colour so as to assimilate it with that of immediately surrounding objects. Like the majority of lizards they are insectivorous, but they are said to be able to live for months without taking food. The dragons or flying lizards of India and the larger Malay island are perhaps the most curious and interesting of living reptiles, owing to their power of passing through the air by means of wing-like membranes, which stretch along each side of the body and are expanded by means of slender bony processes from the first six false ribs. These membranes are folded up close to the body when not in use, and are then almost imperceptible; but when open they form a nearly circular web, the upper surface of which is generally zoned with red or yellow in a highly ornamental manner. By means of this parachute the animal can easily pass from one tree to another for a distance of about thirty feet, descending at first, but as it approaches its destination rising a little so to reach the tree with its head erect. They are very small, being usually not more than two or three inches long, exclusive of the slender tail; and when the wings are expanded in the sunshine they more resemble some strange insects than members of the reptile tribe.

Snakes

Snakes are, fortunately, not so abundant or so obtrusive as lizards, or the tropics would be scarcely habitable. At first, indeed, the traveller is disposed to wonder that he does not see more of them, but he will soon find out that there are plenty; and, if he is possessed by the usual horror or dislike of them, he may think there are too many. In the equatorial zone snakes are less troublesome than in the drier parts of the tropics, although they are probably more numerous and more varied. This is because the country is naturally a vast forest, and the snakes being all adapted to a forest life do not as a rule frequent gardens and come into houses as in India and Australia, where they are accustomed to open and arid places. One cannot traverse the forest, however, without soon coming upon them. The slender green whipsnakes glide among the foliage, and may often be touched before they are seen. The ease and rapidity with which these snakes pass through bushes, almost without disturbing a leaf, is very curious. More dangerous are the green vipers, which lie coiled motionless upon foliage, where their colour renders it difficult to see them. The writer has often come upon them while creeping through the jungle after birds or insects, and has sometimes only had time to draw back when they were within a few inches of his face. It is startling in walking along a forest path to see a long snake glide away from just where you were going to set down your foot; but it is perhaps even more alarming to hear a long-drawn heavy slur-r-r, and just to catch a glimpse of a serpent as thick as your leg and an unknown number of feet in length, showing that you must have passed unheeding within a short distance of where it was lying. The smaller pythons are not, however, dangerous, and they often enter houses to catch and feed upon the rats, and are rather liked by the natives. You wills sometimes be told when sleeping in a native house that there is a large snake in the roof, and that you need not be disturbed in case you should hear it hunting after its prey. These serpents no doubt sometimes grow to an enormous size, but such monsters are rare. In Borneo, Mr. St. John states that he measured one twenty-six feet long, probably the largest ever measured by a European in the East. The great water-boa of South America is believed to reach the largest size. Mr. Bates measured skins twenty-one feet long, but the largest ever met with by a European appears to be that described by the botanist, Dr. Gardiner, in his *Travels in Brazil.* It had devoured a horse, and was found dead, entangled in the branches of a tree overhanging a river, into which it had been carried by a flood. It was nearly forty feet long. These creatures are said to seize and devour full-sized cattle on the Rio Branco; and from what is known of their habits this is by no means improbable.

Frogs and Toads

The only Amphibia that often meet the traveller's eye in equatorial countries are the various kinds of frogs and toads, and especially the elegant tree-frogs. When the rainy season begins, and dried-up pools and ditches become filled with water, there is a strange nightly concert produced by the frogs, some of which croak, others bellow, while many have clanging or chirruping, and not unmusical notes. In roads and gardens one occasionally meets huge toads six or seven inches long; but the most abundant and most interesting of the tribe are those adapted for an arboreal life, and hence called tree-frogs. Their toes terminate in discs, by means of which they can cling firmly to leaves and stems. The majority of them are green or brown, and these usually feed at night, sitting quietly during the day so as to be almost invisible, owing to their colour and their moist shining skins so closely resembling vegetable surfaces. Many are beautifully marbled and spotted, and when sitting on leaves resemble large beetles more than frogs, while others are adorned with bright and staring colours; and these, as Mr. Belt has discovered, have nauseous secretions which render them uneatable, so that they have no need to conceal themselves. Some of these are bright blue, others are adorned with yellow stripes, or have a red body with blue legs. Of the smaller tree-frogs of the tropics there must be hundreds of species still unknown to naturalists.

MAMMALIA

Monkeys

The highest class of animals, the Mammalia, although sufficiently abundant in all equatorial lands, are those which are least seen by the traveller. There is, in fact, only one group—the monkeys—which are at the same time pre-eminently tropical, and which make themselves perceived as one of the aspects of tropical nature. They are to be met with in all the great continents and larger islands, except Australia, New Guinea, and Madagascar, though the latter island possesses the lower allied form of Lemurs; and they never fail to impress the observer with a sense of the exuberant vitality of the tropics. They are pre-eminently arboreal in their mode of life, and are consequently most abundant and varied where vegetation reaches its maximum development. In the East we find that maximum in Borneo, and in the West African forests; while in the West the great forest plain of the Amazon stands pre-eminent. It is near the equator only that the great Anthropoid apes, the gorilla, chimpanzee, and orang-utan are found, and they may be met with by any persevering explorer of the jungle. The gibbons, or long-armed apes, have a wider range in the Asiatic continent and in Malaya, and they are more abundant both in species and individuals. Their plaintive howling notes may often be heard in the forests, and they are constantly to be seen sporting at the summits of the loftiest trees, swinging suspended by their long arms, or bounding from tree to tree with incredible agility. They pass through the forest at a height of a hundred feet or more, as rapidly as a deer will travel along the ground beneath them. Other monkeys of various kinds are more abundant and usually less shy; and in places where firearms are not much used they will approach the houses and gambol in the trees undisturbed by the approach of man. The most remarkable of the tailed monkeys of the East is the proboscis monkey of Borneo, whose long fleshy nose gives it an aspect very different from that of most of its allies.

In tropical America monkeys are even more abundant than in the East, and they present many interesting peculiarities. They differ somewhat in dentition and in other structural features from all Old World apes, and a considerable number of them have prehensile tails, a peculiarity never found elsewhere. In the howlers and spider monkeys the tail is very long and powerful, and by twisting the extremity round a branch the animal can hang suspended as easily as other monkeys can by their hands. It is, in fact, a fifth hand, and is constantly used to pick up small objects from the ground. The most remarkable of the American monkeys are the howlers, whose tremendous roaring exceeds that of the lion or the bull, and is to be heard frequently at morning and evening in the primeval forests. The sound is produced by means of a large, thin, bony vessel in the throat, into which air is forced; and it is very remarkable that this one group of monkeys should possess an organ not found in any other monkey or even in any other mammal, apparently for no other purpose than to be able to make a louder noise than the rest. The only other monkeys worthy of special attention are the marmosets, beautiful little creatures with crests, whiskers, or manes, in outward form resembling squirrels, but with a very small monkey-like face. They are either black, brown, reddish, or nearly white in colour, and are the smallest of the monkey tribe, some of them being only about six inches long exclusive of the tail.

Bats

Almost the only other order of mammals that is specially and largely developed in the tropical zone is that of the Chiroptera or bats, which becomes suddenly much less plentiful when we pass into the temperate regions, and still more rare towards the colder parts of it, although a few species appear to reach the Arctic circle. The characteristics of the tropical bats are their great numbers and variety, their large size, and their peculiar forms or habits. In the East those which most attract the traveller's attention are the great fruit-bats, or flying-foxes as they are sometimes called, from the rusty colour of the the coarse fur and the fox-like shape of the head. These creatures may sometimes be seen in immense flocks which take hours to pass by, and they often devastate the fruit plantations of the natives. They are often five feet across the expanded wings, with the body of a proportionate size; and when resting in the daytime on dead trees, hanging head downwards, the branches look as if covered with some monster fruits. The descendants of the Portuguese in the East use them for food, but all the native inhabitants reject them.

In South America there is a group of bats which are sure to attract attention. These are the so-called vampires or blood-suckers, which abound in most parts of tropical America, and are especially plentiful in the Amazon valley. Their carnivorous propensities were once discredited, but are too well authenticated. Horses and cattle are often bitten, and are found in the morning covered with blood, and repeated attacks weaken and ultimately destroy them. Some persons are especially subject to the attacks of these bats; and as native huts are never sufficiently close to keep them out, these unfortunate individuals are obliged to sleep completely muffled up in order to avoid being made seriously ill or even losing their lives. The exact manner in which the attack is made is not positively known, as the sufferer never feels the wound. The present writer was once bitten on the toe, which was found bleeding in the morning from a small round hole from which the flow of blood was not easily stopped. On another occasion, when his feet were carefully covered up, he was bitten on the tip of the nose, only awaking to find his face streaming with blood. The motion of the wings fans the sleeper into a deeper slumber, and renders him insensible to the gentle abrasion of the skin either by teeth or tongue. This ultimately forms a minute hole, the blood flowing from which is sucked or lapped up by the hovering vampire. The largest South American bats, having wings from two to two and a half feet in expanse, are fruit-eaters like the Pteropi of the East, the true blood-suckers being small or of medium size, and varying in colour in different localities. They belong to the genus Desmodus, and have a tongue with horny papillæ at the end; and it is probably by means of this that they abrade the skin and produce a small round wound. This is the account given by Buffon and Azara, and there seems now little doubt that it is correct.

Beyond these two great types—the monkeys and the hats—we look in vain among the varied forms of mammalian life for any that can be said to be distinctive of the tropics as compared with the temperate regions. Many peculiar groups are tropical, but they are in almost every case confined to limited portions of the tropical zones, or are rare in species or individuals. Such are the lemurs in Africa, Madagascar, and Southern Asia; the tapirs of America and Malaya; the rhinoceroses and elephants of Africa an Asia; the cavies and the sloths of America; the scaly ant-eaters of Africa and Asia; but none of these are sufficiently numerous to come often before the traveller so as to affect his general ideas of the aspects of tropical life, and they are, therefore, out of place in such a sketch of those aspects as we are here attempting to lay before out readers.

Summary of the Aspects of Animal Life in the Tropics

We will now briefly summarise the general aspects of animal life as forming an ingredient in the scenery and natural phenomena of the equatorial regions. Most prominent are the butterflies, owing to their numbers, their size, and their brilliant colours, as well as their peculiarities of form, and the slow and majestic flight of many of them. In other insects, the large size and frequency of protective colours and markings are prominent features, together with the inexhaustible profusion of the ants and other small insects. Among birds the parrots stand forth as the pre-eminent tropical group, as do the apes and monkeys among mammals, the two groups having striking analogies in the prehensile hand and the power of imitation. Of reptiles, the two most prominent groups are the lizards and the frogs; the snakes, though equally abundant, being much less obtrusive.

Animal life is, on the whole, far more abundant and more varied within the tropics than in any other part of the globe, and a great number of peculiar groups are found there which never extend into temperate regions. Endless eccentricities of form and extreme richness of colour are its most prominent features, and these are manifested in the highest degree in those equatorial lands where the vegetation acquires its greatest beauty and its fullest development. The causes of these essentially tropical features are not to be found in the comparatively simple influence of solar light and heat, but rather in the uniformity and permanence with which these and all other terrestrial conditions have acted, neither varying prejudicially throughout the year, nor having undergone any important change for countless past ages. While successive glacial periods have devastated the temperate zones, and destroyed most of the larger and more specialised forms which during more favourable epochs had been developed, the equatorial lands must always have remained thronged with life, and have been unintermittingly subject to those complex influences of organism upon organism which seem the main agents in developing the greatest variety of forms and filling up every vacant place in nature. A constant struggle against the vicissitudes and recurring severities of climate must always have restricted the range of effective animal variation in the temperate and frigid zones, and have checked all such developments of form and colour as were in the least degree injurious in themselves, or which co-existed with any constitutional incapacity to resist great changes of temperature or other unfavourable conditions. Such disadvantages were not experienced in the equatorial zone. The struggle for existence as against the forces of nature was there always less severe; food was there more abundant and more regularly supplied; shelter and concealment were at all times more easily obtained; and almost the only physical changes experienced, being dependent on cosmical or geological revolutions, were so slow that variation and natural selection were always able to keep the teeming mass of organisms in nicely balanced harmony with the changing physical conditions. The equatorial zone, in short, exhibitis to us the result of a comparatively continuous and unchecked development of organic forms; while in the temperate regions there have been a series of periodical checks and extinctions of a more or less disastrous nature, necessitating the commencement of the work of development in certain lines over and over again. In the one, evolution has had a fair change; in the other, it has had countless difficulties thrown in its way. The equatorial regions are then, as regards their past and present life-history, a more ancient world than that represented by the temperate zones, a world in which the laws which have governed the progressive development of life have operated with comparatively little check for countless ages, and have resulted in those infinitely varied and beautiful forms—those wonderful eccentricities of structure, of function, and of instinct—that rich variety of colour, and that nicely balanced harmony of relations, which delight and astonish us in the animal productions of all tropical countries.

IV

HUMMING-BIRDS

AS ILLUSTRATING THE LUXURIANCE OF TROPICAL NATURE

Structure—Colours and Ornaments—Descriptive Names—The Motions and Habits and Humming-birds—Display of Ornaments by the Male—Foods—Nests—Geographical Distribution and Variation—Humming-birds of Juan Fernandez as illustrating Variation and Natural Selection—The Relations and Affinities of Humming-birds—How to Determine doubtful Affinities—Resemblances of Swifts and Humming-birds—Differences between Sun-birds and Humming-birds—Conclusions.

THERE are now about ten thousand different kinds of birds known to naturalists, and these are classed in one hundred and thirty families, which vary greatly in extent, some containing a single species only, while others comprise many hundreds. The two largest families are those of the warblers, with more than six hundred, and the finches with more than five hundred species, spread over the whole globe; the hawks and the pigeons, also spread over the whole globe, number about three hundred and thirty and three hundred and sixty species respectively; while the diminutive humming-birds, confined to one hemisphere, consist of about four hundred different species. They are thus, as regards the number of distinct kinds collected in a limited area, the most remarkable of all the families of birds. It may, however, very reasonably be asked, whether the four hundred species of humming-birds above alluded to are really all distinct—as distinct on the average as the ten thousand species of birds are from each other. We reply that they certainly are perfectly distinct species, which never intermingle; and their differences do not consist in colour only, but in peculiarities of form, of structure, and of habits; so that they have to be classed in more than a hundred distinct genera of systematic groups of species, these genera being really as unlike each other as stonechats and nightingales, or as partridges and blackcocks. The figures we have quoted, as showing the proportion of birds in general to humming-birds, thus represent real facts; and they teach us that these small and in some respects insignificant birds constitute an important item in the animal life of the globe.

Humming-birds are, in many respects, unusually interesting and instructive. They are highly peculiar in form, in structure, and in habits, and are quite unrivalled as regards variety and beauty. Though the name is familiar to every one, few but naturalists are acquainted with the many curious facts in their history, or know how much material they afford for admiration and study. It is proposed, therefore, to give a brief and popular account of the form, structure, habits, distribution, and affinities of this remarkable family of birds, as illustrative of the teeming luxuriance of tropical nature, and as throwing light on some of the most interesting problems of natural history.

Structure

The humming-birds form one compact family named Trochilidæ They are all small birds, the largest known being about the size of a swallow, while the smallest are minute creatures, whose bodies are hardly larger than a humble-bee. Their distinguishing features are excessively short legs and feet, very long and pointed wings, a long and slender bill, and a long extensible tubular tongue; and these characters are found combined in no other birds. The feet are exceedingly small and delicate, often beautifully tufted with down, and so short as to be hardly visible beyond the plumage. The toes are placed as in most birds, three in front and one behind, and have very strong and sharply curved claws; and the feet serve probably to cling to a perch rather than to give any movement to the body. The wings are long and narrow, but strongly formed; and the first quill is the longest, a peculiarity found in hardly any other birds but a few of the swifts. The bill varies greatly in length, but is always long, slender, and pointed, the upper mandible being the widest and lapping over the lower at each side, thus affording complete protection to the delicate tongue, the perfect action of which is essential to the bird's existence. The humming-bird's tongue is very long, and it capable of being greatly extended beyond the beak and rapidly drawn back, by means of muscles which are attached to the hyoid or tongue-bones, and bend round over the back and top of the head to the very forehead, just as in the woodpeckers. The two blades or laminæ of which the tongues of birds usually seem to be formed are here greatly lengthened, broadened out, and each rolled up; so as to form a complete double tube connected down the middle, and with the outer edges in contact but not united. The extremities of the tubes are, however, flat and fibrous. This tubular and retractile tongue enables the bird to suck up honey from the nectaries of flowers, and also to capture small insects; but whether the later pass down the tubes, or are entangled in the fibrous tips and thus draw back into the gullet, is not known. The only other birds with a similar tubular tongue are the sun-birds of the East, which, however, as we shall presently explain, have no affinity whatever with the humming-birds.

Colours and Ornaments

The colours of these small birds are exceedingly varied and exquisitely beautiful. The basis of the colouring may be said to be green, as in parrots; but whereas in the latter it is a silky green, in humming-birds it is always metalic. The majority of the species have some green about them, especially on the back; but in a considerable number rich blues, purples, and various shades of red are the prevailing tints. The greater part of the plumage has more or less of a metallic gloss, but there is almost always some part which has an intenser lustre, as if actually formed of scales of burnished metal. A gorget, covering the greater part of the neck and breast, most commonly displays this vivid colour; but it also frequently occurs on the head, on the back, on the tail-coverts above or below, on the upper surface of the tail, on the shoulders or even the quills. The hue of every precious stone and the lustre of every metal is here represented; and such terms as topaz, amethyst, beryl, emerald, garnet, ruby, sapphire; golden, golden-green, coppery, fiery, glowing, iridescent, refulgent, celestial, glittering, shining, are constantly used to name or describe the different species.

No less remarkable than the colours are the varied development of plumage with which these birds are adorned. The head is often crested in a variety of ways; either a simple flat crest, or with radiating feathers, or diverging into two horns, or spreading laterally like wings, or erect and bushy, or recurved and pointed like that of a plover. The throat and breast are usually adorned with broad scale-like feathers, or these diverge into a tippet, or send out pointed collars, or elegant frills of long and narrow plumes tipped with metallic spots of various colours. But the tail is even a more varied and beautiful ornament, either short and rounded, but pure white or some other strongly contrasted tint; or with short pointed feathers forming a star; or with the three outer feathers on each side long and tapering to a point; or larger, and either square or round, or deeply forked or acutely pointed; or with the two middle feathers excessively long and narrow; or with the tail very long and deeply forked, with broad and richly-coloured feathers; or with the two outer feathers wire-like and having broad spoon-shaped tips. All these ornaments, whether of the head, neck, breast, or tail, are invariably coloured in some effective or brilliant manner, and often contrast strikingly with the rest of the plumage. Again, these colours often vary in tint according to the direction in which they are seen. In some species they must be looked at from above, in others from below; in some form the front, in others behind, in order to catch the full glow of the metallic lustre; hence, when the birds are seen in their native haunts, the colours come and go and change with their motions, so as to produce a startling and beautiful effect.

The bill differs greatly in length and shape, being either straight of gently curved, in some species bent like a sickle, in others turned up like the bill of the avoset. It is usually long and slender, but in one group is so enormously developed that it is nearly the same length as the rest of the bird. The legs, usually little seen, are in some groups adorned with globular tufts of white, brown, or black down, a peculiarity possessed by no other birds. The reader will now be in a position to understand how the four hundred species of humming-birds may be easily distinguished, by the varied combinations of the characters here briefly enumerated, together with many others of less importance. One group of birds will have a short round tail, with crest and long neck-frill; another group a deeply-forked broad tail, combined with glowing crown and gorget; one is both bearded and crested; others have a luminous back and pendent neck-plumes; and in each of these groups the species will vary in combinations of colour, in size, and in the proportions of the ornamental plumes, so as to produce an unmistakable distinctness; while, without any new developments of form or structure, there is room for the discovery of hundreds more of distinct kinds of humming-birds.

Descriptive Names

The name we usually give to the birds of this family is derived from the south of their rapidly-moving wings, a sound which is produced by the largest as well as by the smallest member of the group. The Creoles of Guiana similarly call them Bourdons or hummers. The French term, Oiseau-mouche, refers to their small size; while Colibri is a native name which has come down the Carib inhabitants of the West Indies. The Spaniards and Portuguese call them by more poetical names, such as flower-peckers, flower-kissers, myrtle-suckers—while the Mexican and Peruvian names show a still higher appreciation of their beauties, their meaning being "rays of the sun," "tresses of the day-star," and other such appellations. Even our modern naturalists, while studying the structure and noting the peculiarities of these living gems, have been so struck by their inimitable beauties that they have endeavoured to invent appropriate English names for the more beautiful and remarkable genera. Hence we find in common use such terms as sun-gems, sun-stars, hill-stars, wood-stars, sun-angels, star-throats, comets, coquettes, flame-bearers, sylphs, and fairies; together with many others derived from the character of the tail of the crests.

The Motions and Habits of Humming-Birds

Let us now consider briefly the peculiarities of flight, the motions, the food, the nests, and general habits of the humming-birds, quoting the descriptions of those modern naturalists who have personally observed them. Their appearance, remarks Professor Alfred Newton, is entirely unlike that of any other bird: "One is admiring some brilliant and beautiful flower, when between the blossom and one's eye suddenly appears a small dark object, suspended as it were between four short black threads meeting each other in a cross. For an instant it shows in front of the flower; again another instant, and emitting a momentary flash of emerald and sapphire light, it is vanishing, lessening in the distance, as it shoots away, to a speck that the eye cannot take note of." Audubon observes that the Ruby humming-birds pass through the air in long undulations, but the smallness of their size precludes the possibility of following them with the eye farther than fifty or sixty yards, without great difficulty. A person standing in a garden by the side of a common althæa in bloom, will hear the humming of their wings and see the little birds themselves within a few feet of him one moment, while the next they will be out of sight and hearing. Mr. Gould, who visited North America in order to see living humming-birds while preparing his great work on the family, remarks that the action of the wings reminded him of a piece of mechinery acted upon by a powerful spring. When poised before a flower, the motion is so rapid that a hazy semicircle of indistinctness on each side of the bird is all that is perceptible. Although many short intermissions of rest are taken, the bird may be said to live in the air—an element in which it performs every kind of evolution with the utmost ease, frequently rising perpendicularly, flying backward, pirouetting or dancing off, as it were, from place to place, or from one part of a tree to another, sometimes descending, at others ascending. It often mounts up above the towering trees, and then shoots off like a little meteor at a right angle. At other times it gently buzzes away among the little flowers near the ground; at one moment it is poised over a diminutive weed, at the next it is seen at a distance of forty yards, whither it has vanished with the quickness of thought.

The Rufous Flame-bearer, an exquisite species found on the west coast of North America, is thus described by Mr. Nuttall: "When engaged in collecting its accustomed sweets, in all the energy of life, it seemed like a breathing gem, a magic carbuncle of flaming fire, stretching out its glorious ruff as if to emulate the sun itself in splendour." The Sappho Comet, whose long forked tail barred with crimson and black renders it one of the most imposing of humming-birds, is abundant in many parts of the Andes; and Mr. Bonelli tells us that the difficulty of shooting them is very great from the extraordinary turns and evolutions they make when on the wing; at one instant darting head-long into a flower, at the next describing a circle in the air with such rapidity that the eye, unable to follow the movement, loses sight of the bird until it again returns to the flower which at first attracted it attention. Of the little Vervain humming-bird of Jamaica, Mr. Gosse writes: “I have sometimes watched with much delight the evolutions of this little species at the Moringa-tree.[1] When only one is present, he pursues the round of the blossoms soberly enough. But if two are at the tree, one will fly off, and suspend himself in the air a few yards distant; the other presently starts off to him, and then, without touching each other, they mount upwards with strong rushing wings, perhaps for five hundred feet. They then separate, and each starts diagonally towards the ground like a ball from a rifle, and, wheeling round, comes up to the blossoms again as if it had not moved away at all. The figure of the smaller humming-birds on the wing, their rapidity, their wavering course, and their whole manner of flight, are entirely those of an insect.” Mr. Bates remarks that on the Amazons, during the cooler hours of the morning and form four to six in the afternoon, humming-birds are to be seen whirring about the trees by scores; their motions being unlike those of any other birds. They dart to and fro so swiftly that the eye can scarcely follow them, and when they stop before a flower it is only for a few moments. They poise themselves in an unsteady manner, their wings moving with inconceivable rapidity, probe the flower, and then shoot off to another part of the tree. They do not proceed in that methodical manner which bees follow, taking the flowers seriatim, but skip about from one part of the tree to another in the most capricious way. Mr. Belt remarks on the excessive rapidity of the flight of the humming-bird giving it a sense of security from danger, so that it will approach a person nearer than any other bird, often hovering within two or three yards (or even one or two feet) of one's face. He watched them bathing in a small pool in the forest, hovering over the water, turning from side to side by quick jerks of the tail; now showing a throat of gleaming emerald, now shoulders of glistening amethyst; then darting beneath the water, and rising instantly, throw off a shower of spray from its quivering wings, and again fly up to an overhanging bough and commence to preen its feathers. All humming-birds bathe on the wing, and generally take three or four dips, hovering between times about three or four inches above the surface. Mr. Belt also remarks on the immense numbers of humming-birds in the forests, and the great difficulty of seeing them; and his conclusion is, that in the part of Nicaragua where he was living they equalled in number all the rest of the birds together, if they did not greatly exceed them.

[1] Sometimes called the horse-radish tree. It is the

Moringa pterygosperma, a native of the East Indies, but commonly cultivated in Jamaica. It has yellow flowers.

The extreme pugnacity of humming-birds has been noticed by all observers. Mr. Gosse describes two meeting and chasing each other through the labyrinths of twigs and flowers till, an opportunity occurring, the one would dart with seeming fury upon the other, and then, with a loud rustling of their wings, they would twirl together, round and round, till they nearly came to the earth. Then they parted, and after a time another tussle took place. Two of the same species can hardly meet without an encounter, while in many cases distinct species attack each other with equal fury. Mr. Salvin describes the splendid Eugenes fulgens attacking two other species with as much ferocity as its own fellows. One will knock another off its perch, and the two will go fighting and screaming away at a pace hardly to be followed by the eye. Audubon says they attack any other birds that approach them, and think nothing of assaulting tyrant-shrikes and even birds of prey that come too near their home.

Display of Ornaments by the Male

It is a well-known fact that when male birds possess any unusual ornaments, they take such positions or perform such evolutions as to exhibit them to the best advantage while endeavouring to attract or charm the females, or in rivalry with other males. It is therefore probable that the wonderfully varied decorations of humming-birds, whether burnished breast-shields, resplendent tail, crested head, or glittering back, are thus exhibited; but almost the only actual observation of this kind is that of Mr. Belt, who describes how two males of the Florisuga mellivora displayed their ornaments before a female bird. One would shoot up like a rocket, then, suddenly expanding the snow-white tail like an inverted parachute, slowly descend in front of her, turning round gradually to show off both back and front. The expanded white tail covered more space than all the rest of the bird, and was evidently the grand feature of the performance. Whilst one was descending the other would shoot up and come slowly down expanded.[1]

1 *The Naturalist in Nicaragua*, p. 112.

Food

The food of humming-birds has been a matter of much controversy. All the early writers, down to Buffon, believed that they lived solely on the nectar of flowers; but since that time every close observer of their habits maintains that they feed largely, and in some cases wholly, on insects. Azara observed them on the La Plata in winter, taking insects out of the webs of spiders at a time and place where there were no flowers. Bullock, in Mexico, declares that he saw them catch small butterflies, and that he found many kinds of insects in their stomachs. Waterton made a similar statement. Hundreds and perhaps thousands of specimens have since been dissected by collecting naturalists, and in almost every instance their stomachs have been found full of insects—sometimes, but not generally, mixed with a proportion of honey. Many of them in fact may be seen catching gnats and other small insects just like fly-catchers, sitting on a dead twig over water, darting off for a time in the air, and then returning to the twig. Others come out just at dusk, and remain on the wing, now stationary, now darting about with the greatest rapidity, imitating in a limited space the evolutions of the goatsuckers, and evidently for the same end and purpose. Mr. Gosse also remarks: "All the humming-birds have more or less the habit, when in flight, of pausing in the air and throwing the body and tail into rapid and odd contortions. This is most observable in the Polytmus, from the effect that such motions have on the long feathers of the tail. That the object of these quick turns is the capture of insects, I am sure, having watched one thus engaged pretty close to me. I observed it carefully, and distinctly saw the minute flies in the air which it pursued and caught, and heard repeatedly the snapping of the beak. My presence scarcely disturbed it, if at all."

There is also an extensive group of small brown humming-birds, forming the sub-family Phaëthornithinæ, which rarely or never visit flowers, but frequent the shady recesses of the forest, where they hunt for minute insects. They dart about among the foliage, and visit in rapid succession every leaf upon a branch, balancing themselves vertically in the air, passing their beaks closely over the under-surface of each leaf, and thus capturing, no doubt, any small insects that may lurk there. While doing this, the two long feathers of the tail have a vibrating motion, serving apparently as a rudder to assist them in performing the delicate operation. Others search up and down stems and dead sticks in the same manner, every now and then picking of something, exactly as a bush-shrike or a tree creeper does, with the difference that the humming-bird is constantly on the wing; while the remarkable sickle-bill is said to probe the scale-covered stems of palms and tree-ferns to obtain its insect food.

It is a well-known fact that although humming-birds are easily tamed, they cannot be preserved long in captivity, even in their own country, when fed only on syrup. Audubon states that when thus fed they only live a month or two and die apparently starved; while if kept in a room whose open windows are covered with a fine net, so as to allow small insects to enter, they have been kept for a whole year without any ill effects. Another writer, Mr. Webber, captured and tamed a number of the Ruby-throat in the United States. He found that when fed for three weeks on syrup they drooped, but after being let free for a day or two they would return to the open cage for more of the syrup. Some which had been thus tamed and set free returned the following year, and at once flew straight to the remembered little cup of sweets. Mr. Gosse in Jamaica also kept some in captivity, and found the necessity of giving them insect food; and he remarks that they were very fond of a small ant that swarmed on the syrup with which they were fed. It is strange that, with all this previous experience and information, those who have attempted to bring live humming-birds to this country have fed from exclusively on syrup; and the weakness produced by this insufficient food has no doubt been the chief cause of their death on, or very soon after, arrival. A box of ants would not be difficult to bring as food for them, but even finely-chopped meat or yolk of egg would probably serve, in the absence of insects, to supply the necessary proportion of animal food.

Nests

The nests of the humming-birds are, as might be expected, beautiful objects, some being no larger inside than the half of a walnut shell. These small cup-shaped nests are often placed in the fork of a branch, and the outside is sometimes beautifully decorated with pieces of lichen, the body of the nest being formed of cottony substances and the inside lined with the finest and most silky fibres. Others suspend their nests to creepers hanging over water, or even over the sea; and the Pichincha humming-bird once attached its nest to a strawrope hanging from the roof of a shed. Others again build nests of a hammock-form attached to the face of rocks by spiders' web; while the little forest-haunting species fasten their nests to the points or to the under-sides of palm-leaves or other suitable foliage. They lay only one or two white eggs.

Geographical Distribution and Variation

Most persons know that humming-birds are found only in America; but it is not so generally known that they are almost exclusively tropical birds, and that the few species that are found in the temperate (northern and southern) parts of the continent are migrants, which retire in the winter to the warmer lands near or within the tropics. In the extreme north of America two species are regular summer visitants, one on the east and the other on the west of the Rocky Mountains. On the east the common North American or Ruby-throated humming-bird extends through the United States and Canada, and as far as 57° north latitude, or considerably north of Lake Winnipeg; while the milder climate of the west coast allows the Rufous Flame-bearer to extend its range beyond Sitka to the parallel of 61°. Here they spend the whole summer, and breed, being found on the Columbia River in the latter end of April, retiring to Mexico in the winter. Supposing that those which go farthest north do not return farther south than the borders of the tropics, these little birds must make a journey of full three thousand miles each spring and autumn. The antarctic humming-bird visits the inhospitable shores of Tierra-del-Fuego, where it has been seen visiting the flowers of fuchsias in a snowstorm, while it spends the winter in the warmer parts of Chili and Bolivia.

In the south of California and in the Central United States three or four other species are found in summer; but it is only when we enter the tropics that the number of different kinds becomes considerable. In Mexico there are more than thirty species, while in the southern parts of Central America there are more than double that number. As we go on towards the equator they become still more numerous, till they reach their maximum in the equatorial Andes. They especially abound in the mountainous regions; while the luxuriant forest plains of the Amazons, in which so many other forms of life reach their maximum, are very poor in humming-birds. Brazil, being more hilly and with more variety of vegetation, is richer, but does not equal the Andean valleys, plateaux, and volcanic peaks. Each separate district of the Andes has its peculiar species and often its peculiar genera, and many of the great volcanic mountains possess kinds which are confined to them. Thus, on the great mountain of Pichincha there is a peculiar species found at an elevation of about fourteen thousand feet only; while an allied species on Chimborazo ranges from fourteen thousand feet to the limits of perpetual snow at sixteen thousand feet elevation. It frequents a beautiful yellow-flowered alpine shrub belonging to the Asteraceæ. On the extinct volcano of Chiriqui in Veragua a minute humming-bird, called the little Falme-bearer, has been only found inside the crater. Its scaled gorget is of such a flaming crimson that, as Mr. Gould remarks, it seems to have caught the last spark from the volcano before it was extinguished.

Not only are humming-birds found over the whole extent of America, from Sitka to Tierra-del-Fuego, and from the level of the sea to the snow-line on the Andes, but they inhabit many of the islands at a great distance from the mainland. The West Indian islands possess fifteen distinct species belonging to eight different genera, and these are so unlike any found on the continent that five of these genera are peculiar to the Antilles. Even the Bahamas, so close to Florida, possess two peculiar species. The small group of islands called Tres Marias, about sixty miles from the west coast of Mexico, has a peculiar species. More remarkable are the two humming-birds of Juan Fernandez, situated in the Pacific Ocean, four hundred miles west of Valparaiso in Chili, one of these being peculiar; while another species inhabits the little island Mas-afuera, ninety miles farther west. The Galapagos, though very little farther from the mainland and much more extensive, have no humming-birds; neither have the Falkland islands, and the reason seems to be that both these groups are deficient in forest, and in fact have hardly any trees or large shrubs, while there is a great paucity of flowers and of insect life.

Humming-birds of Juan Fernandez as illustrating Variation and Natural Selection

The three species which inhabit Juan Fernandez and Masafuera present certain peculiarities of great interest. They form a distinct genus, Eustephanus, one species of which inhabits Chili as well as the island of Juan Fernandez. This, which may be termed the Chilian species, is greenish in both sexes, whereas in the two species peculiar to the islands the males are red or reddish-brown, and the females green. The two red males differ very slightly from each other, but the three green females differ considerably; and the curious point it that the female in the smaller and more distant island somewhat resembles the same sex in Chili, while the female of the Juan Fernandez species is very distinct, although the males of the two islands are so much alike. As this forms a comparatively simple case of the action of the laws of variation and natural selection, it will be instructive to see if we can picture to ourselves the process by which the changes have been brought about. We must first go back to an unknown but rather remote period, just before any humming-birds had reached these islands. At that time a species of this peculiar genus, Eustephanus, must have inhabited Chili; but we cannot be sure that it was identically the same as that which is now found there, because we know that species are always undergoing change to a greater or less degree. After perhaps many failures, one or more pairs of the Chilian bird got blown across to Juan Fernandez, and finding the country favourable, with plenty of forests and a fair abundance of flowers and insects, they rapidly increased and permanently established themselves on the island. They soon began to change colour, however, the male getting a tinge of reddish-brown, which gradually deepened into the fine colour now exhibited by the two insular species, while the female, more slowly, changed to white on the under-surface and on the tail, while the breast-spots became more brilliant. When the change of colour was completed in the male, but only partially so in the female, a further emigration westward took place to the small island Mas-afuera, where they also established themselves. Here, however, the change begun in the larger island appears to have been checked, for the female remains to this day intermediate between the Juan Fernandez and the Chilian forms. More recently, the parent form has again migrated form Chili to Juan Fernandez, where it still lives side by side with its greatly changed descendant.[1]. Let us now see how far these facts are in accordance with the general laws of variation, and with those other laws which I have endeavoured to show regulate the development of colour.[1]

[1] In the preceding account of the probable course of events in peopling these islands with humming-birds, I follow Mr. Sclater's paper on the "Land Birds of Juan Fernandez," *Ibis,* 1871, p. 183. In what follows I give my own explanation of the probable causes of the change.

[1] See *Macmillan's Magazine,* September 1867, "On the Colours of Animals and Plants," and chapter v. and vi., *post.*

The amount of variation which is likely to occur in a species will be greatly influenced by two factors—the occurrence of a change in the physical conditions, and the average abundance or scarcity of the individuals composing the species. When from these or other causes variation occurs, it may become fixed as a variety or a race, or may go on increasing to a certain extent, either, from a tendency to vary along certain special lines induced by local or physiological causes, or by the continued survival and propagation of all such varieties as are beneficial to the race. After a certain time a balance will be arrived at, either by the limits of useful variation in this one direction having been reached, or by the species becoming harmoniously adapted to all the surrounding conditions: and without some change in these conditions the specific from may then remain unaltered for a very long time; whence arises the common impression of the fixity of species. Now in a country like Chili, forming part of a great continent very well stocked with all forms of organic life, the majority of the species would be in a state of stable equilibrium; the most favourable variations would have been long ago selected; and the numbers of individuals in each species would be tolerably constant, being limited by the numerous other forms whose food and habits were similar, or which in any way impinged upon its sphere of existence. We may, therefore, assume that the Chilian humming-bird which migrated to Juan Fernandez was a stable form, hardly if at all different from the existing species which is termed Eustephanus galeritus. On the island it met with very changed but highly favourable conditions—an abundant shrubby vegetation and a tolerably rich flora; less extremes of climate than on the mainland; and, most important of all, absolute freedom from the competition of rival species. The flowers and their insect inhabitants were all its own; there were no snakes or mammalia to plunder is nests; nothing to prevent the full enjoyment of existence. The consequence would be, rapid increase and a large permanent population, which still maintains itself; for Mr. Moseley, of the *Challenger* expedition, has informed the writer that humming-birds are extraordinarily abundant in Juan Fernandez, every bush or tree having one or two darting about it. Here, then, we have one of the special conditions which have always been held to favour variation—a great increase in the number of individuals; but, as there was no struggle with allied creatures, there was no need for any modification in form or structure, and we accordingly find that the only important variations which have become permanent are those of size and of colour. The increased size would naturally arise from greater abundance of food with a more equable climate throughout the year; the healthier, stronger, and larger individuals being preserved. The change of colour would depend on molecular changes in the plumage accompanying the increase of size; and the superior energy and vitality in the male, aided by the favourable change in conditions and rapid increase of population, would lead to an increased intensity of colour, the special tint being determined either by local conditions or by inherited tendencies in the race. It is to be noted that the change from green to red is in the direction of the less refrangible rays of change which has been shown to accompany expansion in inorganic growth an development in organic forms.[1] The change of colour in the female, not being urged on by such intense vital activity as in the case of the male, would be much slower, and, owing probably to inherited tendencies, in a different direction. The under-surface of the Chilian bird is ashy with bronzy-green spots on the breast, while the tail is entirely bronze-green. In the Juan Fernandez species the under-surface has become pure white, the breast-spots larger and of a purer golden-green, while the whole inner web of the tail-feathers has become pure white, producing a most elegant effect when the tail is expanded.

[1] See "Colours of Animals," *Macmillan's Magazine* September 1877, pp 394–398, and chapter v., *post.*

We may now follow the two sexes to the remoter island, at a period when the male had acquired his permanent style of colouring, but was not quite so large as he subsequently became; while the change of the female bird had not been half completed. In this small and comparatively barren island (a mere rock, as it is described by some authors) there would be no such constant abundance of food, and therefore no possibility of a large permanent population of humming-birds; while the climate would not differ materially from that of the larger island. Variation would therefore be checked, or might be stopped altogether; and we find the facts exactly correspond to this view. The male, which had already acquired his colour, remains almost undistinguishable form his immedite ancestral form; but he is a little smaller, indicating either that the full size of that form had not been acquired at the period of migration, or that a slight diminution of size has since occurred, owing to a deficiency of food. The female shows also a slight diminution of size, but in other respects is almost exactly intermediate between the Chilian and Juan Fernandez females. The colour beneath is light ashy, the breast-spots are intermediate in size and colour, and the tail-feathers have a large ill-defined white spot on the end of the inner web which has only to be extended along the whole web to produce the exact character which has been acquired in Juan Fernandez. It seems probable, therefore, that the female bird has remained nearly or quite stationary since its migration, while its Juan Fernandez relative has gone on steadily changing in the direction already begun; and the more distant species geographically thus appears to the more nearly related to its Chilian ancestor.

Coming down to a more recent period, we find that the comparatively small and dull-coloured Chilian bird has again migrated to Juan Fernandez; but it at once came into competition with its red descendant, which had firm possession of the soil, and had probably undergone slight constitutional changes exactly fitting it to its insular abode. The new-comer, accordingly, only just manages to maintain its footing; for we are told by Mr. Reed of Santiago that it is by no means common; whereas, as we have seen, the red species is excessively abundant. We may further suspect that the Chilian birds now pass over pretty frequently to Juan Fernandez, and thus keep up the stock; for it must be remembered that whereas, at a first migration, both a male and a female are necessary for colonisation, yet, after a colony is formed, any stray bird which may come over adds to the numbers, and checks permanent variation by crossbreeding.

We find, then, that all the chief peculiarities of the three allied species of humming-birds which inhabit the Juan Fernandez group of islands, may be fairly traced to the action of those general laws which Mr. Darwin and others have shown to determine the variation of animals and the perpetuation of those variations. It is also instructive to note that where the variations of colour and size have been greatest they are accompanied by several lesser variations in other characters. In the Juan Fernandez bird the bill has become a little shorter, the tail feathers somewhat broader, and the fiery cap on the head somewhat smaller; all these peculiarities being less developed or absent in the birds inhabiting Masafuera. These coincident changes may be due, either to what Mr. Darwin has termed correlation of growth, or to the partial reappearance of ancestral characters under more favourable conditions, or to the direct action of changes of climate and of food; but they show us how varied and unaccountable are the changes in specific forms that may be effected in a comparatively short time, and by means of very slight changes of locality.

If now we consider the enormously varied conditions presented by the whole continent of America—the hot, moist, and uniform forest-plains of the Amazon; the open llanos of the Orinoco; the dry uplands of Brazil; the sheltered valleys and forest slopes of the Eastern Andes; the verdant plateaux, the barren paramos, the countless volcanic cones with their peculiar Alpien vegetation; the contrasts of the east and west coasts; the isolation of the West Indian islands, and to a less extent of Central America and Mexico, which we know have been several times separated from South America; and when we further consider that all these characteristically distinct areas have been subject to cosmical and local changes, to elevations and depressions, to diminution and increase of size, to greater extremes and greater uniformity of temperature, to increase or decrease of rainfall; and that with these changes there have been coincident changes of vegetation and of animal life, all affecting in countless ways the growth and development, the forms and colours, of these wonderful little birds—if we consider all these varied and complex influences, we shall be less surprised at their strange forms, their infinite variety, their wondrous beauty. For how many ages the causes above enumerated may have acted upon them we cannot say; but their extreme isolation from all other birds, no less than the abundance and variety of their generic and specific forms, clearly point to a very high antiquity.

The Relations and Affinities of Humming-birds

The question of the position of this family in the class of birds and its affinities or resemblances to other groups is so interesting, and affords such good opportunities for explaining some of the best established principles of classification in natural history in a popular way, that we propose to discuss it at some length, but without entering into technical details.

There is in the Eastern hemisphere, especially in tropical Africa and Asia, a family of small birds called sun-birds, which are adorned with brilliant metallic colours, and which, in shape and general appearance, much resemble humming-birds. They frequent flowers in the same way, feeding on honey and insects; and all the older naturalists placed the two families side by side as undoubtedly allied. In the year 1850, in a general catalogue of birds, Prince Lucien Bonaparte, a learned ornithologist, place the humming-birds next to the swifts, and far removed from the Nectarinidæ or sun-birds; and this view of their position has gained ground with increasing knowledge, so that now all the more advanced ornithologists have adopted it. Before proceeding to point out the reasons for this change of view, it will be well to discuss a few of the general principles which guide naturalists in the solution of such problems.

How to Determine doubtful Affinities

It is now generally admitted that, for the purpose of determining obscure and doubtful affinities, we must examine by preference those parts of an animal which have little or no direct influence on its habits and general economy. The value of an organ, or of any detail of structure, for purposes of classification, is generally in inverse proportion to its adaptability to special uses. And the reason of this is apparent, when we consider that similarities of food and habits are often accompanied by similarities of external form or of special organs, in totally distinct animals. Porpoises, for example, are modified externally so as to resemble fishes; yet they are really mammalia. Some marsupials are carnivorous, and are so like true carnivora that it is only by minute peculiarities of structure that the skeleton of the one can be distinguished from that of the other. Many of the hornbills and toucans have the same general form, and resemble each other in habits, in food, and in their enormous bills; yet peculiarities in the structure of the feet, in the form of the breast-bone, in the cranium, and in the texture and arrangement of the plumage, show that they have no real affinity, the former approaching the kingfishers, the latter the cuckoos. Such structural peculiarities as these have no direct relation to habits; and they are therefore little liable to change, when from any cause a portion of the group may have been driven to adopt a new mode of life. Thus all the Old World apes, however much they may differ in size or habits, and whether we class them as baboons, monkeys, or anthropoids, have the same number of teeth; while the American monkeys all have an additional premolar tooth. This difference can have no relation to the habits of the two groups, because each group exhibits difference of habits greater than often occur between American and Asiatic species; and it thus becomes a valuable character indicating the radical distinctness of the two groups, a distinctness confirmed by other anatomical characters.

On the other hand, peculiarities of organisation which seem specially adapted to certain modes of life are often diminished or altogether lost in a few species of the group, showing their essential unimportance to the type, as well as their small value for classification. Thus the woodpeckers are most strikingly characterised by a very long and highly extensible tongue, with the muscles attached to the tongue-bone prolonged backward over the head so as to enable the tongue to be suddenly darted out; and also by the rigid and pointed tail, which is a great help in climbing up the vertical trunks of trees. But in one group (the Picumni) the tail becomes quite soft, while the tongue remains fully developed; and in another (Meiglyptes) the characteristic tail remains, while the prolonged hyoid muscles have almost entirely disappeared, and the tongue has consequently lost its peculiar extensile power; yet in both these cases the form of the breast-bone and the character of the feet, the skeleton, and the plumage, show that the birds are really woodpeckers; while even the habits and the food are very little altered. In like manner the bill may undergo great changes; as from the short crow-like bill of the true birds-of-paradise to the long slender bills of Epimachinæ, which latter were on that account long classed apart in the tribe of Tenuirostres, or slender-billed birds, but whose entire structure shows them to be closely allied to the paradise-birds. So, the long feathery tongue of the toucans differs from that of every other bird; yet it is not held to overbalance the weight of anatomical peculiarities which show that these birds are allied to the barbets and the cuckoos.

The skeleton, therefore, and especially the sternum or

breast-bone, affords us an almost infallible guide in doubtful cases; because it appears to change its form with extreme slowness, and thus indicates deeper seated affinities than those shown by organs which are in direct connection with the outside world, and are readily modified in accordance with varying conditions of existence. Another, though less valuable guide is afforded, in the case of birds, by the eggs. These often have a characteristic form and colour, and a peculiar texture of surface, running unchanged through whole genera and families which are nearly related to each other, however much they may differ in outward form and habits. Another detail of structure, which has no direct connection with habits and economy, is the manner in which the plumage is arranged on the body. The feathers of birds are by no means set uniformly over their skin, but grow in certain definite lines and patches, which vary considerably in shape and size in the more important orders and tribes, while the mode of arrangement agrees in all which are known to be closely related to each other; and thus the form of the feather-tracts or the "pterylography," as it is termed, of a bird, is a valuable aid in doubtful cases of affinity.

Now, if we apply these three tests to the humming-birds, we find them all pointing in the same direction. The sternum or breast-bone is not notched behind; and this agrees with the swifts, and not with the sun-birds, whose sternum has two deep notches behind, as in all the families of the vast order of Passeres to which the latter belong. The eggs of both swifts and humming-birds are white, only two in number, and resembling each other in texture. And in the arrangement of the feather-tracts the humming-birds approach more nearly to the swifts than they do to any other birds; and altogether differ from the sun-birds, which in this respect, as in so many others, resemble the honey-suckers of Australia and other true passerine birds.

Resemblances of Swifts and Humming-birds

Having this clue to their affinities, we shall find other peculiarities common to these two groups, the swifts and the humming-birds. They have both ten tail-feathers, while the sun-birds have twelve. They have both only sixteen true quill-feathers, and they are the only birds which have so small a number. The humming-birds are remarkable for having, in almost all the species, the first quill the longest of all, the only other birds resembling them in this respect being a few species of swifts; and, lastly, in both groups the plumage is remarkably compact and closely pressed to the body. Yet, with all these points of agreement, we find an extreme diversity in the bills and tongues of the two groups. The swifts have a short, broad, flat bill, with a flat horny-tipped tongue of the usual character; while the humming-birds have a very long, narrow, almost cylindrical bill, containing a tubular and highly extensible tongue. The essential point, however, is, that whereas hardly any of the other characters we have adduced are adaptive, or strictly correlated with habits and economy, this character is preeminently so; for the swifts are pure aerial insect-hunters, and their short, broad bills and wide gape are essential to their mode of life. The humming-birds, on the other hand, are floral insect-hunters, and for this purpose their peculiarly long bills and extensile tongues are especially adapted; while they are at the same time honey-suckers, and for this purpose have acquired the tubular tongue. The formation of such a tubular tongue out of one of the ordinary kind is easily conceivable, as it only requires to be lengthened, and the two laminæ of which it is composed curled in at the sides; and these changes it probably goes through in the young birds.

When on the Amazons I once had a nest brought me containing two little unfledged humming-birds, apparently not long hatched. Their beaks were not at all like those of their parents, but short, triangular, and broad at the base, just the form of the beak of a swallow or swift slightly lengthened. Thinking (erroneously) that the young birds were fed by their parents on honey, I tried to feed them with a syrup made of honey and water, but though they kept their mouths constantly open as if ravenously hungry, they would not swallow the liquid, but threw it out again and sometimes nearly choked themselves in the effort. At length I caught some minute flies, and on dropping one of these into the open mouth it instantly closed, the fly was gulped down and the mouth opened again for more; and each took in this way fifteen or twenty little flies in succession before it was satisfied. They lived thus three or four days, but required more constant care than I could give them. These little birds were in the "swift" stage; they were pure insects-caters, with a bill and mouth adapted for insect-eating only. At that time I was not aware of the importance of the observation of the tongue; but as the bill was so short and the tubular tongue not required, there can be little doubt that the organ was, at that early stage of growth, short and flat, as it is in the birds most nearly allied to them.

Differences between Sun-birds and Humming-birds

In respect of all the essential and deep-seated points of structure, which have been shown to offer such remarkable similarities between the swifts and the humming-birds, the sun-birds of the Eastern hemisphere differ totally from the latter, while they agree with the passerine birds generally, or more particularly with the creepers and honey-suckers. They have a deeply-notched sternum; they have twelve tail-feathers in place of ten; they have nineteen quills in place of sixteen; and the first quill instead of being and longest is the very shortest of all, while the wings are short and round instead of being excessively long and pointed; their plumage is arranged differently; and their feet are long and strong, instead of being excessively short and weak. There remain only the superficial characters of small size and brilliant metallic colours to assimilate them with the humming-birds, and one structural feature—a tubular and somewhat extensile tongue. This, however, is a strictly adaptive character, the sun-birds feeding on small insects and the nectar of flowers, just as do the humming-birds; and it is a remarkable instance of a highly peculiar modification of an organ occurring independently in two widely-separated groups. In the sun-birds the hyoid or tongue-muscles do not extend so completely over the head as they do in the humming-birds, so that the tongue is less extensible; but it is constructed in exactly the same way by the inrolling of the two laminæ of which it is composed.

The tubular tongue of the sun-birds is a special adaptive modification acquired within the family itself and not inherited from a remote ancestral form. This is shown by the amount of variation this organ exhibits in different members of the family. It is most highly developed in the Arachnotheræ, or spider-hunters of Asia, which are sun-birds without any metallic or other brilliant colouring. These have the longest bills and tongues, and the most developed hyoid muscles; they hunt much about the blossoms of palmtrees, and may frequently be seen probing the flowers while fluttering clumsily in the air, just as if they had seen and attempted to imitate the aerial gambols of the American humming-birds. The true metallic sun-birds generally cling about the flowers with their strong feet; and they feed chiefly on minute hard insects, as do many humming-birds. There is, however, one species (Chalcoparia phœnicotis), always classed as a sun-bird, which differs entirely from the rest of the species in having the tongue flat, horny, and forked at the tip; and its food seems to differ correspondingly, for small caterpillars were found in its stomach. More remotely allied, but yet belonging to the same family, are the little flower-peckers of the genus Diceum, which have a short bill and a tongue twice split at the end; and these feed on small fruits, and perhaps on buds and on the pollen of flowers. The little white-eyes (Zosterops), which are probably allied to the last, eat soft fruits and minute insects.

Conclusion

Here, then, we have an extensive group of birds, considerably varied in external form, yet undoubtedly closely allied to each other, one division of which is specially adapted to feed on the juices secreted by flowers and the minute insects that harbour in them; and these alone have a lengthened bill and double tubular tongue, just as in the humming-birds. We can hardly have a more striking example of the necessity of discriminating between adaptive and purely structural characters. The same adaptive character may coexist in two groups which have a similar mode of life, without indicating any affinity between them, because it may have been acquired by each independently to enable it to fill a similar place in nature. In such cases it is found to be an almost isolated character, apparently connecting two groups which otherwise differ radically. Non-adaptive or purely structural characters, on the other hand, are such as have probably been transmitted from a remote ancestor, and thus indicate fundamental peculiarities of growth and development. The changes of structure rendered necessary by modifications of the habits or instincts of the different species have been made to a great extent independently of such characters; and as several of these may always be found in the same animal their value becomes cumulative. We thus arrive at the seeming paradox that the *less* of direct use is apparent in any peculiarity of structure, the *greater* is its value in indicating true, though perhaps remote, affinities; while any peculiarity of an organ which seems essential to its possessor's wellbeing is often of very little value in indicating its affinity for other creatures.

This somewhat technical discussion will, it is hoped, enable the general reader to understand some of the more important principles of the modern or natural classification of animals as distinguished from the artificial system which long prevailed. It will also afford him an easily remembered example of those principles, in the radical distinctness of two families of birds often confounded together,—the sun-birds of the Eastern Hemisphere and the humming-birds of America; and in the interesting fact that the latter are essentially swifts—profoundly modified, it is true, for an aerial and flower-haunting existence, but still bearing in many important peculiarities of structure the unmistakable evidences of a common origin.[1]

[1] Recent researches into the anatomy of the swifts and humming-birds have brought to light so many and such important differences that the above conclusion, founded on comparatively superficial characters, becomes doubtful. Dr. Shufeldt considers that both groups are so isolated that they each require to be classed as a distinct order of birds. But while the swifts are believed to have undoubted though remote affinities with the swallows, it cannot yet be determined whether they have any real affinity with the humming-birds, which latter appear to have no special and unmistakable relationship with any other order or family of birds. See "Studies of the Macrochires, Morphological, and otherwise, with the view of indicating their relationships," etc., by R. W. Shufeldt, M.D., in the *Journal of the Linnean Society,* vol. xx.; *Zoology,* pp. 299, 394: 1889.

V

THE COLOURS OF ANIMALS AND SEXUAL SELECTION[1]

General Phenomena of Colour in the Organic World—Theory of Heat and Light as producing Colour—Changes of Colour in Animals produced by Coloured Light—Classification of Organic Colours—Protective Colours—Warning Colours—Sexual Colours—Normal Colours—The Nature of Colour—How Animal Colours are produced—Colour a normal product of Organisation—Theory of Protective Colours— Theory of Warning Colours—Imitative Warning Colours—The Theory of Mimicry—Theory of Sexual Colours—Colour as a means of Recognition—Colour proportionate to Integumentary Development—Selection by Females not a cause of Colour—Probable use of the Horns of Beetles—Cause of the greater Brilliancy of some Female Insects—Origin of the Ornamental Plumage of Male Birds—Theory of Display of Ornaments by Males—Natural Selection as neutralising Sexual Selection—Greater Brilliancy of some Female Birds—Colour-development as illustrated by Humming-Birds—Theory of Normal Colours—Local causes of Colour-development—The influence of Locality on Colour in Butterflies and Birds—Sense-perception influenced by Colour of the Integuments—Summary on Colour-development in Animals.

1 A first sketch of this essay appeared in *Macmillan's Magazine* of September 1877.

General Phenomena of Colour in the Organic World

THERE is probably no one quality of natural objects from which we derive so much pure and intellectual enjoyment as from their colours. The heavenly blue of the firmament, the glowing tints of sunset, the exquisite purity of the snowy mountains, and the endless shades of green presented by the verdure-clad surface of the earth, are a never-failing source of pleasure to all who enjoy the inestimable gift of sight. Yet these constitute, as it were, but the frame and background of a marvellous and ever-changing picture. In contrast with these broad and soothing tints, we have presented to us in the vegetable and animal worlds an infinite variety of objects adorned with the most beautiful and most varied hues. Flowers, insects, and birds are the organisms most generally ornamented in this way; and their symmetry of form, their variety of structure, and the lavish abundance with which they clothe and enliven the earth, cause them to be objects of universal admiration. The relation of this wealth of colour to our mental and moral nature is indisputable. The child and the savage alike admire the gay tints of flower, bird, and insect; while to many of us their contemplation brings a solace and enjoyment which is both intellectually and morally beneficial. It can then hardly excite surprise that this relation was long thought to afford a sufficient explanation of the phenomena of colour in nature; and although the fact that

Full many a flower is born to blush unseen,
And waste its sweetness on the desert air,

might seem to throw some doubt on the sufficiency of the explanation, the answer was easy,—that in the progress of discovery man would, sooner or later, find out and enjoy every beauty that the hidden recesses of the earth have in store for him. This theory received great support from the difficulty of conceiving any other use or meaning in the colours with which so many natural objects are adorned. Why should the homely gorse be clothed in golden raiment, and the prickly cactus be adorned with crimson bells? Why should our fields be gay with buttercups, and the heather-clad mountains be clad in purple robes? Why should every land produce its own peculiar floral gems, and the alpine rocks glow with beauty, if not for the contemplation and enjoyment of man? What could be the use to the butterfly of its gaily-painted wings, or to the humming-bird of its jewelled breast, except to add the final touches to a world-picture, calculated at once to please and to refine mankind? And even now, with all our recently acquired knowledge of this subject, who shall say that these old-world views were not intrinsically and fundamentally sound; and that, although we now know that colour has "uses" in nature that we little dreamt of, yet the relation of those colours—or rather of the various rays of light—to our senses and emotions may not be another and perhaps more important use which they subserve in the great system of the universe?

We now propose to lay before our readers a general account of the more recent discoveries on this interesting subject; and in doing so it will be necessary first to give an outline of the more important facts as to the colours of organised beings; then to point out the cases in which it has been shown that colour is of use; and lastly, to endeavour to throw some light on its nature and on the general laws of its development.

Among naturalists, colour was long thought to be of little import, and to be quite untrustworthy as a specific character. The numerous cases of variability of colour led to this view. The occurrence of white blackbirds, white peacocks, and black leopards, of white blue-bells, and of white, blue, or pink milk-worts, led to the belief that colour was essentially unstable, that it could therefore be of little or no importance, and belonged to quite a different class of characters from form or structure. But it now begins to be perceived that these cases, though tolerably numerous, are, after all, exceptional; and that colour, as a rule, is a constant character. The great majority of the species, both of animals and plants, are each distinguished by peculiar tints which vary very little, while the minutest markings are often constant in thousands or millions of individuals. All our field buttercups are invariably yellow, and our poppies red, while many of our butterflies and birds resemble each other in every spot and streak of colour through thousands of individuals. We also find that colour is constant in whole genera and other groups of species. The Genistas are all yellow, the Erythrinas all red; many genera of Carabidæ are entirely black; whole families of birds—as the Dendrocolaptidæ—are brown; while among butterflies the numerous species of Lycæna are all more or less blue, those of Pontia white, and those of Callidryas yellow. An extensive survey of the organic world thus leads us to the conclusion that colour is by no means so unimportant or inconstant a character as at first sight it appears to be; and the more we examine it the more convinced we shall become that it must serve some purpose in nature, and that, besides charming us by its diversity and beauty, it must be well worthy of our attentive study, and have many secrets to unfold to us.

Theory of Heat and Light as producing Colour

In commencing our study of the great mass of facts relating to the colours of the organic world, it will be necessary to consider, first, how far the chief theories already proposed will account for them. One of the most obvious and most popular of these theories, and one which is still held, in part at least, by many eminent naturalists, is, that colour is due to some direct action of the heat and light of the sun—thus at once accounting for the great number of brilliant birds, insects, and flowers which are found between the tropics.

But before proceeding to discuss this supposed explanation of the colours of living things, we must ask the preliminary question,—whether it is really the fact that colour is more developed in tropical than in temperate climates in proportion to the whole number of species; and even if we find this to be so, we have to inquire whether there are not so many and such striking exceptions to the rule as to indicate some other causes at work than the direct influence of solar light and heat. As this is a most important branch of the inquiry, we must go into it somewhat fully.

It is undoubtedly the case that there are an immensely greater number or richly-coloured birds and insects in tropical than in temperate and cold countries, but it is by no means so certain that the *proportion* of coloured to obscure species is much or any greater. Naturalists and collectors well know that the majority of tropical birds are dull-coloured; and there are whole families, comprising hundreds of species, not one of which exhibits a particle of bright colour. Such are, for example, the Timaliidæ or babbling thrushes of the eastern, and the Dendrocolaptidæ or tree-creepers of the western hemispheres. Again, many groups of birds which are universally distributed are no more adorned with colour in the tropical than in the temperate zones; such are the thrushes, wrens, goat-suckers, hawks, grouse, plovers, and snipe; and if tropical light and heat have any direct colouring effect, it is certainly most extraordinary that in groups so varied in form, structure, and habits as those just mentioned, the tropical should be in no wise distinguished in this respect from the temperate species.

It is true that brilliant tropical birds mostly belong to groups which are wholly tropical—as the chatterers, toucans, trogons, and pittas; but as there are perhaps an equal number of tropical groups which are wholly dull-coloured, while others contain dull and bright-coloured species in nearly equal proportions, the evidence is by no means strong that tropical light and heat have anything to do with the matter. But there are other groups in which the cold and temperate zones produce finer-coloured species than the tropics. Thus the arctic ducks and divers are handsomer than those of the tropical zone; while the king-duck of temperate America and the mandarin-duck of North China are the most beautifully coloured of the whole family. In the pheasant family we have the gorgeous gold and silver pheasants in North China and Mongolia, and the superb Impeyan pheasant in the temperate North-Western Himalayas, as against the peacock and fire-backed pheasants of tropical Asia. Then we have the curious fact that most of the bright-coloured birds of the tropics are denizens of the forests, where they are shaded from the direct light of the sun, and that they abound near the equator, where cloudy skies are very prevalent; while, on the other hand, places where light and heat are at a maximum have often dull-coloured birds. Such are the Sahara and other deserts, where almost all the living things are sand-coloured; but the most curious case is that of the Galapagos islands, situated under the equator, and not far from South America, where the most gorgeous colours abound, but which are yet characterised by prevailing dull and sombre tints in birds, insects, and flowers, so that they reminded Mr. Darwin of the cold and barren plains of Patagonia rather than of any tropical country. Insects are wonderfully brilliant in tropical countries generally, and any one looking over a collection of South American or Malayan butterflies would scout the idea of their being no more gaily-coloured than the average of European species, and in this he would be undoubtedly right. But on examination we should find that all the more brilliantly-coloured groups were exclusively tropical, and that where a genus has a wide range there is little difference in coloration between the species of cold and warm countries. Thus the European Vanessides, including the beautiful "peacock," "Camberwell beauty," and "red admiral" butterflies, are quite up to the average of tropical colour in the same group; and the remark will equally apply to the little "blues" and "coppers"; while the alpine "apollo" butterflies have a delicate beauty that can hardly be surpassed. In other insects, which are less directly dependent on climate and vegetation, we find even greater anomalies. In the immense family of the Carabidæ or predaceous ground-beetles, the northern forms fully equal, if they do not surpass, all that the tropics can produce. Everywhere, too, in hot countries, there are thousands of obscure species of insects which, if they were all collected, would not improbably bring down the average of colour to much about the same level as that of temperate zones.

But it is when we come to the vegetable world that the greatest misconception on this subject prevails. In abundance and variety of floral colour the tropics are almost universally believed to be pre-eminent, not only absolutely, but relatively to the whole mass of vegetation and the total number of species. Twelve years of observation among the vegetation of the eastern and western tropics has, however, convinced me that this notion is entirely erroneous, and that, in proportion to the whole number of species of plants, those having gaily-coloured flowers are actually more abundant in the temperate zones than between the tropics. This will be found to be not so extravagant an assertion as it may at first appear, if we consider how many of the choicest adornments of our greenhouses and flower-shows are really temperate as opposed to tropical plants. The masses of colour produced by our rhododendrons, azaleas, and camellias, our pelargoniums, calceolarias, and cinerarias—all strictly temperate plants—can certainly not be surpassed, if they can be equalled, by any productions of the tropics.

It may be objected that most of the plants named are choice cultivated *varieties,* far surpassing in colour the original stock, while the tropical plants are mostly unvaried wild *species.* But this does not really much affect the question at issue. For our florists' gorgeous varieties have all been produced under the influence of our cloudy skies, and with even a still further deficiency of light, owing to the necessity of protecting them under glass from our sudden changes of temperature, so that they are themselves an additional proof that tropical light and heat are not needed for the production of intense and varied colour. Another important consideration is, that these cultivated *varieties* in many cases displace a number of wild *species* which are hardly, if at all, cultivated. Thus there are scores of *species* of wild hollyhocks varying in colour almost as much as the cultivated varieties, and the same may be said of the pentstemons, rhododendrons, and many other flowers; and if these were all brought together in well-grown specimens, they would produce a grand effect. But it is far easier, and more profitable for our nurserymen, to grow *varieties* of one or two species, which all require a similar culture, rather than fifty distinct *species*, most of which would require special treatment, the result being that the varied beauty of the temperate flora is even now hardly known, except to botanists and to a few amateurs.

But we may go further, and say that the hardy plants of our cold temperate zone equal, if they do not surpass, the productions of the tropics. Let us only remember such gorgeous tribes of flowers as the roses, pæonies, hollyhocks, and antirrhinums; the laburnum, wistaria, and lilac; the lilies, irises, and tulips; the hyacinths, anemones, gentians, and poppies, and even our humble gorse, broom, and heather; and we may defy any tropical country to produce masses of floral colour in greater abundance and variety. It may be true that individual tropical shrubs and flowers do surpass everything in the rest of the world; but that is to be expected, because the tropical zone comprises a much greater land area than the two temperate zones, while, owing to its more favourable climate, it produces a still larger proportion of species of plants and a greater number of peculiar natural orders.

Direct observation in tropical forests, plains, and

mountains fully supports this view. Occasionally we are startled by some gorgeous mass of colour, but as a rule we gaze upon an endless expanse of green foliage, only here and there enlivened by not very conspicuous flowers. Even the orchids, whose superb blossoms adorn our stoves, form no exception to this rule. It is only in favoured spots that we find them in abundance; the species with small and inconspicuous flowers greatly preponderate; and the flowering season of each kind being of short duration, they rarely produce any marked effect of colour amid the vast masses of foliage which surround them. An experienced collector in the Eastern tropics once told me that although a single mountain in Java had produced three hundred species of Orchideæ, only about 2 per cent of the whole were sufficiently ornamental or showy to be worth sending home as a commercial speculation. The Alpine meadows and rock-slopes, the open plains of the Cape of Good Hope or of Australia, and the flower-prairies of North America, offer an amount and variety of floral colour which can certainly not be surpassed, even if it can be equalled, between the tropics.

It appears, therefore, that we may dismiss the theory that the development of colour in nature is directly dependent on, and in any may proportioned to, the amount of solar heat and light, as entirely unsupported by facts. Strange to say, however, there are some rare and little-known phenomena which prove that in exceptional cases light does directly affect the colours of natural objects, and it will be as well to consider these before passing on to other matters.

Changes of Colour in Animals produced by Coloured Light

A few years ago Mr. T. W. Wood called attention to the curious changes in the colour of the chrysalis of the small cabbage-butterfly (Pontia rapæ) when the caterpillars, just before their change, were confined in boxes lined with different tints. Thus in black boxes they were very dark, in white boxes nearly white; and he further showed that similar changes occurred in a state of nature, chrysalises fixed against a whitewashed wall being nearly white, against a red brick wall reddish, against a pitched pailing nearly black. It has also been observed that the cocoon of the emperor-moth is either white or brown, according to the colours surrounding it. But the most extraordinary example of this kind of change is that furnished by the chrysalis of an African butterfly (Papilio Nireus), observed at the Cape by Mrs. Barber, and described (with a coloured plate) in the *Transactions of the Entomological Society,* 1874, p. 519.

This caterpillar feeds upon the orange tree, and also upon a forest tree (Vepris lanceolata) which has a lighter green leaf; and its colour corresponds with that of the leaves it feeds upon, being of a darker green when it feeds on the orange. The chrysalis is usually found suspended among the leafy twigs of its food-plant, or of some neighbouring tree, but it is probably often attached to larger branches; and Mrs. Barber has discovered that it has the property of acquiring the colour, more or less accurately, of any natural object it may be in contact with. A number of the caterpillars were placed in a case with a glass cover, one side of the case being formed by a red brick wall, the other sides being of yellowish wood. They were fed on orange leaves, and a branch of the bottle-brush tree (Banksia sp.) was also placed in the case. When fully fed, some attached themselves to the orange twigs, others to the bottle-brush branch, and these all changed to green pupæ, but each corresponded exactly in tint to the leaves around it, the one being dark, the other a pale faded green. Another attached itself to the wood, and the pupa became of the same yellowish colour, while one fixed itself just where the wood and brick joined, and became one side red, the other side yellow! These remarkable changes would perhaps not have been credited had it not been for the previous observations of Mr. Wood; but the two support each other, and oblige us to accept them as actual phenomena. It is a kind of natural photography, the particular coloured rays to which the fresh pupa is exposed in its soft, semi-transparent condition effecting such a chemical change in the organic juices as to produce the same tint in the hardened skin. It is interesting, however, to note that the range of colour that can be acquired seems to be limited to those of natural objects to which the pupa is likely to be attached, for when Mrs. Barber surrounded one of the caterpillars with a piece of scarlet cloth no change of colour at all was produced, the pupa being of the usual green tint, but the small red spots with which it is marked were brighter than usual.[1]

1 Mr. E. B. Poulton has since greatly extended these observations, both in pupæ and larvæ, with very remarkable results. See *Proc. of the Royal Society*, No. 243, 1886; *Transactions of the Royal Society*, vol. clxxviii. B., pp. 311–441. These are briefly described in *Darwinism*, p. 197, and more fully in a volume by Mr. Poulton on *The Colours of Animals*, 1890.

Many other cases are known among insects in which the same species acquires a different tint according to its surroundings; this being particularly marked in some South African locusts, which correspond with the colour of the soil wherever they are found. There are also many caterpillars which feed on two or more plants, and which vary in colour accordingly. A number of such changes are quoted by Mr. R. Meldola, in a paper on "Variable Protective Colouring in Insects" (*Proceedings of the Zoological Society of London,* 1873, p. 153); and in some cases it has been shown that green chlorophyll remains unchanged in the tissues of leaf-eating insects, and being discernible through the transparent integument, produces the same colour as that of the food plant.

In all these insects, as well as in the great majority of cases in which a change of colour occurs in other animals, the action is quite involuntary; but among some few of the higher animals the colour of the integument can be modified at the will of the individual, or at all events by a reflex action dependent on sensation. The most remarkable case of this kind occurs with the chameleon, which has the power of changing its colour from dull white to a variety of tints. This singular power has been traced to two layers of movable pigment-cells deeply seated in the skin, but capable of being brought near to the surface. The pigment-layers are bluish and yellowish, and by their contraction or concentration these can be forced upwards either together or separately. When the animal is passive the colour is dirty white, which changes to various tints of bluish, green, yellow, or brown, as more or less of either pigment is forced up and rendered visible. The animal is excessively sluggish and defenceless, and its power of changing its colour so as to harmonise with surrounding objects is essential to its safety. Here too, as with the pupa of Papilio Nireus, colours such as scarlet or blue, which do not occur in the natural environment of the animal, cannot be produced. Somewhat similar changes of colour occur in some prawns and flat-fish, according to the colour of the bottom on which they rest. This is very striking in the chameleon shrimp (Mysis chamæleon), which is gray when on sand, but brown or green when among seaweed of these two colours. Experiment shows, however, that when blinded the change does not occur; so that here too we probably have a voluntary or reflex senseaction.

These peculiar powers of change of colour and adaptation are, however, rare and quite exceptional. As a rule there is no direct connection between the colours of organisms and the kind of light to which they are usually exposed. This is well seen in most fishes, and in such marine animals as porpoises, whose backs are always dark, although this part is exposed to the blue and white light of the sky and clouds, while their bellies are very generally white, although these are constantly subjected to the deep blue or dusky green light from the bottom. It is evident, however, that these two tints have been acquired for concealment and protection. Looking *down* on the dark back of a fish it is almost invisible, while, to an enemy looking *up* from below, the light undersurface would be equally invisible against the light of the clouds and sky. Again, the gorgeous colours of the butterflies which inhabit the depths of tropical forests bear no relation to the kind of light that falls upon them, coming as it does almost wholly from green foliage, dark brown soil, or blue sky; and the bright underwings of many moths, which are only exposed at night, contrast remarkably with the sombre tints of the upper wings, which are more or less exposed to the various colours of surrounding nature.

Classification of Organic Colours

We find, then, that neither the general influence of solar light and heat, nor the special action of variously tinted rays, are adequate causes for the wonderful variety, intensity, and complexity of the colours that everywhere meet us in the animal and vegetable worlds. Let us therefore take a wider view of these colours, grouping them into classes determined by what we know of their actual uses or special relations to the habits of their possessors. This, which may be termed the functional and biological classification of the colours of living organisms, seems to be best expressed by a division into five groups, as follows:—

Animals

1. Protective colours
2. Warning colours.
   a. Of creatures specially protected.
   b. Of defenceless creatures, mimicking *a.*
3. Sexual colours
4. Normal colours.[1]

Plants

5. Attractive colours.

It is now proposed, firstly, to point out the nature of the phenomena presented under each of these heads; then to explain the general laws of the production of colour in nature; and, lastly, to show how far the varied phenomena of animal coloration can be explained by means of those laws, acting in conjunction with the laws of evolution and natural selection.

[1] Many, or perhaps all, of these are now believed to be diversely coloured for purposes of recognition. See *Darwinism,* p. 217.

Protective Colours

The nature of the two first groups, protective and warning colour, has been so fully detailed and illustrated in my chapter on "Mimicry and other Protective Resemblances among Animals," that very little need be added here except a few words of general explanation. Protective colours are exceedingly prevalent in nature, comprising those of all the white arctic animals, the sandy-coloured desert forms, and the green birds and insects of tropical forests. It also comprises thousands of cases of special resemblance—of birds to the surroundings of their nests, and especially of insects to the bark, leaves, flowers, or soil, on or amid which they dwell. Mammalia, fishes, and reptiles, as well as mollusca and other marine invertebrates, present similar phenomena; and the more the habits of animals are investigated, the more numerous are found to be the cases in which their colours tend to conceal them, either from their enemies or from the creatures they prey upon. One of the last-observed and most curious of these protective resemblances has been communicated to me by Sir Charles Dilke. He was shown in Java a pink-coloured Mantis which, when at rest, exactly resembled a pink orchis-flower. The mantis is a carnivorous insect which lies in wait for its prey; and, by its resemblance to a flower, the insects it feeds on would be actually attracted towards it. This one is said to feed especially on better-flies, so that it is really a living trap, and forms its own bait![1]

All who have observed animals, and especially insects, in their native haunts and attitudes, can understand how it is that an insect which in a cabinet looks exceedingly conspicuous, may yet when alive, in its peculiar attitude of repose and with its habitual surroundings, be perfectly well concealed. We can hardly ever tell by the mere inspection of an animal whether its colours are protective or not. No one would imagine the exquisitely beautiful caterpillar of the emperor-moth, which is green with pink star-like spots, to be protectively coloured; yet, when feeding on the heather, it so harmonises with the foliage and flowers as to be almost invisible. Every day fresh cases of protective colouring are being discovered, even in our own country; and it is becoming more and more evident that the need of protection has played a very important part in determining the actual coloration of animals.

[1] These cases form a distinct sub-group of "alluring coloration." See *Darwinism,* p. 210.

Warning Colours

The second class—the warning colours—are exceedingly interesting, because the object and effect of these is, not to conceal the object, but to make it conspicuous. To these creatures it is *useful* to be seen and recognised; the reason being that they have a means of defence which, if known, will prevent their enemies from attacking them, though it is generally not sufficient to save their lives if they are actually attacked. The best examples of these specially protected creatures consist of two extensive families of butterflies, the Danaidæ and Acræidæ, comprising many hundreds of species inhabiting the tropics of all parts of the world. These insects are generally large, are all conspicuously and often most gorgeously coloured, presenting almost every conceivable tint and pattern; they all fly slowly, and they never attempt to conceal themselves; yet no bird, spider, lizard, or monkey (all of which eat other butterflies) ever devours them. The reason simply is that they are not fit to eat, their juices having a powerful odour and taste that is absolutely disgusting to all these animals. Now we see the reason of their showy colours and slow flight. It is good for them to be seen and recognised, for then they are never molested; but if they did not differ in form and colouring from other butterflies, or if they flew so quickly that their peculiarities could not be easily noticed, they would be captured, and though not eaten would be maimed or killed.

As soon as the cause of the peculiarities of these butterflies was clearly recognised, it was seen that the same explanation applied to many other groups of animals. Thus, bees and wasps and other stinging insects are showily and distinctively coloured; many soft and apparently defenceless beetles, and many gay-coloured moths, were found to be as nauseous as the above-named butterflies; other beetles, whose hard and glossy coats of mail render them unpalatable to insect-eating birds, are also sometimes showily coloured; and the same rule was found to apply to caterpillars, all the brown and green (or protectively coloured species) being greedily eaten by birds, while showy kinds which never hide themselves—like those of the magpie-, mullein-, and burnet-moths—were utterly refused by insectivorous birds, lizards, frogs, and spiders (p. 84). Some few analogous examples are found among vertebrate animals. I will only mention here a very interesting case not given in my former work. In his delightful book, entitled *The Naturalist in Nicaragua,* Mr. Belt tells us that there is in that country a frog which is very abundant, which hops about in the day-time, which never hides himself, and which is gorgeously coloured with red and blue. Now frogs are usually green, brown, or earth-coloured, feed mostly at night, and are all eaten by snakes and birds. Having full faith in the theory of protective and warning colours, to which he had himself contributed some valuable facts and observations, Mr. Belt felt convinced that this frog must be uneatable. He therefore took one home, and threw it to his ducks and fowls; but all refused to touch it except one young duck, which took the frog in its mouth, but dropped it directly, and went about jerking its head as if trying to get rid of something nasty. Here the uneatableness of the frog was predicted from its colours and habits, and we can have no more convincing proof of the truth of a theory than such previsions.

The universal avoidance by carnivorous animals of all these specially protected groups, which are thus entirely free from the constant persecution suffered by other creatures not so protected, would evidently render it advantageous for any of these latter which were subjected to extreme persecution to be mistaken for the former; and for this purpose it would be necessary that they should have the same colours, form, and habits. Now, strange to say, wherever there is a large group of directly-protected forms (division *a* of animals with warning colours), there are sure to be found a few otherwise defenceless creatures which resemble them externally so as to be mistaken for them, and which thus gain protection, as it were, on false pretences (division *b* of animals with warning colours). This is what is called "mimicry," and it has already been very fully treated of by Mr. Bates (its discoverer), by myself, by Mr. Trimen, and others. Here it is only necessary to state that the uneatable Danaidæ and Acræidæ are accompanied by a few species of other groups of butterflies (Leptalidæ, Papilios, Diademas, and Moths), which are all really eatable, but which escape attack by their close resemblance to some species of the uneatable groups found in the same locality. In like manner there are a few eatable beetles which exactly resemble species of uneatable groups; and others, which are soft, imitate those which are uneatable through their hardness. For the same reason wasps are imitated by moths, and ants by beetles; and even poisonous snakes are mimicked by harmless snakes, and dangerous hawks by defenceless cuckoos. How these curious imitations have been brought about, and the laws which govern them, have been already discussed. (See p. 54.)

Sexual Colours

The third class comprises all cases in which the colours of the two sexes differ. This diference is very general, and varies greatly in amount, from a slight divergence of tint up to a radical change of coloration. Differences of this kind are found among all classes of animals in which the sexes are separated, but they are much more frequent in some groups than in others. In mammalia, reptiles, and fishes, they are comparatively rare and not great in amount, whereas among birds they are very frequent and very largely developed. So among insects, they are abundant in butterflies, while they are comparatively uncommon in beetles, wasps, and hemiptera.

The phenomena of sexual variations of colour, as well as of colour generally, are wonderfully similar in the two analogous yet totally unrelated groups of birds and butterflies; and as they both offer ample materials, we shall confine our study of the subject chiefly to them. The most common case of difference of colour between the sexes is for the male to have the same general hue as the females, but deeper and more intensified—as in many thrushes, finches, and hawks, and among butterflies in the majority of our British species, In cases where the male is smaller the intensification of colour is especially well pronounced—as in many of the hawks and falcons, and in most butterflies and moths in which the coloration does not materially differ. In another extensive series we have spots or patches of vivid colour in the male, which are represented in the female by far less brilliant tints or are altogether wanting—as exemplified in the gold-crest warbler, the green woodpecker, and most of the orange-tip butterflies (Anthocharis). Proceeding with our survey, we find greater and greater differences of colour in the sexes, till we arrive at such extreme cases as some of the pheasants, the chatterers, tanagers, and birds-of-paradise, in which the male is adorned with the most gorgeous and vivid colours, while the female is usually dull brown, or olive green, and often shows no approximation whatever to the varied tints of her partner. Similar phenomena occur among butterflies; and in both these groups there are also a considerable number of cases in which both sexes are highly but differently coloured. Thus many woodpeckers have the head in the male red, in the female yellow; while some parrots have red spots in the male, replaced by blue in the female, as in Psittacula diopthalma. In many South American Papilios, green spots on the male are represented by red on the female; and in several species of the genus Epicalia, orange bands in the male are replaced by blue in the female, a similar change of colour to that in the small parrot above referred to. For fuller details of the varieties of sexual coloration we refer our readers to Mr. Darwin's *Descent of Man,* chapters x. to xviii., and to chapters iii. iv. and vii. Of the first portion of the present volume.

Normal Colours

The fourth group—of normally coloured animals—includes all species which are brilliantly or conspicuously coloured in both sexes, and for whose particular colours we can assign no function or use.[1] It comprises an immense number of showy birds, such as kingfishers, barbets, toucans, lories, tits, and starlings; among insects most of the largest and handsomest butterflies, innumerable bright-coloured beetles, locusts, dragonflies, and hymenoptera; a few mammalia, as the zebras; a great number of marine fishes; thousands of striped and spotted caterpillars; and abundance of mollusca, star-fish, and other marine animals. Among these we have included some which, like the gaudy caterpillars, have warning colours; but as that theory does not explain the particular colours or the varied patterns with which they are adorned, it is best to include them also in this class. It is a suggestive fact that all the brightly-coloured birds mentioned above build in holes or form covered nest, so that the females do not need that protection during the breeding season which I believe to be one of the chief causes of the dull colour of female birds when their partners are gaily coloured. This subject is fully argued in chapter vi. of the present volume.

Leaving the colours of flowers to be discussed in another chapter, we will now consider how the general facts of colour here sketched out can be explained. We have first to inquire what is colour, and how it is produced; secondly, what is known of the causes of change of colour; and, lastly, what theory best accords with the whole assemblage of facts.

[1] Distinctness of marking for purposes of recognition is probably the use in all cases. See p. 367, and *Darwinism,* p. 217.

The Nature of Colour

The sensation of colour is caused by vibration or undulations of the ethereal medium of different lengths and velocities. The whole body of vibrations caused by the sun is termed radiation, or, more commonly, rays; and consists of sets of waves which vary considerably in their dimensions and rate of recurrence, but of which the middle portion only is capable of exciting in us sensations of light and colour. Beginning with the largest waves, which recur at the longest intervals, we have first those which produce heat-sensations only; as they get smaller and recur quicker, we perceive a dull red colour; and as the waves increase in rapidity and diminish in size, we get successively sensations of orange, yellow, green, blue, indigo, and violet, all fading imperceptibly into each other. Then come more invisible rays, of shorter wavelength and quicker recurrence, which produce, solely or chiefly, chemical effects. The red rays, which first become visible, have been ascertained to recur at the rate of 458 millions of millions times in a second, the length of each wave being 1/36900 of an inch; while the violet rays, which last remain visible, recur 727 millions of millions times per second, and have a wave-length of 1/64516 of an inch. Although the waves recur a different rates, they are all propagated through the ether with the same velocity (192,000 miles per second); just as different musical sounds, which are produced by waves of *air* of different lengths and rates of recurrence, travel at the same speed, so that a tune played several hundred yards off reaches the ear in correct time. There are, therefore, an almost infinite number of different colour-producing undulations, and these may be combined in an almost infinite variety of ways, so as to excite in us the sensation of all the varied colours and tints we are capable of perceiving. When all the different kinds of rays reach us in the proportion in which they exist in the light of the sun, they produce the sensation of white. If the rays which excite the sensation of any one colour are prevented from reaching us, the remaining rays in combination produce a sensation of colour often very far removed from white. Thus green rays being abstracted leave purple light; blue, orange-red light; violet, yellowish-green light, and so on. These pairs are termed complementary colours. And if portions of differently coloured lights are abstracted in various degrees, we have produced all those infinite gradations of colours and all those varied tints and hues which are of such use to us in distinguishing external objects, and which form one of the great charms of our existence. Primary colours would therefore be as numerous as the different wave-lengths of the visible radiations, if we could appreciate all their differences; while secondary or compound colours, caused by the simultaneous action of any combination of rays of different wave-lengths, must be still more numerous.

In order to account for the fact that all colours appear to use to be produced by combinations of three primary colours—red, green, and violet—it is believed that we have three sets of nerve fibers in the retina, each of which is capable of being excited by all rays, but that one set is excited most by the larger or red waves, another by the medium or green waves, and the third set chiefly by the violet or smallest waves of light; and when all three sets are excited together in proper proportions we see white. This view is supported by the phenomena of colour-blindness, which are explicable on the theory that one of these sets of nerve-fibres (usually that adapted to perceive red) has lost its sensibility, causing all colours to appear as if the red rays were abstracted from them.

It is a property of these various radiations that they are unequally refracted or bent in passing obliquely through transparent bodies, the longer waves being least refracted, the shorter most. Hence it becomes possible to analyse white or any other light into its component rays. A small ray of sunlight, for example, which would produce a white spot on a wall, if passed through a prism, is lengthened out into a band of coloured light, exactly corresponding to the colours of the rainbow. Any one colour can thus be isolated and separately examined; and by means of reflecting mirrors the separate colours can be again compounded in various ways, and the resulting colours observed. This band of coloured light is called a *spectrum,* the instrument by which the *spectra* of various kinds of light are examined is called a *spectroscope.* This branch of the subject has, however, no direct bearing on the mode in which the colours of living things are produced, and its has only been alluded to in order to complete our sketch of the nature of colour.

The colour which we perceive in material substances are produced either by the absorption or by the interference of some of the rays which form white light. Pigmental or absorption-colours are the most frequent, comprising all the opaque tints of flowers and insects, and all the colours of dyes and pigments. They are caused by rays of certain wave-lengths being absorbed, while the remaining rays are reflected and give rise to the sensation of colour. When all the colour producing rays are reflected in due proportion, the colour of the object is white; when all are absorbed the colour is black. If blue rays only are absorbed the resulting colour is orange-red; and generally, whatever colour an object appears to us, it is because the complementary colours are absorbed by it. The reason why rays of only certain refrangibilities are reflected, and the rest of the incident light absorbed by each substance, is supposed to depend upon the molecular structure of the body. Chemical action almost always implies change of molecular structure; hence chemical action is the most potent cause of change of colour. Sometimes simple solution in water effects a marvellous change, as in the case of the well-known aniline dyes; the magenta and violet dyes exhibiting, when in the solid form, various shades of golden or bronzy metallic green.

Heat alone often produces change of colour without effecting any chemical change. Mr. Ackroyd has investigated this subject,[1] and has shown that a large number of bodies are changed by heat, returning to their normal colour when cooled, and that this change is almost always in the direction of the less refrangible rays or longer wave-lengths; and he connects the change with the molecular expansion caused by heat. As examples may be mentioned mercuric oxide, which is orange yellow, but which changes to orange, red, and brown when heated; chromic oxide, which is green, and changes to yellow; cinnabar, which is scarlet, and changes to puce; and metaborate of copper, which is blue, and changes to green and greenish yellow.

[1] "Metachromatism, or Colour-Change," *Chemical News,* August 1876.

How Animal Colours are Produced

The colouring matters of animals are very varied. Copper has been found in the red pigment of the wing of the turaco, and Mr. Sorby has detected no less than seven distinct colouring matters in birds' eggs, several of which are chemically related to those of blood and bile. The same colours are often produced by quite different substances in different groups, as shown by the red of the wing on the burnet-moth changing to yellow with muriatic acid, while the red of the red-admiral butterfly undergoes no such change.

These pigmental colours have a different character in animals according to their position in the integument. Following Dr. Hagen's classification, epidermal colours are those which exist in the external chitinised skin of insects, in the hairs of mammals, and, partially, in the feathers of birds. They are often very deep and rich, and do not fade after death. The hypodermal colours are those which are situated in the inferior soft layer of the skin. These are often of lighter and more vivid tints, and usually fade after death. Many of the reds and yellows of butterflies and birds belong to this class, as well as the intensely vivid hues of the naked skin about the heads of many birds. These pigments sometimes exude through the pores, forming an evanescent bloom on the surface.

Interference colours are less frequent in the organic world. They are caused in two ways: either by reflection from the two surfaces of transparent films, as seen in the soap-bubble and in thin films of oil on water; or by fine striæ which produce colours either by reflected or transmitted light, as seen in mother-of-pearl and in finely-ruled metallic surfaces. In both cases colour is produced by light of one wave-length being neutralised, owing to one set of such waves being retarded or shifted so as to be half a wave-length behind the other set, as may be found explained in any treatise on physical optics. The result is, that the complementary colour of that neutralised is seen; and, as the thickness of the film or the fineness of the striæ undergo slight changes, almost any colour can be produced. This is believed to be the origin of many of the glossy or metallic tints of insects, as well as those of the feathers of some birds. The iridescent colours of the wings of dragon-flies are caused by the superposition of two or more transparent lamellæ; while the shining blue of the purple-emperor and other butterflies, and the intensely metallic colours of humming-birds, are probably due to fine striæ

Colour a Normal Product of Organisation

This outline sketch of the nature of colour in the animal world, however imperfect, will at least serve to show us how numerous and varied are the causes which perpetually tend to the production of colour in animal tissues. If we consider that in order to produce white all the rays which fall upon an object must be reflected in nearly the same proportions as they exist in solar light—whereas, if rays of any one or more kinds are absorbed or neutralised, the resultant reflected light will be coloured; and that this colour may be infinitely varied according to the proportions in which different rays are reflected or absorbed—we should expect that white would be, as it really is, comparatively rare and exceptional in nature.[1] The same observation will apply to black, which arises from the absorption of all the different rays. Many of the complex substances which exist in animals and plants are subject to changes of colour under the influence of light, heat, or chemical changes, and we know that chemical changes are continually occurring during the physiological processes which occur in the body during development and growth. We also find that every external character is subject to minute changes, which are generally perceptible to us in closely allied species; and we can therefore have no doubt that the extension and thickness of the transparent lamellæ, and the fineness of the striæ or rugosities of the integuments, must be undergoing constant minute changes; and these changes will very frequently produce changes of colour. These considerations render it probable that colour is a normal and even necessary result of the complex structure of animals and plants; and that those parts of an organism which are undergoing continual development and adaptation to new conditions, and are also continually subject to the action of light and heat, will be the parts in which changes of colour will most frequently appear. Now there is little doubt that the external changes of animals and plants in adaptation to the environment are much more numerous than the internal changes; as seen in the varied character of the integuments and appendages of animals—hair, horns, scales, feathers, etc., etc.—and in plants, the leaves, bark, flowers, and fruits, with their various modifications—as compared with the great uniformity in the texture and composition of their internal tissues; and this accords with the uniformity of the tints of blood, muscle, nerve, and bone throughout extensive groups, as compared with the great diversity of colour of their external organs. It seems a fair conclusion that colour *per se* may be considered to be normal, and to need no special accounting for; while the absence of colour (that is, either *white* or *black*), or the prevalence of certain colours to the constant exclusion of others, must be traced, like other modifications in the economy of living things, to the needs of the species. Or, looking at it in another aspect, we may say that amid the constant variations of animals and plants colour is ever tending to vary and to appear where it is absent; and that natural selection is constantly eliminating such tints as are injurious to the species, or preserving and intensifying such as are useful.

[1] White is produced by the scattering of the various rays in all directions, and is often caused by air-bubbles or transparent globules. See Poulton's *Colours of Animals*, pp. 3–6.

This view is in accordance with the well-known fact of colours which rarely or never appear in the species in a state of nature, continually occurring among domesticated animals and cultivated plants; showing us that the capacity to develop colour is ever present, so that almost any required tint can be produced which may, under changed conditions, be useful, in however small a degree.

Let us now see how these principles will enable us to understand and explain the varied phenomena of colour in nature, taking them in the order of our functional classification of colours.

Theory of Protective Colours

We have seen that obscure or protective tints in their infinitely varied degrees are present in every part of the animal kingdom; whole families or genera being often thus coloured. Now the various brown, earthy, ashy, and other neutral tints are those which would be most readily produced, because they are due to an irregular mixture of many kinds of rays; while pure tints require either rays of one kind only, or definite mixtures in proper proportions of two or more kinds of rays. This is well exemplified by the comparative difficulty of producing definite pure tints by the mixture of two or more pigments; while a haphazard mixture of a number of these will be almost sure to produce browns, olives, or other neutral or dingy colours. An indefinite or irregular absorption of some rays and reflection of others would, therefore, produce obscure tints; while pure and vivid colours would require a perfectly definite absorption of one portion of the coloured rays, leaving the remainder to produce the true complementary colour. This being the case, we may expect these brown tints to occur when the need of protection is very slight or even when it does not exist at all; always supposing that bright colours are not in any way useful to the species. But whenever a pure colour is protective,—as green in tropical forests, or white among arctic snows,—there is no difficulty in producing it, by natural selection action on the innumerable slight variations of tints which are ever occurring. Such variations may, as we have seen, be produced in a great variety of ways, either by chemical changes in the secretions, or by molecular changes in surface structure; and may be brought about by change of food, by the physiological action of light, or by the normal process of generative variation. Protective colours therefore, however curious and complex they may be in certain cases, offer no real difficulties.

Theory of Warning Colours

These differ greatly from the last class, inasmuch as they present us with a variety of brilliant hues, often of the greatest purity, and combined in striking contrasts and conspicuous patterns. Their use depends upon their boldness and visibility, not on the presence of any one colour; hence we find among these groups some of the most exquisitely-coloured objects in nature. Many of the uneatable caterpillars are strikingly beautiful; while the Danaidæ, Heliconidæ, and protected groups of Papilionidæ, comprise a series of butterflies of the most brilliant and contrasted colours. The bright colours of many of the sea-anemones and sea-slugs will probably be found to be in this sense protective, serving as a warning of their uneatableness.[1] On our theory none of these colours offer any difficulty. Conspicuousness being useful, every variation tending to brighter and purer colours was selected; the result being the beautiful variety and contrast we find.

Imitative Warning Colours—the Theory of Mimicry

We now come to those groups which gain protection solely by mistaken for some of these brilliantly coloured but uneatable creatures, and here a difficulty really exists, and to many minds is so great as to be insuperable. It will be well therefore to endeavour to explain how the resemblance in question may have been brought about.

The most difficult case, and the one which may be taken as a type of the whole class, is that of the genus Leptalis (a group of South American butterflies allied to our common white and yellow kings), many of the larger species of which are still white or yellow, and which are all eatable by birds and other insectivorous creatures. But there are also a number of species of Leptalis, which are brilliantly red, yellow, and black, and which, band for band and spot for spot, resemble some one of the Danaidæ or Heliconidæ which inhabit the same district and which are nauseous and uneatable. Now the usual difficulty is, that a slight approach to one of these protected butterflies would be of no use, while a greater sudden variation is not admissible on the theory of gradual change of indefinite slight variations. This objection depends almost wholly on the supposition that, when the first steps towards mimicry occurred, the South American Danaidæ were what they are now; while the ancestors of the Leptalides were like the ordinary white or yellow Pieridæ which they area allied. But the Danaioid butterflies of South America are so immensely numerous and so greatly varied, not only in colour but in structure, that we may be sure they are of vast antiquity and have undergone great modification. A large number of them, however, are still of comparatively plain colours, often rendered extremely elegant by the delicate transparency of the wing membrane, but otherwise not at all conspicuous.

[1] This has since been found to be the case by Professor Herdman *(Trans. Biol. Soc. Liverpool,* vol. iv. p. 150).

Many have only dusky or purplish bands or spots; others have patches of reddish or yellowish brown—perhaps the commonest colour among butterflies; while a considerable number are tinged or spotted with yellow, also a very common colour, and one especially characteristic of the Pieridæ, the family to which Leptalis belongs. We may therefore reasonably suppose that in the early stages of the development of the Danaidæ, when they first began to acquire those nauseous secretions which are now their protection, their colours were somewhat plain; either dusky with paler bands and spots, or yellowish with dark borders, and sometimes with reddish bands or spots. At this time they had probably shorter wings and a more rapid flight, just like the other unprotected families of butterflies. But, so soon as they became decidedly unpalatable to any of their enemies, it would be an advantage to them to the readily distinguished from all the eatable kinds; and as butterflies were no doubt already very varied in colour, while all probably had wings adapted for rather quick or jerking flight, the best distinction might have been found in outline and habits; whence would arise the preservation of those varieties whose longer wings, bodies, and antennæ as well as their slower flight, rendered them noticeable—characters which now distinguish the whole group in every part of the world.

Now it would be at this stage that some of the weaker-flying Pieridæ which happened to resemble some of the Danaidæ around them in their yellow and dusky tints and in the general outline of their wings, would be sometimes mistaken for them by the common enemy, and would thus gain an advantage in the struggle for existence. Admitting this one step to be made, and all the rest must inevitably follow from simple variation and survival of the fittest. So soon as the nauseous butterfly varied in form or colour to such an extent that the corresponding eatable butterfly no longer closely resembled it, the latter would be exposed to attacks, and only those variations would be preserved which kept up the resemblance. At the same time we may well suppose the enemies to become more acute and able to detect smaller differences than at first. This would lead to the destruction of all adverse variations, and thus keep up in continually increasing complexity the outward mimicry which now so amazes us. During the long ages in which this process has been going on, and the Danaidæ have been acquiring those specialities of colour which aid in their preservation, many a Leptalis may have become extinct from not varying sufficiently in the right direction and at the right time to keep up a protective resemblance to its neighbour; and this well accords with the comparatively small number of cases of true mimicry, as compared with the frequency of those protective resemblances to vegetable or inorganic objects whose forms are less definite and colours less changeable. About a dozen other genera of butterflies and moths mimic and Danaidæ in various parts of the world, and exactly the same explanation will apply to all of them. They represent those species of each group which, at the time when the Daniadæ first acquired their protective secretions, happened outwardly to resemble some of them, and which have, by concurrent variation aided by a rigid selection, been able to keep up that resemblance to the present day.[1]

Theory of Sexual Colours

In Mr. Darwin's celebrated work, *The Descent of Man and Selection in Relation to Sex,* he has treated of sexual colour in combination with other sexual characters, and has arrived at the conclusion that all or almost all the colours of the higher animals (including among these insects and all vertebrates) are due to voluntary or conscious sexual selection; and that diversity of colour in the sexes is due, primarily, to the transmission of colour-variations either to one sex only or to both sexes, the difference depending on some unknown law, and not being due to natural selection.

[1] For fuller information on this subject the reader should consult Mr. Bates' original paper, "Contributions to an Insect-fauna of the Amazon Valley," in *Transactions of the Linnean Society,* vol. xxiii. p. 495; Mr. Trimen's paper in vol. xxvi. p. 497; the author's essay on "Mimicry," etc., already referred to; and, in the absence of collections of butterflies, the plates of Heliconidæ and Leptalidæ, in Hewitson's *Exotic Butterflies;* and Felder's *Voyage of the "Novara,"* may be examined.

I have long held this portion of Mr. Darwin's theory to be erroneous, and have argued that the primary cause of sexual diversity of colour was the need of protection, repressing in the female those bright colours which are normally produced in both sexes by general laws; and I have attempted to explain many of the more difficult cases on this principle ("A Theory of Birds' Nests," chap. vi.*ante*). As I have since given much thought to this subject, and have arrived at some views which appear to me to be of considerable importance, it will be well to sketch briefly the theory I now hold, and afterwards show its application to some of the detailed cases adduced in Mr. Darwin's work.

The very frequent superiority of the male bird or insect in brightness or intensity of colour, even when the general coloration is the same in both sexes, now seems to me to be, in great part, due to the greater vigour and activity and the higher vitality of the male. The colours of an animal usually fade during disease or weakness, while robust health and vigour adds to their intensity. This is a most important and suggestive fact, and one that appears to hold universally. In all quadrupeds a "dull coat" is indicative of ill-health or low conditions, while a glossy coat and sparkling eye are the invariable accompaniments of health and vital activity. The same rule applies to the feathers of birds, whose colours are only seen in their purity during perfect health; and a similar phenomenon occurs even among insects, for the bright hues of caterpillars begin to fade as soon as they become inactive preparatory to undergoing their transformation, or if attacked by disease. Even in the vegetable kingdom we see the same thing, for the tints of foliage are deepest, and the colours of flowers and fruits richest, on those plants which are in the most healthy and vigorous condition.

This intensity of coloration becomes most developed in the male during the breeding season, when the vitality is at a maximum. It is also very general in those cases in which the male is smaller than the female, as in the hawks and in most butterflies and months. The same phenomena occur, though in a less marked degree, among mammalia. Whenever there is a difference of colour between the sexes the male is the darker or more strongly marked, and the difference of intensity is most visible during the breeding season (*Descent of Man,* p. 533.) Numerous cases among domestic animals also prove that there is an inherent tendency in the male to special developments of dermal appendages and colour, quite independently of sexual or any other form of selection. Thus—"the hump on the male zebu cattle of India, the tail of fat-tailed rams, the arched outline of the forehead in the males of several breeds of sheep, and the mane, the long hairs on the hind legs, and the dewlap of the male of the Berbura goat"—are all adduced by Mr. Darwin as instances of characters peculiar to the male, yet not derived from any parent ancestral form. Among domestic pigeons the character of the different breeds is often most strongly manifested in the male birds; the wattles of the carriers and the eye-wattles of the barbs are largest in the males, and male pouters distend their crops to a much greater extent than do the females, while the cock fantails often have a greater number of tail-feathers than the females. There are also some varieties of pigeons of which the males are striped or spotted with black, while the females are never so spotted (*Animals and Plants under Domestication*, i. 161); yet in the parent stock of these pigeons there are no differences between the sexes either of plumage or colour, and artificial selection has not been applied to produce them.

The greater intensity of coloration in the male, which may be termed the normal sexual difference, would be further developed by the combats of the males for the possession of the females. The most vigorous and energetic usually being able to rear most offspring, intensity of colour, if dependent on, or correlated with vigour, would tend to increase. But as differences of colour depend upon minute chemical or structural differences in the organism, increasing vigour acting unequally on different portions of the integument, and often producing at the same time abnormal developments of hair, horns, scales, feathers, etc., would almost necessarily lead also to variable distribution of colour, and thus to the production of new tints and markings. These acquired colours would, as Mr. Darwin has shown, be transmitted to both sexes or to one only, according as they first appeared at an early age, or in adults of one sex; and thus we may account for some of the most marked differences in this respect. With the exception of butterflies, the sexes are almost alike in the great majority of insects. The same is the case in mammals and reptiles, while the chief departure from the rule occurs in birds, though even here in very many cases the law of sexual likeness prevails. But in all cases where the increasing development of colour became disadvantageous to the female, it would be checked by natural selection, and thus produce those numerous instances of protective colouring in the female only, which occur most frequently in these two groups, birds and butterflies.

Colour as a means of Recognition

There is also, I believe, a very important purpose and use of the varied colours of the higher animals in the facility it affords for recognition by the sexes or by the young of the same species; and it is this use which probably fixes and determines the coloration in many cases. When differences in the size and form of allied species are very slight, colour affords the only means of recognition at a distance, or while in motion; and such a distinctive character must therefore be of especial value to flying insects which are continually in motion, and encounter each other, as it were, by accident. This view offers us an explanation of the curious fact that among butterflies the females of closely-allied species in the same locality sometimes differ considerably, while the males are much alike; for, and the males are the swiftest and by far the highest fliers, and seek out the females, it would evidently be advantageous for them to be able to recognise their true partners at some distance off. This peculiarity occurs with many species of Papilio, Diadema, Adolias, and Colias; and these are all genera, the males of which are strong on the wing and mount high in the air. In birds such marked differences of colour are not required owning to their higher organisation and more perfect senses, which render recognition easy by means of a combination of very slight differential characters.[1]

[1] For numerous examples of recognition-colours in birds, see *Darwinism,* pp.217–226.

This principle may perhaps, however, account for some anomalies of coloration among the higher animals. Thus, while admitting that the hare and the rabbit are coloured protectively, Mr. Darwin remarks that the latter, while running to its burrow, is made conspicuous to the sportsman, and no doubt to all beasts of prey, by its upturned white tail. But this very conspicuousness while running away, may be useful as a signal and guide to the young, who are thus enabled to escape danger by following the older rabbits, directly and without hesitation, to the safety of the burrow;

and this may be the more important from the semi-nocturnal habits of the animal. If this explanation is correct, and it certainly seems probable, it may serve as a warning of how impossible it is, without exact knowledge of the habits of an animal and a full consideration of all the circumstances, to decide that any particular coloration cannot be protective or in any way useful. Mr. Darwin himself is not free from such assumptions. Thus, he says:—"The zebra is conspicuously striped, and stripes cannot afford any protection on the open plains of South Africa." But the zebra is a very swift animal, and, when in herds, by no means void of means of defence. The stripes therefore *may* be of use by enabling stragglers to distinguish their fellows at a distance, and they *may* be even protective when the animal is at rest among herbage—the only time when it would need protective colouring. Until the habits of the zebra have been observed with the special reference to these points, it is surely somewhat hasty to declare that the stripes "cannot afford any protection."[1]

[1] For further information on this point, see *Darwinism*, p. 220.

Colour proportionate to Integumentary Developement

The wonderful display and endless variety of colour in which butterflies and birds so far exceed all other animals, seems primarily due to the excessive development and endless variations of the integumentary structures of these two groups. No insects have such widely expanded wings in proportion to their bodies as butterflies and moths; in none do the wings vary so much in size and form, and in none are they clothed with such a beautiful and highly-organised coating of scales. According to the general principles of the production of colour already explained, these long continued expansions of membranes and developments of surface-structures must have led to numerous colour-changes, which have been sometimes checked, sometimes fixed and utilised, sometimes intensified, by natural selection, according to the needs of the animal. In birds, too, we have the wonderful clothing of plumage—the most highly organised, the most varied, and the most expanded of all dermal appendages. The endless processes of growth and change during the development of feathers, and the enormous extent of this delicately-organised surface, must have been highly favourable to the production of varied colour-effects, which favourable to the production of varied colour-effects, which, when not injurious, have been merely fixed for purposes for specific identification, but have often been modified or suppressed whenever different tints were needed for purposes of protection.

Selection by Females not a cause of Colour

To conscious sexual selection—that is, the actual choice by the females of the more brilliantly-coloured males or the rejection of those less gaily coloured—I believed very little if any effect is directly due. It is undoubtedly proved that in birds the females do sometimes exert a choice; but the evidence of this fact, collected by Mr. Darwin (*Descent of Man,* chap. xiv.), does not prove that colour determines that choice, while much of the strongest evidence is directly opposed to this view. All the facts appear to be consistent with the choice depending on a variety of male characteristics, with some of which colour is often correlated. Thus it is the opinion of some of the best observers that vigour and liveliness are most attractive, and these are no doubt usually associated with intensity of colour. Again, the display of the various ornamental appendages of the male during courtship may be attractive; but these appendages, with their bright colours or shaded patterns, are due probably to general laws of growth, and to that superabundant vitality which we have seen to be a cause of colour. But there are many considerations which seem to show that the possession of these ornamental appendages and bright colours in the male is not an important character functionally, and that it has not been produced by the action of conscious sexual selection. Amid the copious mass of facts and opinions collected by

Mr. Darwin as to the display of colour and ornaments by the male birds, there is a total absence of any evidence that the females, as a rule, admire or even notice this display. The hen, the turkey, and the pea-fowl go on feeding while the male is displaying his finery; and there is reason to believe that it is his persistency and energy rather than his beauty which wins the day. Again, evidence collected by Mr. Darwin himself, proves that each bird finds a mate under any circumstances. He gives a number of cases of one of a pair of birds shot, and the survivor being always found paired again almost immediately. This is sufficiently explained on the assumption that the destruction of birds by various causes is continually leaving widows and widowers in nearly equal proportions, and thus each one finds a fresh mate; and it leads to the conclusion that permanently unpaired birds are very scarce, so that, speaking broadly, every bird finds a mate and breeds. But this would almost or quite neutralise any effect of sexual selection of colour or ornament, since the less highly-coloured birds would be at little or no disadvantage as regards leaving healthy offspring. If, however, heightened colour is correlated with health and vigour; and if these healthy and vigorous birds provide best for their young, and leave offspring which, being equally healthy and vigorous, can best provide for themselves—which cannot be denied—then natural selection becomes a preserver and intensifier of colour.

Another most important consideration is, that male butterflies rival or even excel themost gorgeous male birds in bright colours and elegant patterns; and among these there is literally not one particle of evidence that the female is influenced by colour, or even that she has any power of choice, while there is much direct evidence to the contrary (*Descent of Man,* p. 318). The weakness of the evidence for conscious sexual selection among these insects is so palpable that Mr. Darwin is obliged to supplement it by the singularly inconclusive argument that, "Unless the female prefer one male to another, the pairing must be left to mere chance, and this does not appear probable" (*l.c.* p. 317). But he has just said: "The males sometimes fight together in rivalry, and many may be seen pursuing or crowding round the same female;" while in the case of the silk-moths, "the females appear not to evince the least choice in regard to their partners." Surely the plain inference from all this is, that males fight and struggle for the almost passive female, and that the most vigorous and energetic, the strongest-winged or the most persevering, wins her. How can there be chance in this? Natural selection would here act, as in birds, in perpetuating the strongest and most vigorous males; and as these would usually be the more highly coloured of their race, the same results would be produced as regards the intensification and variation of colour in the one case as in the other.

Let us now see how these principles will apply to some the cases adduced by Mr. Darwin in support of his theory of conscious sexual selection.

In *Descent of Man,* 2d ed., pp. 307–316, we find an elaborate account of the various modes of colouring of butterflies and moths, proving that the coloured parts are always more or less displayed, and that they have some evident relation to an observer. Mr. Darwin then says: "From the several foregoing facts it is impossible to admit that the brilliant colours of butterflies, and of some few moths, have commonly been acquired for the sake of protection. We have seen that their colours and elegant patterns are arranged and exhibited as if for display. Hence I am led to believe that the females prefer or are most excited by the more brilliant males; for on any other supposition the males would, as far as we can see, be ornamented to no purpose" (*l.c.* p. 316). I am not aware that any one has ever maintained that the brilliant colurs of butterflies have "commonly been acquired for the sake of protection," yet Mr. Darwin has himself referred to cases in which the brilliant colour is so placed as to serve for protection; as, for example, the eye-spots on the hind wings of moths, which are pierced by birds and so save the vital parts of the insects, while the bright patch on the orange-tip butterflies, which Mr. Darwin denies are protective, may serve the same purpose. It is, in fact, somewhat remarkable how very generally the black spots, ocelli, or bright patches of colour are on the tips, margins, or discs of the wings; and as the insects are necessarily visible while flying, and this is the time when they are most subject to attacks by insectivorous birds, the position of the more conspicuous parts at some distance from the body may be a real protection to them. Again, Mr. Darwin admits that the white colour of the male white companions? This seems a more probable supposition than the altogether hypothetical choice of the female, sometimes exercised in favour of and sometimes against every new variety of colour in her partner.

A strictly analogous case is that of the glow-worm, whose light, as originally suggested by Mr. Belt, is admitted to be a warning of its uneatability to insectivorous nocturnal animals. The male, having wings, does not require this protection. In the tropics the number of nocturnal insectivorous birds and bats is very much greater, hence winged species possess the light, as they would otherwise be eaten by mistake for more savoury insects; and it may be that the luminous Elateridæ of the tropics really mimic the true fireflies (Lampyridæ), which are uneatable. This is the more probable, as the Elateridæ, in the great majority of species, have brown or protective colours, and are therefore certainly palatable to insectivorous animals.

Origin of the Ornamental Plumage of Male Birds

We now come to such wonderful developments of plumage and colour as are exhibited by the peacock and the Argus-pheasant; and I may here mention that it was the case of the latter bird, as fully discussed by Mr. Darwin, which first shook my belief in "sexual" or more properly "female" selection. The long series of gradations by which the beautifully shaded ocelli on the secondary wing-feathers of this bird have been produced, are clearly traced out, the result being a set of markings so exquisitely shaded as to represent "balls lying loose within sockets"—purely artificial objects of which these birds could have no possible experience. That this result should have been attained, through thousands and tens of thousands of female birds all preferring those males whose markings varied slightly in this one direction, this uniformity of choice continuing through thousands and tens of thousands of generations, is to me absolutely incredible. And when, further, we remember that those which did not so vary would also, according to all the evidence, find mates and leave offspring, the actual result seems quite impossible of attainment by such means.

Without pretending to solve completely so difficult a problem as that of the origin and uses of the variously coloured plumes and ornaments so often possessed by male birds, I would point out a few facts which seem to afford a clue. And first, the most highly-coloured and most richly-varied markings occur on those parts of the plumage which have undergone the greatest modification, or have acquired the most abnormal development. In the peacock, the tailcoverts are enormously developed, and the "eyes" are situated on the greatly dilated ends of these elongated feathers. In the birds-of-paradise, breast, or neck, or head, or tail-feathers, are greatly developed and highly coloured. The hackles of the cock and the scaly breasts of humming-birds are similar developments; while in the Argus-pheasant the secondary quills are so enormously lengthened and broadened as to have become almost useless for flight. Now it is easily conceivable that during this process of development inequalities in the distribution of colour may have arisen in different parts of the same feathers, and that spots and bands may thus have become broadened out into shaded spots or ocelli, in the way indicated by Mr. Darwin, much as the spots and rings on a soap-bubble increase with increasing tenuity. This is the more probable, because in domestic fowls varieties of colour tend to become symmetrical, quite independently of sexual selection (*Descent of Man,* p. 424). This is one of those crucial facts which, on Mr. Darwin's theory, ought not be happen, and which plainly indicate that symmetrical markings arise from the action of some general laws of colour-development.

If now we accept the evidence of Mr. Darwin's most trustworthy correspondents, that the choice of the female, so far as she exerts any, falls upon the "most vigorous, defiant, and mettlesome male;" and if we further believe, what is certainly the case, that these are as a rule the most brightly coloured and adorned with the finest developments of plumage—we have a real and not a hypothetical cause at work. For these most healthy, vigorous, and beautiful males will have the choice of the finest and most healthy females, will have the most numerous and healthy families, and will be able best to protect and rear those families. Natural selection, and what may be termed male selection, will tend to give them the advantage in the struggle for existence, and thus the fullest plumage and the finest colours will be transmitted, and tend to advance in each succeeding generation.

Theory of Display of Ornaments by Males

The full and interesting account given by Mr. Darwin of the colours and habits of male and female bird *(Descent of Man,* chaps, xiii. and xiv.), proves that in most, if not in all cases, the male birds fully display their ornamental plumage before the females or in rivalry with each other; but on the essential point of whether the female's choice is determined by minute differences in these ornaments or in their colours, there appears to be an entire absence of evidence. In the section on "Preference for particular Males by the Females," the facts quoted show indifference to colour, except that some colour similar to their own seems to be preferred. But in the case of the hen canary who chose a greenfinch in preference to either chaffinch or goldfinch, gay colours had evidently no preponderating attraction. There is some evidence adduced that female birds may, and probably do, choose their mates, but none whatever that the choice is determined by difference of colour; and no less than three eminent breeders informed Mr. Darwin that they "did not believe that the female prefer certain males on account of the beauty of their plumage." Again, Mr. Darwin himself says: "As a general rule colour appears to have little influence on the pairing of pigeons." The oft-quoted case of Sir R. Heron's pea-hens, which preferred an "old pied cock" to those normally coloured, is a very unfortunate one, because pied birds are just those that are not favoured in a state of nature, or the breeds of wild animals would become as varied and mottled as our domestic varieties. If such irregular fancies were not rare exceptions, the production of definite colours and patterns by the choice of the female birds, or in any other way, would be impossible.

There remains, however, to be accounted for, the remarkable fact of the display by the male of each species of its peculiar beauties of plumage and colour—a display which Mr. Darwin evidently considers to be the strongest argument in favour of conscious selection by the female. This display is, no doubt, a very interesting and important phenomenon; but it may, I believe, be satisfactorily explained on the general principles here laid down, without calling to our aid a purely hypothetical choice exerted by the female bird.

At pairing-time the male is in a state of excitement, and full of exuberant energy. Even unornamental birds flutter their wings or spread them out, erect their tails or crests, and thus give vent to the nervous excitability with which they are overcharged. It is not improbable that crests and other erectile feathers may be primarily of use in frightening away enemies, since they are generally erected when angry or during combat. Those individuals who were most pugnacious and defiant, and who brought these erectile plumes most frequently and most powerfully into action, would tend to leave them further developed in some of their descendants. If, in the course of this development, colour appeared—and we have already shown that such development of plumage are a very probable cause of colour—we have every reason to believe it would be most vivid in these most pugnacious and energetic individuals; and as these would always have the advantage in the rivalry for mates (to which advantage the excess of colour and plumage might sometimes conduce), there seems nothing to prevent a progressive development of these ornaments in *all dominant races;* that is, wherever there was such a surplus of vitality, and such complete adaptation to conditions, that the inconvenience or danger produced by such ornaments was so comparatively small as not to affect the superiority of the race over its nearest allies.

But if those portions of the plumage which were originally erected under the influence of anger or fear became largely developed and brightly coloured, the actual display under the influence of jealously or sexual excitement becomes quite intelligible. The males, in their rivalry with each other, would endeavour to excel their enemies as far as voluntary exertion would enable them to do so, just as they endeavour to rival each other in song, even sometimes to the point of causing their own destruction.

Natural Selection as neutralising Sexual Selection

There is also a general argument against Mr. Darwin's views on this question, founded on the nature and potency of "natural" as opposed to "sexual" selection, which appears to me to be of itself almost conclusive as to the whole matter at issue. Natural selection, or the survival of the fittest, acts perpetually and on an enormous scale. Taking the off-spring of each pair of birds as, on the average, only six annually, one-third of these at most will be preserved, while the two-thirds which are least fitted will die. At intervals of a few years, whenever unfavourable conditions occur, five-sixth, nine-tenths, or even a greater proportion of the whole yearly production are weeded out, leaving only the most perfect and best adapted to survive. Now unless these survivors are, on the whole, the most ornamental, this rigid natural selection must neutralise and destroy any influence that may be exerted by female selection. The utmost that can be claimed for the latter is, that a small fraction of the least ornamental do not obtain mates, while a few of the most ornamented may leave more than the average number of offspring. Unless, therefore, there is the strictest correlation between ornament and general perfection, the more brightly coloured or ornamented varieties can obtain no permanent advantage; and if there is (as I maintain) such a correlation, then the sexual selection of colour or ornament, for which there is little or no evidence, becomes needless, because natural selection, which is an admitted *vera causa,* will itself produce all the results.

In the case of butterflies the argument becomes even stronger, because the fertility is so much greater than in birds, and the weeding-out of the unfit takes place, to a great extent, in the egg and larva state. Unless the eggs and larvæ which escaped to produce the next generation were those which would produce the more highly-coloured butterflies, it is difficult to perceive how the slight preponderance of colour sometimes selected by the females should not be wholly neutralised by the extremely rigid selection for other qualities to which the offspring in every stage are exposed. The only way in which we can account for the observed facts is, by the supposition that colour and ornament are strictly correlated with health, vigour, and general fitness to survive. We have shown that there is reason to believe that this is the case, and if so, conscious sexual selection becomes as unnecessary as it would certainly be ineffective.

Greater Brilliancy of some Female Birds

There is one other very curious case of sexual colouring among birds—that, namely, in which the female is decidedly brighter or more strongly marked than the male, as in the fighting quails (Turnix), painted snipe (Rhynchæa), two species of phalarope (Phalaropus), and the common cassowary (Causuarius galeatus). In all these cases it is known that the males take charge of and incubate the eggs, while the females are almost always larger and more pugnacious.

In my "Theory of Birds' Nests" (see p. 132) I imputed this difference of colour to the greater need for protection by the male bird while incubating, to which Mr. Darwin has objected that the difference is not sufficient, and is not always so distributed as to be most effective for this purpose; and he believes that it is due to reversed sexual selection—that is, to the female taking the usual rôle of the male, and being chosen for her brighter tints. We have already seen reason for rejecting this latter theory in every case; and I also admit that Mr. Darwin's criticism is sound, and that my theory of protection is, in this case, only partially, if at all, applicable. But the theory now advanced, of intensity of colour being due to general vital energy, is quite applicable; and the fact that the superiority of the female in this respect is quite exceptional, and is therefore probably not in any case of very ancient date, will account for the difference of colour thus produced being always very slight.

Colour-development as illustrated by Humming-birds

Of the mode of action of the general principles of colour-development among animals, we have an excellent example in the humming-birds. Of all birds these are at once the smallest, the most active, and fullest of vital energy. When poised in the air their wings are invisible owing to the rapidity of their motion, and when startled they dart away with the rapidity of a flash of light. Such active creatures would not be an easy prey to any rapacious bird; and if one at length was captured, the morsel obtained would hardly repay the labour. We may be sure, therefore, that they are practically unmolested. The immense variety they exhibit in structure, plumage, and colour, indicates a high antiquity for the race; while their general abundance in individuals shows that they are a dominant group, well adapted to all the conditions of their existence. Here we find everything necessary for the development of accessory plumes and colour. The surplus vital energy shown in their combats and excessive activity has expended itself in ever-increasing developments of plumage, and greater and greater intensity of colour, regulated only by the need for specific identification which would be especially required in such small and mobile creatures. Thus may be explained those remarkable differences of colour between closely-allied species, one having a crest like the topaz, while in another it resembles the sapphire. The more vivid colours and more developed plumage of the males, I am now inclined to think, may be almost wholly due to their greater vital energy, and to those general laws which lead to such superior developments even in domestic breeds; but in some cases the need of protection by the female while incubating, to which I formerly imputed the whole phenomenon, may have suppressed a portion of the ornament which she would otherwise have attained.

The extreme pugnacity of humming-birds has been noticed by all observes, and it seems to be to some extent proportioned to the degree to colour and ornament in the species. Thus Mr. Salvin observes of Eugenes fulgens, that it is "a most pugnacious bird," and that "hardly any species shows itself more brilliantly on the wing." Again of Campylopterus hemileucurus: "the pugnacity of this species is remarkable. It is very seldom that two males meet without an aerial battle;" and "the large and showy tail of this humming-bird makes it one of the most conspicuous on the wing." Again, the elegant frill-necked Lophornis ornatus "is very pugnacious, erecting its crest, throwing out its whiskers, and attacking every humming-bird that may pass within its range of vision;" and of another species, L. magnificus, it is said that "it is so bold that the sight of man creates no alarm." The beautifully-coloured Thaumastura cora "rarely permits any other humming-bird to remain in its neighbourhood, but wages a continual and terrible war upon them." The magnificent bar-tail, Cometes sparganurus, one of the most imposing of all the humming-birds, is extremely fierce and pugnacious, "the males chasing each other through the air with surprising perseverance and acrimony." These are all the species I find noticed as being especially pugnacious, and every one of them is exceptionally coloured or ornamental species are so described, although many of them are common and well observed species. It is also to be noticed that the remarkable pugnacity of these birds is not confined to one season or even to birds of the same species, as in usual in sexual combats, but extends to any other species that may be encountered, while they are said even to attack birds of prey that approach too closely to their nests. It must be admitted that these facts agree well with the theory that colour and ornament are due to surplus vital energy and a long course of unchecked development. We have also direct evidence that the males are more active and energetic than the females. Mr. Gosse says that the whirring made by the male Polytmus humming-bird is shriller that that produced by the female; and he also informs us that the male flies higher and frequents mountains, while the female keeps to the lowlands.[1]

[1] Some other cases are noticed at p. 317. For some further developments and illustrations of the theory of sexual colour, see *Darwinism,* chap. x.

Theory of Normal Colours

The remaining kinds of animal colours, those which can neither be classed as protective, warning, nor sexual, are for the most part readily explained on the general principles of the development of colour which we have now laid down. It is a most suggestive fact that in cases where colour is required only as warning, as among the uneatable caterpillars, we find, not one or two glaring tints only, but every kind of colour disposed in elegant patterns, and exhibiting almost as much variety and beauty as among insects and birds. Yet here, not only is sexual selection out of the question, but the need for recognition and identification by others of the same species seems equally unnecessary. We can then only impute this variety to the normal production of colour in organic forms when fully exposed to light and air and undergoing great and rapid developmental modification. Among more perfect animals, where the need for recognition has been added, we find intensity and variety of colour at its highest pitch among the South American butterflies of the families Helicondiæ and Danaidæ, as well as among the Nymphalidæ and Erycinidæ, many of which obtain the necessary protection in other ways. Among birds also, wherever the habits are such that no special protection is needed for the females, and where the species frequent the depths of tropical forests, and are thus naturally protected from the swoop of birds of prey, we find almost equally intense coloration, as in the trogons, barbets, and gapers.

Local Causes of Colour-development

Another real, though as yet inexplicable cause of diversity of colour is to be found in the influence of locality. It is observed that species of totally distinct groups are coloured alike in one district, while in another district the allied species all undergo the same change of colour. Cases of this kind have been adduced by Mr. Bates, by Mr. Darwin, and by myself, and I will here give the more curious and important examples which I have been able to collect.[1]

[1] These were first given in my Address to the Biological Section of the British Association at Glasgow in 1876.

The Influence of Locality on Colour in Butterflies and Birds

Our first example is from tropical Africa, where we find two unrelated groups of butterflies belonging to two very distinct families (Nymphalidæ and Papilionidæ) characterised by a prevailing blue-green colour not found in any other continent.[2] Again, we have a group of African Pieridæ which are white or pale yellow with a marginal row of bead-like black spots; and in the same country one of the Lycænidæ

(Leptena erastus) is coloured so exactly like these that it was at first described as a species of Pieris. None of these four groups are known to be in any way specially protected, so that the resemblance cannot be due to protective mimicry.

[2] Romaleosoma and Euryphene (Nymphaildæ), Papilio zalmoxis and several species of the Nireus-group (Papilionidæ).

In South America we have far more striking cases, for in the three subfamilies Danainæ, Acræinæ, and Heliconiinæ, all of which are specially protected, we find identical tints and patterns reproduced, often in the greatest detail, each peculiar type of coloration being characteristic of distinct geographical subdivision of the continent. Nine very distinct genera are implicated in these parallel changes—Lycorea, Ceratinia, Mechanitis, Ithomia, Melinæa, Tithorea, Acræa, Heliconius, and Eueides, groups of three or four (or even five) of them appearing together in the same livery in one district, while in an adjoining district most or all of them undergo a simultaneous change of coloration or of making. Thus in the genera Ithomia, Mechanitis, and Heliconius, we have species with yellow apical spots in Guiana, all represented by allied species with white apical spots in South Brazil. In Mechanitis, Melinæa, and Heliconius, and sometimes in Tithorea, the species of the Southern Andes

(Bolivia and Peru) are characterised by an orange and black livery, while those of the Northern Andes (New Granada) are almost always orange-yellow and black. Other changes of a like nature, which it would be tedious to enumerate, but which are very striking when specimens are examined, occur in species of the same groups inhabiting these same localities, as well as Central America and the Antilles. The resemblance thus produced between widely different insects is sometimes general, but often so close and minute that only a critical examination of structure can detect the difference between them. Yet all are alike protected by the nauseous secretion which renders them unpalatable to birds.[1]

[1] The above cases have now been satisfactorily explained as a modified form of mimicry. See *Darwinism*, pp. 249–257.

In another series of genera (Catagramma, Callithea, and Agrias), all belonging to the Nymphalidæ, we have the most vivid blue ground, with broad bands of orange, crimson, or a different tint of blue or purple, exactly reproduced in corresponding, yet unrelated species, occurring in the same locality; yet, as none of these groups are known to be specially protected, this can hardly be true mimicry. A few species of two other genera in the same country (Eunica and Siderone) also reproduce the same colours, but with only a general resemblance in the markings. Yet again, in tropical

America we have species of Apatura which, sometimes in both sexes, sometimes in the female only, exactly imitate the peculiar markings of another genus (Heterochroa) confined to America; here, again, neither genus is protected, and the similarity must be due to unknown local causes.

But it is among island that we find some of the most striking examples of the influence of locality on colour, generally in the direction of paler, but sometimes of darker and more brilliant hues, and often accompanied by an unusual increase of size. Thus in the Moluccas and New Guniea we have several Papilios (P. euchenor, P. ormenus, and P. tydeus) distinguished from their allies by a much paler colour, especially in the females, which are almost white. Many species of Danais (forming the subgenus Ideopsis) are also very pale. But the most curious are the Euplœas, which in the larger islands are usually of rich dark colours, while in the small islands of Banda, Ké, and Matabello at least three species not nearly related to each other (E. hoppferi, E. euripon, and E. assimilata) are all broadly banded or suffused with white, their allies in the larger islands being in each case very much darker. Again, in the genus Diadema, belonging to a distinct family, three species from the small Aru and Ké islands (D. deois, D. hewitsonii, and D. polymena) are all more conspicuously white-marked that their representatives in the larger islands. In the beautiful genus Cethosia, a species from the small island of Waigiou (C. cyrene) is the whitest of the genus. Prothoë is represented by a blue species in the continental island of Java, while those inhabiting the ancient insular groups of the Moluccas and New Guinea are all pale yellow or white. The genus Drusilla, almost confined to these islands, comprises many species which are all very pale; while in the small island of Waigiou is found a very distinct genus, Hyantis, which, though differing completely in the neuration of the wings, has exactly the same pale colours and large ocellated spots as Drusilla.

Equally remarkable is the increase of size in some islands. The small island of Amboina produces larger butterflies than any of the much larger islands which surround it. This is the case with at least a dozen butterflies belonging to many distinct genera,[1] so that it is impossible to attribute the fact to other than some local influence. In Celebes, as I have elsewhere pointed out,[2] we have a peculiar form of wing and much larger size running through a whole series of distinct butterflies; and this seems to take the place of any speciality in colour.

In a very small collection of insects recently brought from Duke-of-York island (situated between New Britain and New Ireland)are several of remarkably white or pale coloration. A species of Euplæa is the whitest of all known species of that extensive genus; while a beautiful diurnal moth is much whiter than its ally in the larger island of New

Guinea. There is also a magnificent longicorn beetle almost entirely of an ashy white colour.[3]

From the Fiji islands we have comparatively few butterflies; but there are several species of Diadema of unusually pale colours, some almost white.

The Philippine island seem to have the peculiarity of developing metallic colours. We find there at least three species of Euplæa[4] not closely related, and all of more intense metallic luster than their allies in other islands. Here also we have one of the large yellow Ornithopteræ (O. magellanus), whose hind wings glow with a intense opaline lustre not found in any other species of the entire group; and an Adolias[5] is larger and of more brilliant metallic colouring than any other species in the archipelago. In these islands also we find the extensive and wonderful genus of weevils (Pachyrhynchus), which in their brilliant metallic colouring surpass anything found in the whole eastern hemisphere, if not in the whole world.

[1] Ornithoptera priamus, O. helena, Papilio deiphobus, P. ulysses, P. gambrisius, P. codrus, Iphias Ieucippe, Euplæa prothoë, Hestia idea, Athyma jocaste, Diadema pandarus, Nymphalis pyrrhus, N. euryalus, Drusilla jairus.

[2] *Contributions to the Theory of Natural Selection,* pp. 168–173.

[3] These insects are described and figured in the *Proceedings*

of the Zoological Society for 1877, p. 139. Their names are Euplæa browni, Alcides aurora, and Batocera browni.

[4] Euplæa hewitsonii, E. diocletiana, E. lætifica.

[5] Adolias calliphorus.

In the Andaman islands, in the Bay of Bengal, there are a considerable number of peculiar species of butterflies differing slightly from those on the continent, and generally in the direction of paler or more conspicuous colouring. Thus two species of Papilio which on the continent have the tails black, in their Andaman representatives have them either red or white-tipped.[1] Another species[2] is richly blue-banded where its allies are black; while three species of distinct genera of Nymphalidæ[3] all differ from their allies on the continent in being of excessively pale colours as well as of somewhat larger size.

In Madagascar we have the very large and singularly white-spotted Papilio antenor, while species of three other genera[4] are very white or conspicuous as compared with their continental allies.

Passing to the West Indian islands and Central America (which latter country has formed a group of islands in very recent times), we have similar indications. One of the largest of the Papilios inhabits Jamaica,[5] while another, the largest of its group, is found in Mexico.[6] Cuba has two of the same genus whose colours are of surpassing brilliancy;[7] while the fine genus Clothilda—confined to the Antilles and Central America—is remarkable for its rich and showy colouring.

[1] Papilio rhodifer (near P. doubledayi), and Papilio charicles (near P. memnon).

[2] Papilio mayo.

[3] Euplæa andamanensis, Cethosia biblis, Cyrestis cocles.

[4] Danais nossima, Melanitis massoura, Diadema dexithea.

[5] Papilio homerus.

[6] P. daunus.

[7] P. gundlachianus, P. villiersi.

Persons who are not acquainted with the important structural differences that distinguish these various genera of butterflies can hardly realise the importance and the significance of such facts as I have now detailed. It may be well, therefore, to illustrate them by supposing parallel cases to occur among the Mammalia. We might have, for example, in Africa, the gnus, the elands, and the buffaloes, all coloured and marked like zebras, stripe for stripe over the whole body exactly corresponding. So the hares, marmots, and squirrels of Europe might be all red with black feet, while the corresponding species of Central Asia were all yellow with black heads. In North America we might have raccoons, squirrels, and opossums, in parti-coloured livery of white and black, so as exactly to resemble the skunk of the same country; while in South America they might be black with a yellow throat-patch, so as to resemble with equal closeness the tayra of the Brazilian forests. Were such resemblances to occur in anything like the number and with the wonderful accuracy of imitation met with among the Lepidoptera, they would certainly attract universal attention among naturalists, and would lead to the exhaustive study of the influence of local causes in producing such starling results.

One somewhat similar case does indeed occur among the Mammalia, two singular African animals, the Aard-wolf (Proteles) and the hyæna-dog (Lycaon), both strikingly resembling hyænas in their general form as well as in their spotted markings. Belonging as they all do to the Carnivora, though to three distinct families, it seems quite an analogous case to those we have imagined; but as the Aard-wolf and the hyæna-dog are both weak animals compared with the hyæna, the resemblance may be useful, and in that case would come under the head of mimicry. This seems the more probable because, as a rule, the colours of the Mammalia are protective, and are too little varied to allow of the influence of local causes producing any well-marked effects.

When we come to birds, however, the case is different, for although they do not exhibit such distinct marks of the influence of locality as to butterflies—probably because the causes which determine colour are in their case more complex—yet there are distinct indications of some effect of the kind, and we must devote some little time to their consideration.

One of the most curious cases is that of the parrots of the West Indian islands and Central America, several of which have white heads or foreheads, occurring in two distinct genera,[1] while none of the more numerous parrots of South America are so coloured. In the small island of Dominica we have a very large and richy-coloured parrot (Chrysotis augusta) corresponding to the large and richy-coloured buttefly (Papilio homerus) of Jamaica.

[1] Pionus albifrons and Chrysotis senilis (C. America), Chrysotis sallæi (Hayti).

The Andaman islands are equally remarkable, at least six of the peculiar birds differing from their continental allies in being much lighter, a sometimes with a large quantity of pure white in the plumage,[1] exactly corresponding to what occurs among the butterflies.

In the Philippines this is not so marked a feature; yet we have here the only known white-breasted kingcrow (Dicrurus mirabilis); the newly discovered Eurylæmus steerii, wholly white beneath; three species of Diceum, all white beneath; several species of Parus, largely white-spotted; while many of the pigeons have light ashy tints. The birds generally, however, have rich dark colours, similar to those which prevail among the butterflies.

In Celebes we have a swallow-shrike and a peculiar small crow allied to the jackdaw,[2] whiter than any of their allies in the surrounding islands, but otherwise the colours of the birds call for no special remark.

In Timor and Flores we have white-headed pigeons,[3] and a long-tailed flycatcher almost entirely white.[4]

In Duke-of-York island, east of New Guinea, we find that the four new species figured in the *Proceedings of the Zoological Society* for 1877 are *all* remarkable for the unusual quantity of white in their plumage. They consist of a flycatcher, a diceum, wood-swallow, and a ground-pigeon,[5] all equaling if not surpassing their nearest allies in whiteness, although some of these, from the Philippines, Moluccas, and Celebes, are sufficiently remarkable in this respect.

In the small Lord Howe's island we have the recently extinct white rail (Notornis alba), remarkably contrasting with its allies in the larger islands of New Zealand.

[1] Kittacincla albiventris, Geocichla albigularis, Sturnia andamanensis, Hyloterpe grisola var., Ianthænas palumboides, Osmotreron chloroptera.

[2] Artamus monachus, Corvus advena.

[3] Ptilopus cinctus, P. albocinctus.

[4] Tchitrea affinis, var.

[5] Monarcha verticalis, Diceum eximium, Artamus insignis, Phlogænas johannæ

We cannot, however, lay any stress on isolated examples of white colour, since these occur in most of the great continents; but where we find a series of species of distinct genera all differing from their continental allies in a whiter coloration, as in the Andaman islands, Duke-of-York island, and the West Indies, and, among butterflies, in the smaller Moluccas, the Andamans, and Madagascar, we cannot avoid the conclusion that in these insular localities some general cause is at work.

There are other cases, however, in which local influences seem to favour the production or preservation of intense crimson or a very dark coloration. Thus in the Moluccas and New Guinea alone we have bright red parrots belonging to two distinct families,[1] and which therefore most probably have been independently produced or preserved by some common cause. Here, too, and in Australia we have black parrots and pigeons;[2] and it is a most curious and suggestive fact that in another insular sub-region—that of Madagascar and the Mascarene islands—these same colours reappear in the same two groups.[3]

Sense-perception influenced by Colour of the Integuments

Some very curious physiological facts bearing upon the presence or absence of white colours in the higher animals have lately been adduced by Dr. Ogle.[4] It has been found that a coloured or dark pigment in the olfactory region of the nostrils is essential to perfect smell, and this pigment is rarely deficient except when the whole animal is pure white. In these cases the creature is almost without smell or taste. This, Dr. Ogle believes, explains the curious case of the pigs in Virginia adduced by Mr. Darwin, white pigs being killed by a poisonous root with does not affect black pigs. Mr. Darwin imputed this to a constitutional difference accompanying the dark colour, which rendered what was poisonous to the white-coloured animals quite innocuous to the black. Dr. Ogle, however, observes that there is no proof that the black pigs eat the root, and he believes the more probable explanation to be that it is distasteful to them; while the white pigs, being deficient in smell and taste, eat it and are killed. Analogous facts occur in several distinct families. White sheep are killed in the Tarentino by eating Hypericum crispum, while black sheep escape; white rhinoceroses are said to perish from eating Euphorbia candelabrum; and white horses are said to suffer from poisonous food where coloured ones escape. Now it is very improbable that a constitutional immunity from poisoning by so many distinct plants should, in the case of such widely different animals, be always correlated with the same difference of colour; but the facts are readily understood if the senses of smell and taste are dependent on the presence of a pigment which is deficient in wholly white animals. The explanation, has however, been carried a step further, by experiments showing that the absorption of odours by dead matter, such as clothing, is greatly affected by colour, black being the most powerful absorbent, then blue, red, yellow, and lastly white. We have here a physical cause for the sense-inferiority of totally white animals which may account for their rarity in nature, for few, if any, wild animals are wholly white. The head, the face, or at least the muzzle or the nose, are generally black; the ears and eyes are also often black; and there is reason to believe that dark pigment is essential to good hearing, as it certainly is to perfect vision. We can therefore understand why white cats with blue eyes are so often deaf, a peculiarity we notice more readily than their deficiency of smell or taste.

[1] Lorius, Eos (Trichoglossidæ), Eclectus (Palæornithidæ).

[2] Microglossus, Calyptorhynchus, Turacæna.

[3] Coracopsis, Alectrænas.

[4] *Medico-Chirurgical Transactions,* vol. liii. (1870).

If, then, the prevalence of white coloration is generally

associated with some deficiency in the acuteness of the most important senses, this colour becomes doubly dangerous, for it not only reason its possessor more conspicuous to its enemies, but a the same time makes it less ready in detecting the presence of danger. Hence, perhaps, the reason why white appears more frequently in islands, where competition is less severe and enemies less numerous and varied. Hence, also, a reason why *albinoism*, although freely occurring in captivity, never maintains itself in a wild state, while *melanism* does. The peculiarity of some islands in having all their inhabitants of dusky colours (as the Galapagos) may also perhaps be explained on the same principles, for poisonous fruits may there abound which weed out all white or light-coloured varieties owing to their deficiency of smell and taste. We can hardly believe, however, that this would apply to white-coloured butterflies; and this may be a reason why the effect of an insular habitat is more marked in these insects than in birds or mammals.[1]

It is even possible that this relation of sense-acuteness with colour may have had some influence on the development of the higher human races. If light tints of the skin were generally accompanied by some deficiency in the senses of smell, hearing, and vision, the white could never complete with the darker races so long as man was in a very low or savage condition, and wholly dependent for existence on the acuteness of his senses. But as the mental faculties became more fully developed and more important to his welfare than mere sense-acuteness, the lighter tints of skin and hair and eyes would cease to be disadvantageous whenever they were accompanied by superior brain-power. Such variations would then be preserved; and thus may have arisen the Xanthochroic race of mankind, in which we find a high development of intellect accompanied by a slight deficiency in the acuteness of the senses as compared with the darker forms.

[1] In *Darwinism*, pp. 229, 230, I have suggested an explanation of most of the facts of colour in islands as due to the lesser need of protection.

Summary on Colour-development in Animals

Let us now sum up the conclusions at which we have arrived as to the various modes in which colour is produced or modified in the animal kingdom.

The various causes of colour in the animal world are, molecular and chemical change of the substance of their integuments, or the action on it of heat, light, or moisture. It is also produced by interference of light in superposed transparent lamellæ, or by excessively fine surface-striæ.

These elementary conditions for the production of colour are found everywhere in the surface-structures of animals, so that its presence must be looked upon as normal, its absence as exceptional.

Colours are fixed or modified in animal by natural selection for various purposes; obscure or imitative colours for concealment; gaudy colours as a warning; and special markings, either for easy recognition by strayed individuals females, or young, or to divert attack from a vital part, as in the large brilliantly-marked wings of some butterflies and moths.

[1] In *Darwinism,* pp. 229, 230, I have suggested an explanation of most of the facts of colour in islands as due to the lesser need of protection.

Colours are produced or intensified by processes of development, either where the integument or its appendages undergo great extension or modification, or where there is a surplus of vital energy, as in male animals generally, and more especially at the breeding season.

Colours are also more or less influenced by a variety of causes, such as the nature of the food, the photographic or physiological action of light, and also by some unknown local action, probably dependent on chemical peculiarities in the soil or vegetation.

These various causes have acted and reacted in a variety of ways, and have been modified by conditions dependent on age or on sex, on competition with new forms, or on geographical or climatic changes. In so complex a subject, for which experiment and systematic inquiry have done so little, we cannot expect to explain every individual case, or solve every difficult; but it is believed that all the great features of animal coloration and many of the details becomes explicable on the principles we have endeavoured to lay down.

It will perhaps be considered presumptuous to put forth this sketch of the subject of colour in animals as a substitute for one of Mr. Darwin's most highly elaborated theories—that of voluntary or perceptive sexual selection; yet I venture of think that it is more in accordance with the whole of the facts, and with the theory of natural selection itself; and I would ask such of my readers as may be sufficiently interested in the subjects, to read again chapters xi. to xvi. of the *Descent of Man,* and consider the whole subject from the point of view here laid down. The explanation of almost all the ornaments and colours of birds and insects as having been produced by the perceptions and choice of the females, has, I believe, staggered many evolutionists, but has been provisionally accepted because it was the only theory that even attempted to explain the facts. It may perhaps be a relief to some of them, as it has been to myself, to find that the phenomena can be conceived as dependent on the general laws of development, and on the action of "natural selection," which theory will, I venture to think, be relieved from an abnormal excrescence and gain additional vitality by the adoption of the views here imperfectly set forth.[1]

[1] These views have been restated and enforced by much fresh illustration and argument in *Darwinism,* chap. x.

Although we have arrived at the conclusion that tropical light and heat can in no sense be considered as the cause of colour, there remains to be explained the undoubted fact that all the more intense and gorgeous tints are manifested by the animal life of the tropics; while in some groups, such as butterflies and birds, there is a marked preponderance of highly-coloured species. This is probably due to a variety of causes, some of which we can indicate, while others remain to be discovered. The luxuriant vegetation of the tropics throughout the entire year affords so much concealment that colour may there be safely developed to a much greater extent than in climates where the trees are bare in winter, during which season the struggle for existence is most severe, and even the slightest disadvantage may prove fatal. Equally important, probably, has been the permanence of favourable conditions in the tropics, allowing certain groups to continue dominant for long periods, and thus to carry out in on unbroken line whatever developments of plumage or colour may once have acquired an ascendancy. Changes of climatal conditions, and pre-eminently the glacial epoch, probably led to the extinction of a host of highly-developed and finely-coloured insects and birds in temperate zones, just as we know that it led to the extinction of the larger and more powerful mammalia which formerly characterised the temperate zone in both hemispheres; and this view is supported by the fact that it is amongst those groups only which are now exclusively tropical that all the more extraordinary developments of ornament and colour are found. The obscure local causes of colour to which we have referred will also have acted most efficiently in regions where the climatal condition remained constant, and where migration was unnecessary; while whatever direct effect may be produced by light or heat will necessarily have acted more powerfully within the tropics. And lastly, all these causes have been in action over an actually greater area in tropical than in temperate zones; while, estimated potentially, in proportion to their life-sustaining power, the lands which enjoy a practically tropical climate (extending as they do considerably beyond the geographical tropics) are very much larger than the temperate regions of the earth.

Combining the effects of all these various causes, we are quite able to understand the superiority of the tropical parts of the globe, not only in the abundance and variety of their forms of life, but also as regards the ornamental appendages and vivid coloration which these forms present.

VI

THE COLOURS OF PLANTS AND THE ORIGIN OF THE COLOUR-SENSE

Source of Colouring Matter in Plants—Protective Coloration and Mimicry in Plants—Attractive Colours of Fruits—Protective Colours of Fruits—Attractive Colours of Flowers—Attractive Odours in Flowers—Attractive Grouping of Flowers—Why Alpine Flowers are so beautiful—Why Allied Species of Flowers differ in size and beauty—Absence of Colour in Wind-fertilised Flowers—The same Theory of Colour applicable to Animals and Plants—Relation of the Colours of Flowers and their Geographical Distribution—Recent Views as to Direct Action of Light on the Colours of Flowers and Fruits—On the Origin of the Colour-sense: Supposed increase of Colour-perception within the Historical Period—Concluding Remarks on the Colour-sense.

Source of Colouring Matter in Plants

THE colouring of plants is neither so varied nor so complex as that of animals, and its explanation accordingly offers fewer difficulties. The colours of foliage are, comparatively, little varied, and can be traced in almost all cases to a special pigment termed chlorophyll, to which is due the general green colaour of leaves; but the recent investigations of Mr. Sorby and others have shown that chlorophyll is not a simple green pigment, but that it really consists of at least seven distinct substances, varying in colour from blue to yellow and orange. These differ in their proportions in the chlorophyll of different plants; they have different chemical reactions; they are differently affected by light; and they give distinct spectra. Mr. Sorby further states that scores of different colouring matters are found in the leaves and flowers of plants, to some of which appropriate names have been given, as erythrophyll, which is red; and phaiophyll, which is brown; and many of these differ greatly from each other in their chemical composition. These inquiries are at present in their infancy, but as the original term chlorophyll seems scarcely applicable under the present aspect of the subject, it would perhaps be better to introduce the analogous word chromophyll as a general term for the colouring matters of the vegetable kingdom.

Light has a much more decided action on plants than

on animals. The green colour of leaves is almost wholly dependent on it; and although some flowers will become fully coloured in the dark, others are decidedly affected by the absence of light, even when the foliage is fully exposed to it. Looking therefore at the numerous colouring matters which are developed in the tissues of plants, the sensitiveness of these pigments to light, the changes they undergo during growth and development, and the facility with which new chemical combinations are effected by the physiological processes of plants as shown by the endless variety in the chemical constitution of vegetable products, we have no difficulty in comprehending the general causes which aid in producing the colours of the vegetable world, or the extreme variability of those colours. We may therefore here confine ourselves to an inquiry into the various uses of colour in the economy of plants, and this will generally enable us to understand how it has become fixed and specialised in the several genera and species of the vegetable kingdom.

Protective Coloration and Mimicry in Plants

In animals, as we have seen, colour is greatly influenced by the need of protection form, or of warning to, their numerous enemies, and by the necessity for identification and easy recognition. Plants rarely need to be concealed, and obtain protection either by their spines, their hardness, their hairy covering, or their poisonous secretions. A very few cases of what seem to be true protective colouring do, however, exist, the most remarkable being that of the "stone mesembryanthemum" of the Cape of Good Hope, which, in form and colour, closely resembles the stones among which it grows; and Dr. Burchell, who first discovered it, believes that the juicy little plant thus generally escapes the notice of cattle and wild herbivorous animals. Mr. J. P. Mansel Weale also noticed that many plants growing in the stony Karoo have their tuberous roots above the soil, and these so perfectly resemble the stones among which they grow that, when not in leaf, it is almost impossible to distinguish them (*Nature,* vol. iii. p. 507). A few cases of what seems to be protective mimicry have also been noted, the most curious being that of three very rare British fungi, found by Mr. Worthington Smith, each in company with common species which they so closely resembled that only a minute examination could detect the difference. One of the common species is stated in botanical works to be "bitter and nauseous," so that it is not improbable that the rare kind may escape being eaten by being mistaken for an uneatable species, though itself palatable. Mr. Mansel Weale also mentions a labiate plant, the Ajuga ophrydis, of South Africa, as strikingly resembling an orchid. This may be a means of attracting insects to fertilise the flower in the absence of sufficient nectar or other attraction in the flower itself; and the supposition is rendered more probable by this being the only species of the genus Ajuga in South Africa. Many other cases of resemblances between very distinct plants have been noticed—as that of some Euphorbias to Cacti; but these very rarely inhabit the same country or locality, and it has not been proved that there is in any of these cases the amount of inter-relation between the species which is the essential feature of the protective "mimicry" that occurs in the animal world.

The different colours exhibited by the foliage of plants and the changes it undergoes during growth and decay, appear to be due to the general laws already sketched out, and to have little if any relation to the special requirements of each species. But flowers and fruits exhibit definite and well-pronounced tints, often varying from species to species, and more or less clearly related to the habits and functions of the plant. With the few exceptions already pointed out, these may be generally classed as *attractive* colours.

Attractive Colours of Fruits

The seeds of plants require to be dispersed so as to reach places favourable for germination and growth. Some are very minute and are carried abroad by the wind, or they are violently expelled and scattered by the bursting of the containing capsules. Others are downy or winged, and are carried long distances by the gentlest breeze, or they are hooked and stick to the fur of animals. But there is a large class of seeds which cannot be dispersed in either of these ways, and they are mostly contained in eatable fruits. These fruits are devoured by birds or beasts, and the hard seeds pass through their stomachs undigested, and, owing probably to the gentle heat and moisture to which they have been subjected, in a condition highly favourable for germination. The dry fruits or capsules containing the first two classes of seeds are rarely, if ever, conspicuously coloured, whereas the eatable fruits almost invariably acquire a bright colour as they ripen, while at the same time they become soft and often full of agreeable juices. Our *red* haws and hips, our *black* elderberries, our *blue* sloes and whortleberries, our *white* mistletoe and snowberry, and our *orange* sea-buckthorn, are examples of the colour-sign of edibility; and in every part of the world the same phenomenon is found. Many such fruits are poisonous to man and to some animals, but they are harmless to others; and there is probably nowhere a brightly coloured pulpy fruit which does not serve as food for some species of bird or mammal.

Protective Colours of Fruits

The nuts and other hard fruits of large forest-trees, though often greedily eaten by animals, are not rendered attractive to them by colour, because they are not intended to be eaten. This is evident, for the part eaten in these cases is the seed itself, the destruction of which must certainly be injurious to the species. Mr. Grant Allen, in his ingenious work on *Physiological Æsthetics*, well observes that the colours of all such fruits are protective—green when on the tree, and thus hardly visible among the foliage, but turning brown as they ripen and fall on the ground, as filberts, chestnuts, walnuts, beechnuts, and many others. It is also to be noted that many of these are specially though imperfectly protected, some by a prickly coat as in the chestnuts, or by a nauseous covering as in the walnut; and the reason why the protection is not carried farther is probably because it is not needed, these trees producing such vast quantities of fruit, that, however many are eaten, more than enough are always left to produce young plants. In the case of the attractively coloured fruits, it is curious to observe how the *seeds* are always of such a nature as to escape destruction when the fruit itself is eaten. They are generally very small and comparatively hard, as in the strawberry, gooseberry, and fig; if a little larger, as in the grape, they are still harder and less eatable; in the fruit of the rose or (hip) they are disagreeably hairy; in the orange tribe excessively bitter. When the seeds are larger, softer, and more eatable, they are protected by an excessively hard and stony covering, as in the plum and peach tribe; or they are enclosed in a tough horny core, as with crabs and apples. These last are much eaten by swine, and are probably crushed and swallowed without bruising the core or the seeds, which pass through their bodies undigested. These fruits may also be swallowed by some of the larger frugivorous birds, just as nutmegs are swallowed by pigeons for the sake of the mace which encloses the nut, and which by its brilliant red colour is an attraction as soon as the fruit has split open, which it does upon the tree.

There is, however, one curious case of an attractively coloured seed which has no soft eatable covering. The Abrus precatoria, or "rosary bean," is a leguminous shrub or small tree growing in many tropical countries, whose pods curl up and split open on the tree, displaying the brilliant red seeds within. It is very hard and glossy, and is said to be, as no doubt it is, "very indigestible." It may be that birds, attracted bay the bright colour of the seeds, swallow them, and that they pass through their bodies undigested, and so get dispersed. If so it would be a case among plants analogous to mimicry among animals—ana appearance of edibility put on to deceive birds for the plant's benefit. Perhaps it succeeds only with young and inexperienced birds, and it would have a better chance of success, because such deceptive appearance are very rare among plants.

The smaller plants whose seeds simply drop upon the ground, as in the grasses, sedges, composites, umbelliferæ, etc., always have dry and obscurely coloured capsules and small brown seeds. Others whose seeds are ejected by the bursting open of their capsules, as with the oxalis and many of the caryophyllaceæ, scrophulariaceæ, etc., have their seeds very small and rarely or never edible.

It is to be remarked that most of the plants whose large-seeded nuts cannot be eaten without destroying their germinating power—as the oaks, beeches, and chestnuts—are trees of large size which bear great quantities of fruit, and that they are long lived and have a wide geographical range. They belong to what are called dominant groups, and are thus able to endure having a large proportion of their seeds destroyed with impunity. It is a suggestive fact that they are among the most ancient of known dicotyledonous plants—oaks and beeches going back to the Cretaceous period with little change of type, so that it is not improbable that they may be older than any fruit-eating mammal adapted to feed upon their fruits. The *attractive* coloured fruits on the other hand, having so many special adaptations to dispersal by birds and mammals, are probably of more recent origin.[1] The apple and plum tribes are not known earlier than the Miocene period; and although the record of extinct vegetable life is extremely imperfect, and the real antiquity of these groups is no doubt very much greater, it is not improbable that the *comparative* antiquity of the fruit-bearing and nut-bearing trees may remain unchanged by further discoveries, as has almost always happened as regards the comparative antiquity of animal groups.

[1] I owe this remark to Mr. Grant Allen, author of *Physiological æsthetics.*

Attractive Colours of Flowers

The colours of flowers serve to render them visible and recognisable by insects, which are attracted by secretions of nectar or pollen. During their visits for the purpose of obtaining these products, insects involuntarily carry the pollen of one flower to the stigma of another, and thus effect cross-fertilisation, which, as Mr. Darwin was the first to demonstrate, immensely increases the vigour and fertility of the next generation of plants. This discovery has led to the careful examination of great numbers of flowers, and the result has been that the most wonderful and complex arrangements have been found to exist, all having for their object to secure that flowers shall not be self-fertilised perpetually, but that pollen shall be carried, either constantly or occasionally, from the flowers of one plant to those of another. Mr. Darwin himself first worked out the details in orchids, primulas, and some other groups, and hardly less curious phenomena have since been found to occur even among some of the most regularly-formed flowers. The arrangement, length, and position of all the parts of the flower is now found to have a purpose, and not the least remarkable portion of the phenomenon is the great variety of ways in which the same result is obtained. After the discoveries with regard to orchids, it was to be expected that the irregular, tubular, and spurred flowers should present various curious adaptations for fertilisation by insect-agency. But even among the open, cup-shaped, and quite regular flowers, in which it seemed inevitable that the pollen must fall on the stigma and produce constant self-fertilisation, it has been found that this is often prevented by a physiological variation—the anthers constantly emitting their pollen either a little earlier or a little later than the stigmas of the same flower, or of other flowers on the same plant, were in the best state to receive it; and as individual plants in different stations, soils, and aspects differ somewhat in the time of flowering, the pollen of one plant would often be conveyed by insects to the stigmas of some other plant in a condition to be fertilised by it. This mode of securing cross-fertilisation seems so simple and easy that we can hardly help wondering why it did not always come into action, and so obviate the necessity for those elaborate, varied, and highly complex contrivances found perhaps in the majority of coloured flowers. The answers to this of course is, that *variation* sometimes occurred most freely in one part of a plant's organisation and sometimes in another, and that the benefit of cross-fertilisation was so great that *any* variation that favoured it was preserved, and then formed the starting-point of a whole series of further variations, resulting in those marvellous adaptations for insect fertilisation which have given much of their variety, elegance, and beauty to the floral world. For details of these adaptations we must refer the reader to the works of Darwin, Lubbock, Herman Müller, and others. We have here only to deal with the part played by colour, and by those floral structures in which colour is most displayed.

Attractive Odours in Flowers

The sweet odours of flowers, like their colours, seem to have been developed as an attraction or guide to insect fertilisers, and the two phenomena are often complementary to each other. Thus, many inconspicuous flowers, like the mignonette and the sweet-violet, can be distinguished by their odours before they attract the eye, and this may often prevent their being passed unnoticed; while very showy flowers, and especially those with variegated or spotted petals, are seldom sweet. White, or very pale flowers, on the other hand, are often excessively sweet, as exemplified by the jasmine and clematis; and many of these are only scented at night, as is strikingly the case with the night-smelling stock, our butterfly orchis (Habenaria chlorantha), the greenish-yellow Daphne pontica, and many others. These white flowers are mostly fertilised by night-flying moths, and those which reserve their odours for the evening probably escape the visits of diurnal insects, which would consume their nectar without effecting fertilisation. The absence of odour in showy flowers, and its preponderance among those that are white, may be shown to be a fact by an examination of the lists in Mr. Mongredien's work on hardy trees and shrubs.[1] He gives a list of about 160 species with showy flowers, and another list of sixty species with fragrant flowers; but only twenty of these latter are included among the showy species, and these are almost all white flowered. Of the sixty species with fragrant flowers, more than forty are white, and a number of others have greenish, yellowish, or dusky and inconspicuous flowers. The relation of white flowers to nocturnal insects is also well shown by those which, like the evening primroses, only open their large white blossoms after sunset, while most of the yellow species remain open all day. The red Martagon lily has been observed by Mr. Herman Müller to be fertilised by the humming-bird hawk moth, which flies in the morning and afternoon, when the colours of this flower, exposed to the nearly horizontal rays of the sun, glow with brilliancy, and when it also becomes very sweet-scented.

[1] *Trees and Shrubs for English Plantations,* by Augustus Mongredien. Murray, 1870.

Attractive Grouping of Flowers

To the same need of conspicuousness the combination of so many individually small flowers into heads and bunches is probably due, producing such broad masses as those of the elder, the guilder-rose, and most of the Umbelliferæ or such elegant bunches as those of the lilac, laburnum, horse chestnut, and wisteria. In other cases minute flowers are gathered into dense heads, as with Globularia, Jasione, clover, and all the Compositæ; and among the latter the outer flowers are often developed into a ray, as in the sunflowers, the daisies, and the asters, forming a starlike compound flower, which is itself often produced in immense profusion.

Why Alpine Flowers are so beautiful

The beauty of Alpine flowers is almost proverbial. It consists either in the increased size of the individual flowers as compared with the whole plant, in increased intensity of colour, or in the massing of small flowers into dense cushions of bright colour; and it is only in the higher Alps, above the limit of forests and upwards towards the perpetual snow-line, that these characteristics are fully exhibited. This effort at conspicuousness under adverse circumstances may be traced to the comparative scarcity of winged insects in the higher regions, and to the necessity for attracting them from a distance. Amid the vast slopes of débris and the huge masses of rock so prevalent in higher mountain regions, patches of intense colour can alone make themselves visible and serve to attract the wandering butterfly from the valleys. Mr. Herman Müller's careful observations have shown that in the higher Alps bees and most other groups of winged insects are almost wanting, while butterflies are tolerably abundant; and he has discovered that in a number of cases where are lowland flower is adapted to be fertilised by bees, its Alpine ally has had its structure so modified as to be adapted for fertilisation only by butterflies.[1] But bees are always (in the temperate zone) far more abundant than butterflies, and this will be another reason why flowers specially adapted to be fertilised by the latter should be rendered unusually conspicuous. We find, accordingly, the yellow primroses and cowslips of the plains replaced by pink and magenta-coloured Alpine species; the straggling wild pinks of the lowlands by the masses of large flowers in such mountain species as Dianthus alpinus and D. glacialis; the saxifrages of the high Alps with bunches of flowers a foot long as in Saxifraga longifolia and S. cotyledon, or forming spreading masses of flowers as in S. oppositifolia; while the soapworts, silenes, and louseworts are equally superior to the allied species of the plains.

[1] *Nature,* vol. xi. pp. 32, 110.

Why Allied Species of Flowers differ in Size and Beauty

Again, Dr. Müller has discovered that when there are showy and inconspicuous species in the same genus of plants, there is often a corresponding difference of structure, those with large and showy flowers being quite incapable of self-fertilisation, and thus depending for their very existence on the visits of insects, while the others are able to fertilise themselves should insects fail to visit them. We have examples of this difference in Malva sylvestris, Epilobium angustifolium, Polygonum bistorta, and

Geranium pratense—which have all large or showy flowers, and must be fertilised by insects—as compared with Malva rotundifolia, Epilobium parviflorum, Polygonum aviculare, and Geranium pusillum, which have small or inconspicuous flowers, and are so constructed that if insects should not visit them they are able to fertilise themselves.[2]

[2] *Ib.*, vol. ix. p. 164.

Absence of Colour in Wind-fertilised Flowers

As supplementing these curious facts, showing the relation of colour in flowers to the need of the visits of insects to fertilise them, we have the remarkable, and, on any other theory, utterly inexplicable circumstance that in all the numerous cases in which plants are fertilised by the agency of the wind they never have specially coloured floral envelopes. Such are our pines, oaks, poplars, willows, beeches, and hazel, our nettles, grasses, sedges, and many others. In some of these the male flowers are very conspicuous, as the catkins of the willows, and these secrete honey and attract numerous insects at a season when there are few other flowers, and thus secure cross-fertilisation. Sedges and grasses are also occasionally visited by insects.

The same Theory of Colour applicable to Animals and Plants

It may be thought that this absence of colour where it is not wanted is opposed to the view maintained in the earlier part of the preceding chapter, that colour is normal and is constantly tending to appear in natural objects. It must be remembered, however, that the green colour of foliage, due to chlorophyll, prevails throughout the greater part of the vegetable kingdom, and has, almost certainly, persisted through long geological periods. It has thus acquired a fixity of character which cannot be readily disturbed; and, as a matter of fact, we find that colour rarely appears in plants except in association with a considerable modification of leaf-texture, such as occurs in the petals and coloured sepals of flowers. Wind-fertilised plants never have such specially organised floral envelopes, and, in most cases, are entirely without a calyx or corolla. The connection between modification of leaf-structure and colour is further seen in the greater amount and variety of colour in irregular than in regular flowers. The latter, which are least modified, have generally uniform or but slightly varied colours, while the former, which have undergone great modification, present an immense range of colour and marking, culminating in the spotted and variegated flowers of such groups as the Scrophularineæ and Orchideæ. The same laws as to the conditions of a maximum production of colour are thus found to obtain both in plants and animals.

Relation of the Colours of Flowers and their Geographical Distribution

The adaptation of flowers to be fertilised by insects—often to such an extent that the very existence of the species depends upon it—has had an important influence on the distribution of plants and the general aspects of vegetation. The seeds of a particular species may be carried to another country, may find there a suitable soil and climate, may grow and produce flowers; but if the insect which alone can fertilise it should not inhabit that country, the plant cannot maintain itself, however frequently it may be introduced or however vigorously it may grow. Thus may probably be explained the poverty in flowering-plants and the great preponderance of ferns that distinguishes many oceanic islands, as well as the deficiency of gaily-coloured flowers in others. New Zealand is, in proportion to its total number of flowering-plants, exceedingly poor in handsome flowers, and it is correspondingly poor in insects, especially in bees and butterflies, the two groups which so greatly aid in fertilisation. In both these aspects it contrasts strongly with Southern Australia and Tasmania in the same latitudes, where there is a profusion of gaily-coloured flowers and an exceeding rich insect-fauna. Another case is presented by the Galapagos islands, which, though situated on the equator off the west coast of South America, and with a tolerably luxuriant vegetation in the damp mountain zone, yet produce hardly a single conspicuously-coloured flower; and this is correlated with, and no doubt dependent on, an extreme poverty of insect life, not one bee and only a single butterfly having been found there.

Again, there is reason to believe that some portion of the large size and corresponding showiness of tropical flowers is due to their being fertilised by very large insects and even by birds. Tropical sphinx-moths often have their probosces nine or ten inches long, and we find flowers whose tubes or spurs reach about the same length, while the giant bees, and the numerous flower-sucking birds, aid in the fertilisation of flowers whose corollas or stamens are proportionately large.

Recent Views as to direct Action of Light on the Colours of Flowers and Fruits

The theory that the brilliant colours of flowers and fruits are due to the direct action of light has been supported by a recent writer by examples taken from the arctic instead of from the tropical flora. In the arctic regions vegetation is excessively rapid during the short summer, and this is held to be due to the continuous action of light throughout the long summer days. "The farther we advance towards the north the more the leaves of plants increase in size, as if to absorb a greater proportion of the solar rays. M. Grisebach says that during a journey in Norway he observed that the majority of deciduous trees had already, at the 60th degree of latitude, larger leaves than in Germany, while M. Ch. Martins has made a similar observation as regards the leguminous plants cultivated in Lapland."[1] The same writer goes on to say that all the seeds of cultivated plants acquire a deeper colour the farther north they are grown, white haricots becoming brown or black, and white wheat becoming brown, while the green colour of all vegetation becomes more intense. The flowers also are similarly changed: those which are white or yellow in central Europe becoming red or orange in Norway. This is what occurs in the Alpine flora, and the cause is said to be the same in both—the greater intensity of the sunlight. In the one the light is more persistent, in the other more intense because it traverses a thinner atmosphere.

Admitting the facts as above stated to be in themselves correct, they do no by any means establish the theory founded on them; and it is curious that Grisebach, who has been quoted by this writer for the fact of the increased size of the foliage, gives a totally different explanation of the more vivid colours of arctic flowers. He says: "We see flowers become larger and more richly coloured in proportion as, by the increasing length of winter, insects become rarer, and their co-operation in the act of fecundation is exposed to more uncertain chances" (*Vegetation du Globe,* vol. i. p. 61—French translation). This is the theory here adopted to explain the colours of Alpine plants, and we believe there are many facts that will show it to be the preferable one. The statement that the white and yellow flowers of temperate Europe become red or golden in the arctic regions must, we think, be incorrect. By roughly tabulating the colours of the plants given by Sir Joseph Hooker[2] as permanently arctic, we find among fifty species with more or less conspicuous flowers, twenty-five white, twelve yellow, eight purple or blue, three lilac, and two red or pink, showing a very similar proportion of white and yellow flowers to what obtains farther south.

[1] *Revue des Deux Mondes,* 1877—"La Vegetation dans les hautes Latitudes," par M. Tisserand.

[2] "On the Distribution of Arctic Plants," *Linn. Trans.* vol. xxiii. (1862).

We have, however, a remarkable flora in the southern hemisphere, which affords a crucial test of the theory of greater intensity of light being the direct cause of brilliantly-coloured flowers. The Auckland and Campbell's islands, south of New Zealand, are in the same latitude as the middle and the south of England, and the summer days are therefore no longer than with us. The climate, though cold, is very uniform, and the weather "very rainy and stormy." It is evident, then, that there can be no excess of sunshine above what we possess, yet in a very limited flora there are a number of flowers which—Sir Joseph Hooker states—are equal in brilliancy to those of the arctic flora. These consist of brilliant gentians, handsome veronicas, large and magnificent Compositæ with purple flowers, bright ranunculi, showy Umbelliferæ, and the golden-flowered Chrysobactron Rossii, one of the finest of the Asphodeleæ.[1] All these find plants, it must be remembered, are peculiar to these islands, and have therefore been developed under the climatal conditions that prevail there; and as we have no reason to suppose that these conditions have undergone any recent change, we may be quite sure that an excess of light has had nothing to do with the development of these exceptionally bright and handsome flowers. Unfortunately we have no information as to the insects of these islands, but from their scarcity in New Zealand we can hardly expect them to be otherwise than very scarce. There are, however, two species of honey-sucking birds (Prosthemadera and Anthornis), as well as a small warbler (Myiomoira); and we may be pretty sure that the former at least visit these large and handsome flowers, and so effect their fertilisation. The most abundant tree on the islands is a species of Metrosideros, and we know that trees of this genus are common in the Pacific islands, where they are almost certainly fertilised by the same family of Meliphagidæ or honey-sucking birds.

[1] Coloured figures of all these plants are given in the *Flora Antarctica,* vol. i.

I have now concluded this sketch of the general phenomena of colour in the organic world. I have shown reasons for believing that its presence, in some of its infinitely-varied hues, is more probable than its absence, and that variation of colour is an almost necessary concomitant of variation of structure, of development, and of growth. It has also been shown how colour has been appropriated and modified both in the animal and vegetable worlds for the advantage of the species in a great variety of ways, and that there is no need to call in the aid of any other laws than those of organic development and "natural selection" to explain its countless modifications. From the point of view here taken, it seems at once improbable and unnecessary that the lower animals should have the same delicate appreciation of the infinite variety and beauty, of the delicate contrasts and subtle harmonies of colour, which are possessed, by the more intellectual races of mankind, since even the lower human races do not possess it. All that seems required in the case of animals is a perception of *distinctness* or *contrast* of colours; and the dislike of so many creatures to scarlet may perhaps be due to the rarity of that colour in nature, and to the glaring contrast it offers to the sober greens and browns which form the general clothing of the earth's surface, though it may also have a direct irritating effect on the retina.

The general view of the subject now given must convince us that, so far from colour being—as it has sometimes been thought to be—unimportant, it is intimately connected with the very existence of a large proportion of the species of the animal and vegetable worlds. The gay colours of the butterfly and of the Alpine flower which it unconsciously fertilises while seeking for its secreted honey, are each beneficial to its possessor, and have been shown to be dependent on the same class of general laws as those which have determined the form, the structure, and the habits of every living thing. The complex laws and unexpected relations which we have seen to be involved in the production of the special colours of flower, bird, and insect must give them an additional interest for every thoughtful mind; while the knowledge that, in all probability, each style of coloration, and sometimes the smallest details, have a meaning and a use must add a new charm to the study of nature.

ON THE ORIGIN OF THE COLOUR-SENSE

Throughout the preceding discussion we have accepted the subjective phenomena of colour—that is, our perception of varied hues and the mental emotions excited by them—as ultimate facts needing no explanation. Yet they present certain features well worthy of attention, a brief consideration of which will form a fitting sequel to the present essay.

The perception of colour seems, to the present writer, the most wonderful and the most mysterious of our sensations. Its extreme diversities and exquisite beauties seem out of proportion to the causes that are supposed to have produced them, or the physical needs to which they minister. If we look at pure tints of red, green, blue, and yellow, they appear so absolutely contrasted and unlike each other, that it is almost impossible to believe (what we nevertheless know to be the fact) that the rays of light producing theses very distinct sensations differ only in wave-length and rate of vibration, and that there is from one to the other a continuous series and gradation of such vibrating waves. The positive diversity we see in them must then depend upon special adaptations in ourselves; and the question arises, For what purpose have our visual organs and mental perceptions become so highly specialised in this respect?

When the sense of sight was first developed in the animal kingdom, we can hardly doubt that what was perceived was light only, and its more or less complete withdrawal. As the sense became perfected, more delicate gradations of light and shade would be perceived, and there seems no reason why a visual capacity might not have been developed as perfect as our own, or even more so in respect of light and shade, but entirely insensible to differences of colour, except in so far as these implied a difference in the quantity of light. The world would in that case appear somewhat as we see it in good stereoscopic photographs; and we all know how exquisitely beautiful such pictures are, and how completely they give us all requisite information as to form, surface-texture, solidity, and distance, and even to some extent as to colour, for almost all colours are distinguishable in a photograph by some differences of tint, and it is quite conceivable that visual organs might exist which would differentiate what we term colour by delicate gradations of some one characteristic neutral tint. Now such a capacity of vision would be simple as compared with that which we actually possess, which, besides distinguishing infinite gradations of the *quantity* of light, distinguishes also, by a totally distinct set of sensations, gradations of *quality*, as determined by differences of wave-lengths or rate of vibration. At what grade in animal development this new and more complex sense first began to appear we have no means of determining. The fact that the higher vertebrates, and even some insects, distinguish what are to us diversities of colour by no means proves that their *sensations* of colour bear any resemblance whatever to ours. An insect's capacity to distinguish red from blue or yellow may be (and probably is) due to perceptions of a totally distinct nature, and quite unaccompanied by any of that sense of enjoyment or even of radial distinctness which pure colours excite in us. Mammalia and birds, whose structure and emotions are so similar to our own, do probably receive somewhat similar impressions of colour; but we have no evidence to show that they experience pleasurable emotions from colour itself when not associated with the satisfaction of their wants or the gratification of their passions.

The primary necessity which led to the development of the sense of colour was probably the need of distinguishing objects much alike in form and size, but differing in important properties, such as ripe and unripe, or eatable and poisonous fruits, flowers with honey or without, the sexes of the same or of closely allied species. In most cases the strongest contrast would be the most useful, especially as the colours of the objects to be distinguished would form but minute spots or points when compared with the broad masses of tint of sky, earth, or foliage against which they would be set.

Throughout the long epochs in which the sense of sight was being gradually developed in the higher animals, their visual organs would be mainly subjected to two groups of rays—the green from vegetation, and the blue from the sky.

The immense preponderance of these over all other groups of rays would naturally lead the eye to become specially adapted for their perception; and it is quite possible that at first these were the only kinds of light-vibrations which could be perceived at all. When the need for differentiation of colour arose, rays of greater and of smaller wave-lengths would necessarily be made use of to excite the new sensations required, and we can thus understand why green and blue form the central portion of the visible spectrum, and are the colours which are most agreeable to us in large surfaces; while at its two extremities we find yellow, red, and violet—colours which we best appreciate in smaller masses, and when contrasted with the other two, or with light neutral tints. We have here probably the foundations of a natural theory of harmonious colouring, derived from the order in which our colour-sensations have arisen and the nature of the emotions with which the several tints have been always associated. The agreeable and soothing influence of green light may be in part due to the green rays having little heating power; but this can hardly be the chief cause, for the blue and violet, though they contain less heat, are not generally felt to be so cool and sedative. But when we consider how dependent are all the higher animals on vegetation, and that man himself has been developed in the closest relation to it, we shall find, probably, a sufficient explanation. The green mantle with which the earth is overspread caused this one colour to predominate over all others that meet our sight, and to be almost always associated with the satisfaction of human wants. Where the grass is greenest, and vegetation most abundant and varied, there has man always found his most suitable dwelling-place. In such spots hunger and thirst are unknown, and the choicest productions of nature gratify the appetite and please the eye. In the greatest heats of summer, coolness, shade, and moisture are found in the green forest glades, and we can thus understand how our visual apparatus has become especially adapted to receive pleasurable and soothing sensation from this class of rays.

Supposed increase of Colour-perception within the Historical Period

Some writers believe that our power of distinguishing colours has increased even in historical times. The subject has attracted the attention of German philologists, and I have been furnished by a friend with some notes from a work of the late Lazarus Geiger, entitled, *Zur Entwickelungsgeschichte der Menschheit* (Stuttgart, 1871). According to this writer it appears that the *colour* of grass and foliage is never alluded to as a beauty in the Vedas or the Zendavesta, though these productions are continually extolled for other properties. Blue is described by terms denoting sometimes green, sometimes black, showing that it was hardly recognised as a distinct colour. The *colour* of the sky is never mentioned in the Bible, the Vedas, the Homeric poems, or even in the Koran. The first distinct allusion to it known to Geiger is in an Arabic work of the ninth century. "Hyacinthine locks" are black locks, and Homer calls iron "violet-coloured." Yellow was often confounded with green, but, along with red, it was one of the earliest colours to receive a distinct name. Aristotle names three colours in the rainbow—red, yellow, and green. Two centuries earlier Xenophanes had described the rainbow as purple, reddish, and yellow. The Pythagoreans admitted four primary colours—white, black, red, and yellow; the Chinese the same, with the addition of green.

Simultaneously with the first publication of this essay in *Macmillan's Magazine,* there appeared in the *Nineteenth Century* an article by Mr. Gladstone on the colour-sense, chiefly as exhibited in the poems of Homer. He shows that the few colour-terms used by Homer are applied to such different objects that they cannot denote colours only, as we perceive and differentiate them, but seem more applicable to different intensities of light and shade. Thus, to give on example, the word *porphureos* is applied to clothing, to the rainbow, to blood, to a cloud, to the sea, and to death; and no one meaning will suit all these applications except comparative darkness. In other cases the same thing has many different epithets applied to it according to its different aspects or conditions; and as the colours of objects are generally indicated in ancient writings by comparative rather than by abstract terms,—as wine-colour, fire-colour, bronze-colour, etc.,—it becomes still more difficult to determine in any particular case what colour was really meant. Mr. Gladstone's general conclusion is, that the archaic man had a positive perception only of degrees of light and darkness, and that in Homer's time he had advanced to the imperfect discrimination of red and yellow, but no further; the green of grass and foliage or the blue of the sky being never once referred to.

These curious facts cannot, however, be held to prove so recent an origin for colour-sensations as they would at first sight appear to do, because we have seen that both flowers and fruits have become diversely coloured in adaptation to the visual powers of insects, birds, and mammals. Red being a very common colour of ripe fruits which attract birds to devour them and thus distribute their seeds, we may be sure that the contrast of red and green is to them very well marked. It is indeed just possible that birds may have a more advanced development of the colour-sense than mammals, because the teeth of the latter commonly grind up and destroy the seeds of the larger fruits and nuts which they devour, and which are not usually coloured; but the irritating effect of bright colours on some of them does not support this view. It seems most probable, therefore, that man's *perception* of colour in the time of Homer was little if any inferior to what it is now, but that, owing to a variety of causes, no precise *nomenclature* of colours had become established. One of these causes probably was, that the colours of the objects of most importance, and those which were most frequently referred to in songs and poems, were uncertain and subject to variation. Blood was light or dark red, or when dry, blackish; iron was gray or dark or rusty; bronze was shining or dull; foliage was of all shades of yellow, green, or brown; and horses or cattle had no one distinctive colour. Other objects, as the sea, the sky, and wine, changed in tint according to the light, the time of day, and the mode of viewing them; and thus colour, indicated at first by reference to certain coloured objects, had no fixity. Things which had more definite and purer colours—as certain species of flowers, birds, and insects—were probably too insignificant or too much despised to serve as colour-terms; and even these often vary, either in the same or in allied species, in a manner which would render their use unsuitable. Colour-names, being abstractions, must always have been a late development in language, and their comparative unimportance in an early state of society and of the arts would still further retard their appearance; and this seems quite in accordance with the various facts set forth by Mr. Gladstone and the other writers referred to. The fact that colour-blindness is so prevalent even now is, however, an indication that the fully-developed colour-sense is not of primary importance to man. If it had been so, natural selection would long ago have eliminated the disease itself, and its tendency to recur would hardly be so strong as it appears to be.

Concluding Remarks on the Colour-sense

The preceding considerations enable us to comprehend both why a perception of difference of colour has become developed in the higher animals, and also why colours require to be presented or combined in varying proportions in order to be agreeable to us. But they hardly seem to afford a sufficient explanation either of the wonderful contrasts and total unlikeness of the sensations produced in us by the chief primary colours, or of the exquisite charm and pleasure we derive from colour itself, as distinguished from variously-coloured objects, in the case of which association of ideas comes into play. It is hardly conceivable that the material *uses* of colour to animals and to ourselves required such very distinct and powerfully-contrasted sensations; and it is still less conceivable that a sense of delight in colour *per se* should have been necessary for our utilisation of it.

The emotions excited by colour and by music alike

seen to rise above the level of a world developed on purely utilitarian principles.

THE ANTIQUITY AND ORIGIN OF MAN1

Indications of Man's Extreme Antiquity—Antiquity of Intellectual Man—Sculptures on Easter Island—North American Earthworks—The Great Pyramid—Conclusion.

MANY now living remember the time (for it is little more than twenty years ago) when the antiquity of man, as now understood, was universally discredited. Not only theologians, but even geologists, then taught us that man belonged altogether to the existing state of things; that the extinct animals of the Tertiary period had finally disappeared, and that the earth's surface had assumed its present condition, before the human race first came into existence. So prepossessed were even scientific men with this idea—which yet rested on purely negative evidence, and could not be supported by any arguments of scientific value—that numerous facts which had been presented at intervals for half a century, all tending to prove the existence of man at very remote epochs, were silently ignored; and, more than this, the detailed statements of three distinct and careful observes, confirming each other, were rejected by a great scientific Society as too improbable for publication, only because they proved (if they were true) the coexistence of man with extinct animals.[2]

[1] This formed part of the author's address to the Biological Section of the British Association at Glasgow in 1876.

[2] In 1854 (?) a communication from the Torquay Natural History Society confirming previous accounts by Mr. Godwin-Austen, Mr. Vivian, and the Rev. Mr. M'Enery, that worked flints occurred in Kent's Hole with remains of extinct species, was rejected as too improbable for publication. See Lubbock's *Prehistoric Times, 2d ed., p. 306.*

But this state of belief in opposition to facts could not long continue. In 1859 a few of our most eminent geologists examined for themselves into the alleged occurrence of flint implements in the gravels of the north of France, which had been made public fourteen years before, and found them strictly correct. The caverns of Devonshire were about the same time carefully examined by equally eminent observers, and were found fully to bear out the statements of those who had published their results eighteen years before. Flint implements began to be found in all suitable localities in the south of England, when carefully searched for, often in gravels of equal antiquity with those of France. Caverns giving evidence of human occupation at various remote periods were explored in Belgium and the south of France—

lake-dwelling were examined in Switzerland—refuse-heaps in Denmark—and thus a whole series of remains have been discovered carrying back the history of mankind from the earliest historic periods to a long distant past.

The antiquity of the races thus discovered cannot be measured in years; but it may be approximately determined by the successively earlier and earlier stages of civilisation through which we can trace them, and by the changes in physical geography and of animal and vegetable life that have since occurred. As we go back metals soon disappear, and we find only tools and weapons of stone and of bone. The stone weapons get ruder and ruder; pottery, and then the bone implements, cease to occur; and in the earliest stages we find only chipped flints of rude design, though still of unmistakably human workmanship. In like manner domestic animals disappear as we go backward; and though the dog seems to have been the earliest, it is doubtful whether the makers of the ruder flint implements of the gravels possessed even this. Still more important as a measure of time are the changes in the distribution of animals, indicating changes of climate, which have occurred during the human period. At a comparatively recent epoch in the record of prehistoric times we find that the Baltic was far salter than it is now and produced abundance of Oysters, and that Denmark was covered with pine forests inhabited by Capercailzies, such as now only occur farther north in

Norway. A little earlier we find that reindeer were common even in the south of France; and still earlier this animal was accompanied by the mammoth and woolly rhinoceros, by the arctic glutton, and by huge bears and lions of extinct species. The presence of such animals implies a change of climate; and both in the caves and gravels we find proofs of a much colder climate than now prevails in Western Europe. Even more remarkable are the changes of the earth's surface which have been effected during man's occupation of it. Many extensive valleys in England and France are believed by the best observers to have been deepened at least a hundred feet; caverns now far out of the reach of any stream must for a long succession of years have had streams flowing through them, at least in times of floods; and this often implies that vast masses of solid rock have since been worn away. In Sardinia land has risen at least 300 feet since men lived there who made pottery and probably used fishing-nets;[1] while in Kent's Cavern remains of man are found buried beneath two separate beds of stalagmite, each having a distinct texture, and each covering a deposit of cave-earth having well-marked differential characters, while each contains a distinct assemblage of extinct animals.

Such, briefly, are the results of the evidence that has been rapidly accumulating for about fifteen years, as to the antiquity of man; and it has been confirmed by so many discoveries of a like nature in all parts of the globe, and especially by the comparison of the tools and weapons of prehistoric man with those of modern savages (so that the use of even the rudest flint implements has become quite intelligible), that we can hardly wonder at the vast revolution effected in public opinion. Not only is the belief in man's vast and still unknown antiquity universal among men of science, but it is hardly disputed by any well-informed theologian; and the present generation of science-students must, we should think, be somewhat puzzled to understand what there was in the earliest discoveries that should have aroused such general opposition, and been met with such universal incredulity.

[1] Lyell's *Antiquity of Man, 4th ed., p. 115.*

But the question of the mere "Antiquity of Man" almost sank into insignificance at a very early period of the inquiry, in comparison with the far more momentous and more exciting problem of the development of man from some lower animal form, which the theories of Mr. Darwin and of Mr. Herbert Spencer soon showed to be inseparably bound up with it. This has been, and to some extent still is, the subject of fierce conflict; but the controversy as to the fact to such development is now almost at an end, since one of the most talented representatives of Catholic theology, and an anatomist of high standing—Professor Mivart—fully adopts it as regards physical structure, reserving his opposition for those parts of the theory which would deduce man's whole intellectual and moral nature from the same source and by a similar mode of development.

Never, perhaps, in the whole history of science or philosophy has so great a revolution in thought and opinion been effected as in the twelve years from 1859 to 1871, the respective dates of publication of Mr. Darwin's *Origin of Species* and *Descent of Man.* Up to the commencement of this period the belief in the independent creation or origin of the species of animals and plants, and the very recent appearance of man upon the earth, were, practically, universal. Long before the end of it these two beliefs had utterly disappeared, not only in the scientific world, but almost equally so among the literary and educated classes generally. The belief in the independent origin of man held its ground somewhat longer; but the publication of Mr. Darwin's great work gave even that its deathblow, for hardly any one capable of judging of the evidence now doubts the derivative nature of man's bodily structure as a whole, although many believe that his mind, and even some of his physical characteristics, may be due to the action of other forces than have acted in the case of the lower animals.

We need hardly be surprised, under these circumstances, if there has been a tendency among men of science to pass from one extreme to the other; from a profession (so few years ago) of total ignorance as to the mode of origin of all living things, to a claim to almost complete knowledge of the whole progress of the universe, from the first speck of living protoplasm up to the highest development of the human intellect. Yet this is really what we have seen in the last sixteen years. Formerly difficulties were exaggerated, and it was asserted that we had not sufficient knowledge to venture on any generalisations on the subjects. Now difficulties are set aside, and it is held that our theories are so well established and so far-reaching that they explain and comprehend all nature. It is not long ago (as I have already reminded you) since *facts* were contemptuously ignored, because they favoured our now popular views; at the present day it seems to me that facts which oppose them hardly receive due consideration. And as opposition is the best incentive to progress, and it is not well even for the best theories to have it all their own way, I propose to direct your attention to a few such facts, and to the conclusions that seem fairly deducible from them.

Indications of Man's Extreme Antiquity

It is a curious circumstance that, notwithstanding the attention that has been directed to the subject in every part of the world, and the numerous excavations connected with railways and mines, which have offered such facilities for geological discovery, no advance whatever has been made for a considerable number of years in detecting the time or mode of man's origin. The Palæolithic flint weapons first discovered in the north of France more than thirty years ago are still the oldest undisputed proofs of man's existence; and amid the countless relics of a former world that have been brought to light, no evidence of any one of the links that must have connected man with the lower animals has yet appeared.

It is, indeed, well known that negative evidence in geology is of very slender value; and this is, no doubt, generally the case. The circumstances here are, however, peculiar, for many converging lines of evidence show that, on the theory of development by the same laws which have determined the development of the lower animals, man must be immensely older than any traces of him yet discovered. As this is a point of great interest we must devote a few moments to its consideration.

1. The most important difference between man and

such of the lower animals as most nearly approach him is undoubtedly in the bulk and development of his brain, as indicated by the form and capacity of the cranium. We should therefore anticipate that these earliest races, who were contemporary with the extinct animals and used rude stone weapons, would show a marked deficiency in this respect. Yet the oldest known crania (those of the Engis and Cro-Magnon caves) show no marks of degradation. The former does not present so low a type as that of most existing savages, but is (to use the words of Professor Huxley) "a fair average human skull, which might have belonged to a philosopher, or might have contained the thoughtless brains of a savage." The latter are still more remarkable, being unusually large and well-formed. Dr. Pruner-Bey states that they surpass the average of modern European skulls in capacity, while their symmetrical form, without any trace of prognathism, compares favourably not only with those of the foremost savage races, but with many civilised nations of modern times.

One or two other crania of much lower type, but of less antiquity than this, have been discovered; but they in no way invalidate the conclusion which so highly developed a form at so early a period implies, viz. that we have as yet made a hardly perceptible step towards the discovery of any earlier stage in the development of man.

2. This conclusion is supported and enforced by the nature of many of the works of art found even in the oldest cave-dwellings. The flints are of the old chipped type, but they are formed into a large variety of tools and weapons—such as scrapers, awls, hammers, saws, lances, etc., implying a variety of purposes for which these were used, and a corresponding degree of mental activity and civilisation. Numerous articles of bone have also been found, including well-formed needles; implying that skins were sewn together, and perhaps even textile materials woven into cloth. Still more important are the numerous carvings and drawings representing a variety of animals, including horses, reindeer, and even a mammoth, executed with considerable skill on bone, reindeer-horns, and mammoth-tusks. These, taken together, indicate a state of civilisation much higher than that of the lowest of our modern savages, while they are quite compatible with a considerable degree of mental advancement, and lead us to believe that the crania of Engis and Cro-Magnon are not exceptional, but fairly represent the characters of the race. If we further remember that these people lived in Europe under the unfavourable conditions of a sub-arctic climate, we shall be inclined to agree with Dr. Daniel Wilson that it is far easier to produce evidences of deterioration than of progress, in instituting a comparison between the contemporaries of the mammoth and later prehistoric races of Europe or savage nations of modern times.[1]

3. Yet another important line of evidence as to the extreme antiquity of the human type has been brought prominently forward by Professor Mivart.[2] He shows, by a careful comparison of all parts of the structure of the body, that man is related not to any one, but almost equally to many of the existing apes—to the orang, the chimpanzee, the gorilla, and even to the gibbons, in a variety of ways; and these relations and differences are so numerous and so diverse that, on the theory of evolution, the ancestral form which ultimately developed into man must have diverged from the common stock whence all these various forms and their extinct allies originated. But so far back as the Miocene deposits of Europe we find the remains of apes allied to these various forms, and especially to the gibbons; so that in all probability the special line of variation which led up to man branched off at a still earlier period. And these early forms, being the initiation of a far higher type, and having to develop by natural selection into so specialised and altogether distinct a creature as man, must have risen at a very early period into the position of a dominant race, and spread in dense waves of population over all suitable portions of the great continent—for this, on Mr. Darwin's hypothesis, is essential to developmental progress through the agency of natural selection.

[1] *Prehistoric Man,* 3d ed., vol. i. p. 117.

[2] *Man and Apes,* pp. 171–193.

Under these circumstances we might certainly expect to find some relics of these earlier forms of man along with those of animals, which were presumably less abundant. Negative evidence of this kind is not very weighty, but still it has *some* value. It has been suggested that as apes are mostly tropical, and anthropoid apes are now confined almost exclusively to the vicinity of the equator, we should expect the ancestral forms of man to have inhabited these same localities—West Africa and the Malay islands. But this objection is hardly valid, because existing anthropoid apes are wholly dependent on a perennial supply of easily accessible fruits, which is only found near the equator; while not only had the south of Europe an almost tropical climate in Miocene times, but we must suppose even the earliest ancestors of man to have been terrestrial and omnivorous, since it must have taken ages of slow modification to have produced the perfectly erect form, the short arms, and the wholly non-prehensile foot,[1] which so strongly differentiate man from the arboreal apes.

The conclusion which I think we must arrive at is, that if man has been developed from a common ancestor with all existing apes, *and by no other agencies than such as have affected their development,* then he must have existed, in something approaching his present form, during the Tertiary period—and not merely existed, but predominated in numbers, wherever suitable conditions prevailed. If, then, continued researches in all parts of Europe and Asia fail to bring to light any proofs of his presence, it will be at least a presumption that he came into existence at a much later date, and by a much more rapid process of development. In that case it will be a fair argument that, just as he is in his mental and moral nature, his capacities and aspirations, so infinitely raised above the brutes, so his origin is due, in part, to distinct and higher agencies than such as have affected their development.

[1] The common statement of travellers as to savages having great prehensile power in the toes has been adopted by some naturalists as indicating an approach to the apes. But this notion is founded on a complete misconception. Savages pick up objects with their feet, it is true, but always by a lateral motion of the toes, which we should equally possess if we never wore shoes or stockings. In no savage have I ever seen the slightest approach to opposability of the great toe, which is the essential distinguishing feature of apes; not have I ever seen it stated that any variation in this direction has been detected in the anatomical structure of the foot of the lower races.

Antiquity of Intellectual Man

There is yet another line of inquiry bearing upon this subject to which I wish to call your attention. It is a somewhat curious fact that, while all modern writers admit the great antiquity of man, most of them maintain the very recent development of his intellect, and will hardly contemplate the possibility of men equal in mental capacity to ourselves having existed in prehistoric times. This question is generally assumed to be settled by such relics as have been preserved of the manufactures of the older races, showing a lower and lower state of the arts; by the successive disappearance in early times of iron, bronze, and pottery; and by the ruder forms of the older flint implements. The weakness of this argument has been well shown by Mr. Albert Mott in his very original but little-known presidential address to the Literary and Philosophical Society of Liverpool in 1873. He maintains that "our most distant glimpses of the past are still of a world peopled as now with men both civilised and savage," and "that we have often entirely misread the past by supposing that the outward signs of civilisation must always be the same, and must be such as are found among ourselves." In support of this view he adduces a variety of striking facts and ingenious arguments, a few of which I will briefly summarise.

Sculptures on Easter Island

On one of the most remote islands of the Pacific—Easter island—2000 miles from South America, 2000 from the Marquesas, and more than 1000 from the Gambier islands, are found hundreds of gigantic stone images, now mostly in ruins. They are often forty feet high, while some seem to have been much larger, the crowns on their heads, cut out of a red stone, being sometimes ten feet in diameter, while even the head and neck of one is said to have been twenty feet high.[1] These images once all stood erect on extensive stone platforms.

[1] *Journ, of Roy. Geog. Soc.*, 1870, pp. 177, 178.

The island containing these remarkable works of art has only an area of about thirty square miles, or considerably less than Jersey. Now, as one of the smallest images (eight feet high) weighs four tons, the largest must weigh over a hundred tons, if not much more; and the existence of such vast works implies a large population, abundance of food, and an established government. Yet how could these coexist on a mere speck of land wholly cut off from the rest of the world? Mr. Mott maintains that these facts necessarily imply the power of regular communication with larger islands or a continent, the arts of navigation, and a civilisation much higher than now exists in any part of the Pacific. Very similar remains in other islands scattered widely over the Pacific add weight to this argument.

North American Earthworks

The next example is that of the ancient mounds and earthworks of the North American continent, the bearing of which is even more significant. Over the greater part of the extensive Mississippi valley, four well-marked classes of these earthworks occur. Some are camps, or works of defence, situated on bluffs, promontories, or isolated hills; others are vast inclosures in the plains and lowlands, often of geometric forms, and having attached to them roadways or avenues often miles in length; a third are mounds corresponding to our tumuli, often seventy to ninety feet high, and some of them covering acres of ground; while a fourth group consists of representations of various animals modelled in relief on a gigantic scale, and occurring chiefly in an area somewhat to the north-west of the other classes, in the plains of Wisconsin.

The first class—the camps or fortified inclosures—resemble in general features the ancient camps of our own island, but far surpass them in extent. Fort Hill, in Ohio, is surrounded by a wall and ditch a mile and a half in length, part of the way cut through solid rock. Artificial reservoirs for water were made within it, while at one extremity, on a more elevated point, a keep is constructed with its separate defences and water-reservoirs. Another, called Clark's Work, in the Scioto valley, which seems to have been a fortified town, incloses an area of 127 acres, the embankments measuring three miles in length, and containing not less than three million cubic feet of earth. This area incloses numerous sacrificial mounds and symmetrical earthworks, in which many interesting relics and works of art have been found.

The second class—the sacred inclosures—may be compared for extent and arrangement with Avebury or Karnak, but are in some respects even more remarkable. One of these at Newark, Ohio, covers an area of several miles, with its connected groups of circles, octagons, squares, ellipses, and avenues on a grand scale, and formed by embankments from twenty to thirty feet in height. Other similar works occur in different parts of Ohio; and by accurate survey it is found, not only that the circles are true, though some of them are one-third of a mile in diameter, but that other figures are truly square, each side being over 1000 feet long; and, what is still more important, the dimensions of some these geometrical figures, in different parts of the country and seventy miles apart, are identical. Now this proves the use, by the builders of these works, of some standard measures of length; while the accuracy of the squares, circles, and, in a less degree, of the octagonal figures, shows a considerable knowledge of rudimentary geometry and some means of measuring angles. The difficulty of drawing such figures on a large scale is much greater than any one would imagine who has not tried it; and the accuracy of these is far beyond what is necessary to satisfy the eye. We must, therefore, impute to the builders the wish to make these figures as accurate as possible, and this wish is a greater proof of habitual skill and intellectual advancement than even the ability to draw such figures. If, then, we take into account this ability and this love of geometric truth, and further consider the dense population and civil organisation implied by the construction of such extensive systematic works, we must allow that these ancient people had reached the earlier stages of a civilisation of which no traces existed among the savage tribes who alone occupied the country when first visited by Europeans.

The animal mounds are of comparatively less importance for our present purpose, as they imply a somewhat lower grade of advancement; but the sepulchral and sacrificial mounds exist in vast numbers, and their partial exploration has yielded a quantity of articles and works of art which throw some further light on the peculiarities of this mysterious people. Most of these mounds contain a large concave hearth or basin of burnt clay, of perfectly symmetrical form, on which are found deposited more or less abundant relics, all bearing traces of the action of fire. We are therefore only acquainted with such articles as are practically fire-proof, or have accidentally escaped combustion. These consist of bone and copper implements and ornaments, disks and tubes; pearl, shell, and silver beads, more or less injured by the fire; ornaments cut in mica; ornamental pottery; and numbers of elaborate carvings in stone, mostly forming pipes for smoking.[1] The metallic articles are all formed by hammering, but the execution is very good; plates of mica are found cut into scrolls and circles; the pottery, of which very few remains have been found; is far superior to that of any of the Indian tribes, since Dr. Wilson is of opinion that it must have been formed on a wheel, as it is often of uniform thickness throughout (sometimes not more than one-sixth of an inch), polished, and ornamented with scrolls and figures of birds and flowers in delicate relief. But the most instructive objects are the sculptured stone pipes, representing not only various easily recognisable animals, but also human heads, so well executed that they appear to be portraits. Among the animals, not only are such native forms as the panther, bear, otter, wolf, beaver, raccoon, heron, crow, turtle, frog, rattlesnake, and many others well represented, but also the manatee, which perhaps then ascended the Mississippi as it now does the Amazon, and the toucan, which could hardly have been obtained nearer than Mexico. The sculptured heads are especially remarkable, because they present to us the features of an intellectual and civilised people. The nose in some is perfectly straight, and neither prominent nor dilated; the mouth is small, and the lips thin; the chin and upper lip are short, contrasting with the ponderous jaw of the modern Indian, while the cheek-bones present no marked prominence. Other examples have the nose somewhat projecting at the apex in a manner quite unlike the features of any American indigenes; and although there are some which show a much coarser face, it is very difficult to see in any of them that close resemblance to the Indian type which these sculptures have been said to exhibit. The few authentic crania from the mounds present corresponding features, being far more symmetrical and better developed in the frontal region than those of any American tribes, although somewhat resembling them in the occipital outline;[1] while one was described by its discover (Mr. W. Marshall Anderson) as a "beautiful skull, worthy of a Greek."

[1] Woven cloth, apparently of flax or hemp, as well as gauges supposed to have been used to regulate the thickness of the thread, have also been found in several of the mounds of Ohio (Foster's *Prehistoric Races of the United States,* 1873, pp. 225–229).

The antiquity of this remarkable race may perhaps not be very great as compared with the prehistoric man of Europe, although the opinion of some writers on the subject seems affected by that "parsimony of time" on which the late Sir Charles Lyell so often dilated. The mounds are well over-grown with dense forest, and one of the large trees was estimated to be 800 years old, while other observers consider the forest growth to indicate an age of at least 1000 years. But it is well known that it requires several generations of trees to pass away before the growth on a deserted clearing comes to correspond with that of the surrounding virgin forest, while this forest, once established, may go on growing for an unknown number of thousands of years. The 800 or 1000 years estimates from the growth of existing vegetation is a minimum which has no bearing whatever on the actual age of these mounds; and we might almost as well attempt to determine the time of the glacial epoch from the age of the pines or oaks which now grow on the moraines.

The important thing for us, however, is that when North America was first settled by Europeans, the Indian tribes inhabiting it had no knowledge or tradition of any preceding race of higher civilisation than themselves. Yet we find that such a race existed—that they must have been populous and have lived under some established government; while there are signs that they practised agriculture largely, as, indeed, they must have done to have supported a population capable of executing such gigantic works in such cast profusion; for it is stated that the mounds and earthworks of various kinds in the State of Ohio alone amount to between eleven and twelve thousand. In their habits, customs, religion, and arts, they differed strikingly from all the Indian tribes; while their love of art and of geometric forms, and their capacity for executing the latter upon so gigantic a scale, render it probable that they were a really civilised people, although the form their civilisation took may have been very different from that of later peoples, subject to very different influences and the inheritors of a longer series of ancestral civilisations. We have here, at all event, a striking example of the transition, over an extensive country, from comparative civilisation to comparative barbarism, the former leaving no tradition and hardly and trace of its influence on the latter.

[1] Wilson's *Prehistoric Man,* 3d ed., vol. ii. pp. 123–130.

As Mr. Mott well remarks: "Nothing can be more striking than the fact that Easter island and North America both give the same testimony as to the origin of the savage life found in them, although in all circumstances and surroundings the two cases are so different. If no stone monuments had been constructed in Easter island, or mounds containing a few relics saved from fire, in the United States, we might never have suspected the existence of these ancient peoples."

He argues, therefore, that it is very easy for the records of an ancient nation's life entirely to perish or to be hidden from observation. Even the arts of Nineveh and Babylon were unknown only a generation ago, and we have only just discovered the facts about the mound-builders of North America.

But other parts of the American continent exhibit parallel phenomena. Recent investigations show that in Mexico, Central America, and Peru, the existing race of Indians has been preceded by a distinct and more civilised race. This is proved by the sculptures of the ruined cities of Central America, by the more ancient terra-cottas and paintings of Mexico, and by the oldest portrait-pottery of Peru. All alike show markedly non-Indian features, while they often closely resemble modern European types. Ancient crania, too, have been found in all these countries, presenting very different characters from those of any of the existing indigenous races of America.[1]

[1] Wilson's *Prehistoric Man,* 3d ed., vol. ii. pp. 125, 144.

The Great Pyramid

There is one other striking example of a higher phase of development in science and the arts being succeeded by a lower phase, which is in danger of being forgotten because it has been made the foundation of theories which seem wild and fantastic, and are probably in great part erroneous. I allude to the Great Pyramid of Egypt, whose form, dimensions, structure, and uses have recently been the subject of elaborate works by Professor Piazzi Smyth. Now the admitted facts about the pyramid are so interesting and so apposite to the subject we are considering, that I beg to recall them to your attention. Most of you are aware that this pyramid has been carefully explored and measured by successive Egyptologists, and that the dimensions have lately become capable of more accurate determination owing to the discovery of some of the original casing-stones, and the clearing away of the earth from the corners of the foundation, showing the sockets in which the corner-stones fitted. Professor Smyth devoted many months of work with the best instruments in order to fix the dimensions and angles of all accessible parts of the structure: and he has carefully determined these by a comparison of his own and all previous measures, the best of which agree pretty closely with each other. The results arrived at are—

1. That the pyramid is truly square, the sides being equal and the angles right angles.

2. That the four sockets on which the four first stones of the corners rested are truly on the same level.

3. That the directions of the sides are accurately to the four cardinal points.

4. That the vertical height of the pyramid bears the same proportion to its circumference at the base as the radius of a circle does to its circumference.

Now all these measures, angles, and levels are accurate, not as an ordinary surveyor or builder could make them, but to such a degree as requires the very best modern instruments and all the refinements of geodetical science to discover any error at all. In addition to this we have the wonderful perfection of the workmanship in the interior of the pyramid, the passages and chambers being lined with huge blocks of stones fitted with the utmost accuracy, while every part of the building exhibits the highest structural science.

In all these respects this largest pyramid surpasses every other in Egypt. Yet it is universally admitted to be the oldest, and also the oldest historical building in the world.

Now these admitted facts about the Great Pyramid are surely remarkable and worthy of the deepest consideration. They are facts which, in the pregnant words of the late Sir John Herschel, "according to received theories ought not to happen," and which, he tells us, should therefore be kept ever present to our minds, since "they belong to the class of facts which serve as the clue to new discoveries." According to modern theories, the higher civilisation is ever a growth and an outcome from a preceding lower state; and it is inferred that this progress is visible to us throughout all history and in all material records of human intellect. But here we have a building which marks the very dawn of history, which is the oldest authentic monument of man's genius and skill, and which, instead of being far inferior, is very much superior to all which followed it. Great men are the products of their age and country, and the designer and constructors of this wonderful monument could never have arisen among an unintellectual and half-barbarous people. So perfect a work implies many preceding less perfect works which have disappeared. It marks the culminating point of an ancient civilisation, of the early stages of which we have no trace or record whatever.

Conclusion

The three cases to which I have now adverted (and there are many others) seem to require for their satisfactory interpretation a somewhat different view of human progress from that which is now generally accepted. Taken in connection with the great intellectual power of the ancient Greeks—which Mr. Galton believes to have been far above that of the average of any modern nation—and the elevation, at once intellectual and moral, displayed in the writings of Confucius, Zoroaster, and the Vedas, they point to the conclusion that, while in material progress there has been a tolerably steady advance, man's intellectual and moral development reached almost its highest level in a very remote past. The lower, the more animal, but often the more energetic types have, however, always been far the more numerous; hence such established societies as have here and there arisen under the guidance of higher minds have always been liable to the swept away be the incursions of barbarians. Thus in almost every part of the globe there may have been a long succession of partial civilisations, each in turn succeeded by a period of barbarism; and this view seems supported by the occurrence of degraded types of skull along with such "as might have belonged to a philosopher," at a time when the mammoth and the reindeer inhabited southern France.

Nor need we fear that there is not time enough for the rise and decay of so many successive civilisations as this view would imply, for the opinion is now gaining ground among geologists that palæolithic man was really preglacial, and that the great gap (marked alike by a change of physical conditions and of animal life) which in Europe always separates him from his neolithic successor, was caused by the coming on and passing away of the great ice age.

If the views now advanced are correct, many, perhaps most, of our existing savages are the successors of higher races; and their arts, often showing a wonderful similarity in distant continents, may have been derived from a common source among more civilised peoples.

VIII

THE ANTIQUITY OF MAN IN NORTH AMERICA[1]

Ancient Shell Mounds—Man Coeval with Extinct Mammalia—Man in the Glacial Period—Palæolithic Implements in North America—The Auriferous Gravels of California—Fossil Remains under the Ancient Lava Beds—Works of Art in the Auriferous Gravels—Human Remains in the Auriferious Gravels—Concluding Remarks on the Antiquity of Man.

OVER a considerable portion of the northern hemisphere the remains of man, or his works, have been found in association with bones of the extinct mammalia which characterised the Glacial epoch, and no evidence has been obtained that man at that time differed more from modern savages than they do among themselves. The facts which prove this antiquity were, when first put forth, doubted, neglected, or violently opposed, and it is now admitted that such opposition was due to prejudice alone, and in every case led to the rejection of important scientific truths. Yet after nearly thirty years' experience we find that an exactly similar prejudice prevails, even among geologists, against all evidence which carries man one little step farther back into pre-Glacial or Pliocene times, although if there is any truth whatever in the doctrine of evolution as applied to man, and if we are not to adopt the exploded idea that the Palæolithic men were specially created just when the flood of ice was passing away, they *must* have had ancestors who *must* have existed in the Pliocene period, if not earlier. Is it then so improbable that some trace of man should be discovered at this period, that each particle of evidence as it arises must be attacked with all the weapons of doubt, accusation, and ridicule, which for so many years crushed down the truth with regard to Palæolithic man? One would think, as Jeremy Bentham said of another matter, that is was "wicked or else unwise" to accept any evidence for facts which are yet so inherently probably that the entire absence to evidence for their existence ought to be felt to be the greatest stumbling-block.

[1] This article appeared in the *Nineteenth Century*, Nov. 1887.

No better illustration of this curious prejudice can be given than the way in which some recent discoveries of stone implements in deposits of considerable antiquity in India are dealt with. These implements are of quartzite, and are of undoubtedly human workmanship. They were found in the

Lower Laterite formation, which is said to have undergone great denudation and to be undoubtedly very ancient. Old stone circles of a great but unknown antiquity are formed of it. It is also stated that the distinction between the Tertiary and post-Tertiary is very difficult in India, and the age of these Laterite beds cannot be determined either by fossils, which are absent, or by superposition. Yet we are informed, "The presence of Palæolithic implements *proves* that the rock is of post-Tertiary origin."[1] Here we have the origin of man taken as fixed and certain, so certain that his remains may be used to *prove* the age of a doubtful deposit! Nor do these indications of great antiquity stand alone, for in the Nerbudda fluviatile deposits Mr. Hackel has found stone weapons *in situ* along with eleven species of *extinct* fossil mammalia.

Believing myself that the existence of man in the Tertiary epoch is a *certainly,* and the discovery of his remains or works in deposits of that age to be decidedly *probable,* I hold it to be both wise and scientific to accept all evidence of his existence before the Glacial epoch which would be held satisfactory for a later period, and when there is any little doubt, to give the benefit of the doubt in favour of the find rather than against it. I hold further that it is equally sound doctrine to give some weight to cumulative evidence; since, when a thing is not improbable in itself, it surely adds much to the argument in its favour that facts which tend to prove it come from many different and independent sources—from those who are quite ignorant of the interest that attaches to their discovery, as well as from trained observers who are fully aware of the importance of every additional fact and the weight of each fresh scrap of evidence. Having by the kindness of Major Powell, the able Director of the United States Geological Survey, been able to look into the evidence recently obtained bearing on this question in the North American continent, I believe that a condensed account of it will certainly prove of interest to English readers.

[1] *Manual of the Geology of India,* p. 370.

The most certain tests of great antiquity, even though they afford us no accurate scale of measurement, are furnished by such natural changes as we know occur very slowly. Changes in the distribution of animals or plants, modifications of the earth's surface, the extinction of some species and the introduction of others, are of this nature, and they are the more valuable because during the entire historical period changes of this character are either totally unknown or of very small amount. Let us then see what changes of this kind have occurred since man inhabited the North American continent.

Ancient Shell Mounds

The shell heaps of the Damariscotta River, in Maine, are remarkable for their number and extent. The largest of these stretches for about half a mile along the shore, and is often six or seven feet, and in one place twenty-five feet, in thickness. They consist almost exclusively of oyster shells of remarkable size, frequently having a length of eight or ten inches, and sometimes reaching twelve or fourteen inches. They contain fragments of bones of edible animals, charcoal, bone implements, and some fragments of pottery. The surface is covered to a depth of several inches with vegetable mould, and large trees grow on them, some more than a century old. The special feature to which we now call attention is "that at the present time oysters are only found in very small numbers, too small to make it an object to gather them; and we were credibly informed that they have not been found in larger quantities since the settlement in the neighbourhood. It cannot be supposed that the immense accumulations now seen on the shores of Salt Bay could have been made unless oysters had existed in very large numbers in the adjoining waters."[1] Here we have evidence of an important change in the distribution of a species of mollusc since the banks were formed.

[1] *Second Annual Report of Trustees of Peabody Museum,* p. 18.

On the St. John's river, Florida, are enormous heaps largely composed of two freshwater shells, Ampullaria depressa and Paludina multilineata, which cover acres of ground, and are often six or eight feet thick. Professor Wyman, who explored these heaps, remarks: "It seems incredible to one who searches the waters of the St. John's and its lakes at the present time, that the two small species of shells above mentioned could have been obtained in such vast quantities as are seen brought together in these mounds, unless at the times of their formation the shells existed more abundantly than now, or the collection of them extended through very long periods of time. When it is borne in mind that the shell heaps afford the only suitable surface for dwellings, being most commonly built in swamps, or on lands liable to be annually overflowed by the rise of the river, they appear to be necessarily the result of the labours of a few living on a limited area at one time. At present it would be a very difficult matter to bring together in a single day enough of these shells for the daily meals of an ordinary family." [1]

On the Lower Mississippi, at Grand Lake, are shell banks of great extent which are now fifteen miles inland; while Nott and Gliddon describe similar banks on the Alabama River fifty miles inland, and they believe that Mobile Bay must have extended so far at the time the shells were collected. These beds are often covered with vegetable mould from one to two feet thick, and on this grow large forest trees. Equally indicative of long occupation and great antiquity is the enormous shell mound at San Pablo, on the bay of San Francisco, which is nearly a mile long and half a mile wide, and more than twenty feet thick. Numerous Indian skeletons and mummies have been found in it, showing that it had been subsequently used as a place of burial. Some mounds in Florida have growing on them enormous live oaks from thirteen to twenty-six feet in circumference at five feet from the ground, some of which are estimated to be about 600 years old, indicating the minimum age possible for the heaps, but not necessarily approaching to their real age.

[1] *Fifth Annual Report of Peabody Museum,* p. 22.

The extensive shell heaps of the Aleutian islands have been carefully examined and reported on by Mr. Dall, and are found to exhibit some remarkable and probably unique peculiarities. Complete sections were made across several of these, and they were found to consist of a series of distinct layers, each marked by some well-defined characteristics. In the upper layers only are there any mammalian remains, and these may be divided into three subdivisions. In the upper bed there are found seals, walruses, etc., aquatic and land birds, the arctic fox and dog, with well-made weapons and implements, awls, whetstones, needles, and lamps. In the next layer the dog and fox are absent, as are remains of large whales; and in the lower mammalian layer there are seals and small cetacea only, but no birds or land animals, and the weapons found are ruder. We then come to a considerable layer in which there are no mammalian remains whatever, but only fish-bones and molluscan shells, with rude knives, lance heads, etc. Belows this is a bottom deposit consisting no weapons, tools, or implements of any kind, except towards the surface of the layer, where a few hammer stones are found, round pebbles with an indentation on each side for the finger and thumb. Echinus' eggs are now eaten raw by the Aleuts, and it is the only eatable part of the animal. It takes forty or fifty full-sized echini for a meal. Some of the heaps cover five acres, and from a careful estimate founded on experiments, and taking the probable numbers of a colony which could have lived on such a spot, Mr. Dall calculates that it would take about 2200 years to form such an accumulation. A similar estimate applied to the upper layers brings the time required for the accumulation of the entire series to 3000 years, but that is on the supposition that they were formed continuously. This, however, was evidently not the case. Each layer indicates a change of inhabitants with different habits and in a somewhat different phase of civilisation, and each such change may imply the lapse of a long period, during which the site was abandoned and no accumulation went on. These shell heaps may, therefore, carry us back to a very remote antiquity.

Man Coeval with Extinct Mammalia

We next come to remains of man or his works found in association with the bones of extinct mammalia. The great mastodon skeleton in the British Museum found by Dr. Koch in the Osage valley, Missouri, had stone arrow-heads and charcoal found near it, but the fact was at the time received with the same incredulity as all other evidences of the antiquity of man. This animal was found at a depth of twenty feet, under seven alternate layers of loam, gravel, clay, and peat, with a forest of old trees on the surface, and one of the arrow-heads lay under the thigh-bone of the mastodon and in contact with it. About the same date (1859) Dr. Holmes communicated to the Philadelphia Academy of Natural Sciences his discovery of fragments of pottery in connection with bones of the mastodon and megatherium on the Ashley river of South Carolina.

Such cases as these remove all improbability from the celebrated Natchez man, a portion of a human pelvis from the loess of the Mississippi, which contains bones of the mastodon, megalonyx, horse, bison, and other extinct animals. This bone was stated by Sir Charles Lyell "to be quite in the same state of preservation and of the same black colour as the other fossils." Dr. Joseph Leidy agrees with this statement, yet he and Professor C. G. Forshey maintain that it is "more probable" that the human bone fell down the cliff from some Indian grave near the surface. Sir Charles Lyell well remarks that "had the bone belonged to any other recent mammal, such a theory would never have been resorted to." The admitted identity of the state of preservation and appearance of the human and animal bones is certainly not consistent with the view that the one is recent, the other ancient; the one artificially buried near the surface, the other in a natural deposit thirty feet below the surface.

Of a similar character to the above is the basket-work mat found in a rock-salt deposit fifteen to twenty feet below the surface in Petit Anse island, Louisiana, two feet above which were fragments of tusks and bones of an elephant. The slat is said to be very pure, extending over an area of 5000 acres, and the formation of such a deposit requires a considerable change of physical conditions from those now existing, an thus of itself implies great antiquity.[1]

The indications of the great antiquity of American man are now supported by such a mass of evidence of the same character that all the improbability supposed at first to attach to them has been altogether removed. As an illustration of this evidence I need only refer here to the Report on the Loess of Nebraska, by an experienced geologist, Dr. Samuel Aughey, who states that this deposit, which is now believed by the best American geologists to be of Glacial origin, and which covers enormous areas, contains throughout its entire extent many remains of mastodons and elephants, and that he himself had found an arrow and a spear-head of flint at depths of fifteen and twenty feet in the deposit. One of these was thirteen feet below a lumber vertebra of Elephas americanus.

Man in the Glacial Period

We now take a decided step backwards in time, to relics of human industry within or at the close of the Glacial period itself. About twenty years ago a well was sunk through the drift at Games, a few miles south of Lake Ontario, and at a depth of seventeen feet there were found lying on the solid rock three large stone enclosing a space within which were about a dozen charred sticks, thus closely resembling the cooking fires usually made by savages. Mr. G. K. Gilbert, of the U.S. Geological Survey, obtained the information from the intelligent farmer who himself found it, and after a close examination of the locality and the drift deposit in its relation to the adjacent lakes, comes to the conclusion that the hearth must have been used "near the end of the second Glacial period," and at the time of the separation of Lake Ontario from Lake Erie. When Mr. Gilbert gave an account of his researches on this matter at the meeting of the Washington Anthropological Society, 16th November 1886, two other gentlemen reported finds of similar character. Mr. Murdock, of the Point Barrow Station, near the extreme north-west corner of the continent, in making an excavation for an earth thermometer, found an Eskimo snow-goggle beneath more than twenty feet of frozen gravel and earth capped by a foot of turf. This being near the shores of the Arctic Sea may be a comparatively recent beach-formation and of no very great antiquity; but the remaining discovery was more important. Mr. W. J. M'Gee, a gentleman who has specially studied the Glacial and post-Glacial formations for the U.S. Geological Survey, described the finding by himself of a spearhead in the quaternary deposits of the Walker River Cañon, Nevada. These beds consist of several feet of silt and loose material at the top, then a layer of calcareous tufa lying upon twenty to thirty feet of white marl, containing remains of extinct mammalia, and resting unconformably upon somewhat similar beds of earlier date. The spear-head was found with its point just projecting from the face of the marl about twenty-six feet below the surface. Before removing the implement, he carefully studied the whole surroundings, and finally came to the conclusion that it had been embedded in the marl during its formation. The beds were deposited by the ancient Lake Lahonton. They have been thoroughly investigated by able geologists, and have been referred to the close of the Glacial period, or about the same time as the hearth described by Mr. Gilbert. The spear-head is three and a half inches in length, finely made, and well preserved.

[1] *Foster's Prehistoric Races of the United States,* p. 56.

About a hundred miles north-west of St. Paul, in Central Minnesota, a thin deposit has been discovered containing numerous quartzite implements. They occur at a depth of from twelve to fifteen feet in an old river terrace of modified drift, and the deposit marks an ancient land surface on which the implements are found, and which must have been deposited at about the close of the last Glacial epoch.[1] Mr. N. H. Winchell, State geologist of Minnesota, has found similar chips and implements in the upper part of the same deposit; and also human bones in the eastern terrace bluffs at Minneapolis, in a formation of about the same age as the above.

[1] "Vestiges of Glacial in Minnesota," by F. E. Babbitt, *Proc. of Am. Assoc.,* vol. Xxxii. 1883.

The same writer reports a still more remarkable discovery of a fragment of a human lower jaw in the red clay and boulder drift, but resting immediately on the limestone rock. This red clay belongs to the first or oldest Glacial period, and we thus have tie proofs of man's existence carried back not only to the end of the Glacial epoch, but perhaps to its very commencement.[1]

[1] *Annual Report of the State Geologist of Minnesota,* 1877, p. 60.

Palæolithic Implements in North America

We now come to the very interesting discoveries of Dr. Charles c. Abbott, of Trenton, New Jersey. In the extensive deposits of gravel in the valley of the Delaware, fresh surfaces of which are continually exposed in the cliffs on the river's banks, he has found large numbers of rude stone implements, almost identical in size and general form with the well-known palælithic implements of the valley of the Somme. These have been found at depths of from five to over twenty feet from the surface, in perfectly undisturbed soil, and that they are characteristic of this particular deposit is shown by the fact that they are found nowhere else in the same district. Large boulders, some of very great size, are found throughout the deposit, and in one case Dr. Abbott found a well-chipped spear-shaped implement immediately beneath a stone weighing at least half a ton. Professor N. S. Shaler, of Cambridge, Massachusetts, after examining the locality and himself obtaining some implements *in situ,* says, "I am disposed to consider these deposits as formed in the sea near the foot of the retreating ice-sheet when the sub-Glacial rivers were pouring out the vast quantity of water and waste that clearly were released during the breaking up of the great ice-time." Dr. Abbott, however, adduces facts which seem to prove that some part of the deposit at all events was sub-aerial, for he states that the very large boulders often have immediately under them a foot or more of soil between the lower surface of the stone and the gravel, and that this layer often extends some distance laterally, showing that if formed a land surface on which the boulders rested, and which was subsequently removed by water action, except where thus protected. At any rate we may accept Professor Shaler's conclusion: "If these remains are really those of man, they prove the existence of inter-Glacial man on this part of our shore." That the implements *are* of human workmanship is quite certain, and the fact stated by Professor Shaler himself, that "they are made of a curious granular argillite, the like of which I do not know in the place," is an additional proof of it. The further fact that the remains of man himself have been discovered in the same deposit completes the demonstration. First a human cranium was found of peculiar characteristics, being small, long, and very thick; then a tooth; and, lastly, a portion of a human under jaw, found at a depth of sixteen feet from the surface, near where a fragment of mastodon tusk had been found some years before. In recording this last discovery the curator of the Peabody Museum remarks: "To Dr. Abbott alone belongs the credit of having worked out the problem of the antiquity of man on the Atlantic coast," so that this gentleman appears to stand in a somewhat similar relation to this great question in America as did Boucher de Perthes in Europe. His researches are recorded in the first, second, and third volumes of the Reports of the Peabody Museum.

The interesting series of researches now briefly recorded has led us on step by step through the several stages of the quaternary at least as far back as the first great Glacial period, thus corresponding to the various epochs of Neolithic and Palæolithic man in Europe, terminating in the Suffolk flints, claimed to be pre-Glacial by Mr. Skertchley, or the earliest traces of human occupancy in Kent's Cavern, of which Mr. Pengelly states that "he is compelled to believe that the earliest men of Kent's Hole were inter-Glacial if not pre-Glacial." It now remains to adduce the evidence which carries us much farther back, and demonstrates the existence of man in Pliocene times. This evidence is derived from the works of art and human crania found in the auriferous gravels of California, and in order to appreciate duly its weight and importance, it is necessary to understand something of the physical characteristics of the country and the nature of the gravels themselves, with their included fossils, since both these factors combine to determine their geological age.

The Auriferous Gravels of California

The great lateral valleys of the Sierra Nevada are characterised by enormous beds of gravel, sometimes in thick deposits on the sides or filling up the whole bed of the valley, at other times forming detached hills or even mountains of considerable size. These gravel deposits are often covered with a bed of hard basalt or lava, having a generally level but very rugged surface, and hence possessing, when isolated, a very peculiar form, to which the name "table mountain" is often given. These tabular hills are sometimes 1000 or even 1500 feet high, and the basaltic capping varies from fifty to 200 feet thick. The gravels themselves are frequently interstratified with a fine white clay and sometimes with layers of basalt.

Geological exploration of the district clearly exhibits the origin of this peculiar conformation of the surface. At some remote period the lower lateral valleys of the Sierra Nevada became gradually filled with deposits of gravel brought down from the higher and steeper valleys. During the time this was going on there were numerous volcanic eruptions in the higher parts of the range, sending out great showers of ashes, which formed the beds now consolidated into pipe-clay or cement, while occasional lava streams produced intercalating layers of basalt. After this had gone on for a long period, and the valleys had in many places been filled up with débris to the depth of many hundred feet, there was a final and very violent eruption, causing outflows of lava, which ran down many of the valleys, filled the river beds, and covered up a considerable portion of the gravel deposits. These lava streams, some of which may be now traced for a length of twenty miles, of course flowed down the lower or middle portion of each valley, so that any part of the gravel remaining uncovered would be that most remote from the river bed towards one or other side of the valley. This gravel, being now the lowest ground as well as that most easily denuded, would of course be eaten away by the torrents and mark the commencement of new river beds, which thenceforth went on deepening their channels and forming new valleys which undermined and carried away some of the gravel, but always left steep slopes and cliffs wherever the lava flow protected the surface from the action of the rains. Hence it happens that the existing rivers are often in very different directions from the old ones, and sometimes cut across them, and thus isolated table mountains have been left rising up out of the surrounding plain or valley. What was once a single lave stream now forms several detached hills, the tops of which can be seen to form parts of one gently inclined plane, the surface of the original lava flow, now 1000 feet or more above the adjacent valleys. The American and Yuba valleys have been lowered from 800 to 1500 feet, while the Stanislaus river gorge has cut through one of these basalt-covered hills to the depth of 1500 feet.

While traveling by stage, in the summer of 1887, from Stockton to the Yosemite valley, I passed through this very district, and was greatly impressed by the indications of vast change in the surface of the country since the streams of lava flowed down the valleys. In the Stanislaus valley the numerous "table mountains" were very picturesque, often running out into castellated headlands or exhibiting long ranges of rugged black cliffs. At one spot the road passed through the ancient river-bed, clearly marked by its gravel, pebbles, and sand, but now about three or four hundred feet above the present river. We also often saw rock surfaces of metamorphic slates far above the present river-bed, thus proving that the original bed-rocks of the valley, as well as the lava and gravels, have been cut away to a considerable depth since the epoch of the lava flows. The ranges of "table mountains," now separated by deep valleys more than 1000 feet below them, could easily be seen, by their perfect agreement of slope and level, to have once formed part of an enormous lava stream spread over a continuous surface of gravel and rock.

Fossil Remains under the Ancient Lava Beds

These great changes in the physical conditions and in the surface features of the country alone imply a great lapse of time, but they are enforced and rendered even more apparent by the proofs of change in the flora and fauna afforded by the fossils, which occur in some abundance both in the gravels and volcanic clays. The animal remains found beneath the basaltic cap are very numerous, and are all of extinct species. They belong to the genera rhinoceros, elotherium, felis, canis, bos, tapirus, hipparion, equus, elephas, mastodon, and auchenia, and form an assemblage entirely distinct from those that now inhabit any part of the North American continent. Besides these we have a tolerably abundant series of vegetable remains, well preserved in the white clays formed from the volcanic ash. These comprise forty-nine species of deciduous trees and shrubs, all distinct from those now living, while not a single coniferous leaf or fruit has been found, although pines and firs are now the prevalent trees all over the sierra. Professor Lesquereaux, who has described these plants, considers them to be of Pliocene age with some affinities to Miocene; while Professor Whitney, the State geologist of California, considers that the animal remains indicate at least a similar antiquity.

These abundant animal and vegetable remains have mostly been discovered in the process of gold-mining, the gravel and sand of the old river-beds preserved under the various flows of basalt being especially rich in gold. Numerous shafts have been sunk and underground tunnels excavated in the auriferous gravels and clays, and the result has been the discovery not only of extinct animals and plants, but of works of art and human remains. The former have been found in nine different counties in the same gravels in which the extinct animals occur, while in no less than five widely separate localities, underneath the ancient lava flows, remains of man himself have been discovered. In order to show the amount of this evidence, and to enable us to appreciate the force or weakness of the objections with which, as usual, it has been received, a brief enumeration of these discoveries will be made. We will begin with the works of art as being the most numerous.

Works of Art in the Auriferous Gravels

In Tuolumne County from 1862 to 1865 stone mortars and platters were found in the auriferous gravel along with bones and teeth of mastodon ninety feet below the surface, and a stone muller was obtained in a tunnel driven under Table Mountain. In 1870 a stone mortar was found at a depth of sixty feet in gravel under clay and "cement," as the hard clay with vegetable remains (the old volcanic ash) is called by the miners. In Calaveras Country from 1860 to 1869 many mortars and other stone implements were found in the gravels under lava beds, and in other auriferous gravels and clays at a depth of 150 feet. In Amador County stone mortars have been found in similar gravel at a depth of forty feet. In Placer County stone platters and dishes have been found in auriferous gravels from ten to twenty feet below the surface. In Nevada County stone mortars and ground discs have been found from fifteen to thirty feet deep in the gravel. In Butte County similar mortars and pestles have been found in the lower gravel beneath lava beds and auriferous gravel; and many other similar finds have been recorded. It must be noted that the objects found are almost characteristic of California, where they are very abundant in graves or on the sites of old settlements, having been used to pound up acorns, which formed an important part of the food of the Indians. They occur literally by hundreds, and are so common that they have little value. It seems therefore absurd to suppose that in scores of cases, over a wide area of country and over a long series of years, gold-miners should have taken the trouble to carry down into their mines or mix with their refuse gravel these articles, of whose special scientific interest in the places where found they have no knowledge whatever. It is further noted that many of these utensils found in the old gravels are coarse and rudely finished as compared with those of more recent manufacture found on the surface. The further objection has been made that there is too great a similarity between these objects and those made in comparatively recent times. But the same may be said of the most ancient arrow and spear heads and those made by modern Indians. The use of the articles has in both cases been continuous, and the objects themselves are so necessary and so comparatively simple, that there is no room for any great modification of form.

Human Remains in the Auriferous Gravels

We will now pass on to the remains of man himself. In the year 1857 a fragment of a human skull with mastodon débris was brought up from a shaft in Table Mountain, Tuolumne County, from a depth of 180 feet below the surface. The matter was investigated by Professor Whitney, the State geologist, who was satisfied that the specimen had been found in the "pay gravel," beneath a bed three feet thick of cement with fossil leaves and branches, over which was seventy feet of clay and gravel. The most remarkable discovery, however is that known as the Calaveras skull. In the year 1866 some miners found in the cement, in close proximity to a petrified oak, a curious rounded mass of earthy and stony material containing bones, which they put on the side, thinking it was a curiosity of some kind.

Professor Wyman, to whom it was given, had great difficulty in removing the cemented gravel and discovering that it was really a human skull nearly entire. Its base was embedded in a conglomerate mass of ferruginous earth, water-worn volcanic pebbles, calcareous tufa, and fragments of bones, and several bones of the human foot and other parts of the skeleton were found wedged into the internal cavity of the skull. Chemical examination showed the bones to be in a fossilised condition, the organic matter and phosphate of lime being replaced by carbonate. It was found beneath four beds of lava, and in the fourth bed of gravel from the surface; and Professor Whitney, who after-wards secured the specimen for the State Geological Museum, has no doubt whatever of its having been found as described.

In Professor Whitney's elaborate Report on the Auriferous Gravels of the Sierra Nevada, from which most of the preceding sketch is taken, he arrives at the conclusion that the whole evidence distinctly proves "that man existed in California previous to the cessation of volcanic activity in the Sierra Nevada, to the epoch of greatest extension of the glaciers in that region, and to the erosion of the present rivercañons and valleys, at a time when the animal and vegetable creations differed entirely from what they are now, and when the topographical features of the State were extremely unlike those exhibited by the present surface." He elsewhere states that the animal and vegetable remains of these deposits prove them to be of "at least as ancient a date as the European Pliocene."

Professor Whitney enumerates two other cases in which human bones have been discovered in the auriferous gravel, and in one of them the bones were found by an educated observer, Dr. Boyce, M.D., under a bed of basaltic lava eight feet thick; but these are of but little importance when compared with the preceding cases, as to which we have such full and precise details. The reason why these remarkable discoveries should have been made in California rather than in any other part of America is sufficiently apparent if we consider the enormous amount of excavation of the Pliocene gravels in the long-continued prosecution of gold-mining, and also the probability that the region was formerly, as now, characterised by a milder climate, and a more luxuriant perennial vegetation, and was thus able to support a comparatively dense population even in those remote times. Admitting that man did inhabit the Pacific slope at the time indicated, the remains appear to be of such a character as might be anticipated, and present all the characteristics of genuine discoveries.

Concluding Remarks on the Antiquity of Man

Even these Californian remains do not exhaust the proofs of man's great antiquity in America, since we have the record of another discovery which indicates that he may, possibly, have existed at an even more remote epoch. Mr. E. L. Berthoud has described the finding of stone implements of a rude type in the Tertiary gravels of the Crow Creek, Colorado. Some shells were obtained from the same gravels, which were determined by Mr. T. A. Conrad to be species which are "certainly not later than Older Pliocene, or possibly Miocene." The account of this remarkable discovery, published in the *Proceedings of the Academy of Natural Sciences of Philadelphia*, 1872, is not very clear or precise, and it is much to be wished that some competent geologist would examine the locality. But the series of proofs of the existence of man by the discovery of his remains or his works going back step by step to the Pliocene period, which have been now briefly enumerated, takes away from this alleged discovery the extreme improbability which would be held to attach to it at the time when it was made.

It is surely now time that this extreme scepticism as to any extension of the human period beyond that reached by Boucher de Perthes, half a century ago, should give way to the ever-increasing body of facts on the other side of the question. Geologists and anthropologists must alike feel that there is a great, and at present inexplicable, chasm intervening between the earliest remains of man and those of his animal predecessors—that the entire absence of the "missing link" is a reproach to the doctrine of evolution; yet with strange inconsistency they refuse to accept evidence which in the case of any extinct or living animal, other than man, would be at least provisionally held to be sufficient, but follow in the very footsteps of those who blindly refused even to examine into the evidence adduced by the earlier discoverers of the antiquity of man, and thus play into the hands of those who can adduce his recent origin and unchangeability as an argument against the descent of man from the lower animals. Believing that the whole bearing of the comparative anatomy of man and of the anthropoid apes, together with the absence of indications of any essential change in his structure during the quaternary period, lead to the conclusion that he *must* have existed, as man, in Pliocene times, and that the intermediate forms connecting him with the higher apes probably lived during the early Pliocene or the Miocene period, it is urged that all such discoveries as those described in the present article are in themselves probable and such as we have right to expect. If this be the case, the proper way to treat evident as to man's antiquity is to place it on record, and admit it provisionally wherever it would be held adequate in the case of other animals; not, as is too often now the case, ignore it as unworthy of acceptance or subject its discoverers to indiscriminate accusations of being either impostors themselves or the victims of impostors. Error is sure to be soon detected, and its very detection is often a valuable lesson. But facts once rejected are apt to remain long buried in obscurity, and their non-recognition may often act as a check to further progress. It is in the hope of inducing a more healthy public opinion on this interesting and scientifically important question that this brief record of the evidences of man's antiquity in North America has been compiled.

IX

THE DEBT OF SCIENCE OF DARWIN[1]

The Century before Darwin—The Voyage of the *Beagle*—The Journal of Researches—Studies of Domestic Animals—Studies of Cultivated and Wild Plants—Researches on the Cowslip, Primrose, and Loosestrife—The Struggle for Existence—Geographical Distribution and Dispersal of Organisms—The Descent of Man and Later Works—Estimate of Darwin's Life-Work.

THE great man recently taken from us had achieved an amount of reputation and honour perhaps never before accorded to a contemporary writer on science. His name has given a new word to several languages, and his genius is acknowledged wherever civilisation extends. Yet the very greatness of his fame, together with the number variety, and scientific importance of his works, has caused him to be altogether misapprehended by the bulk of the reading public. Every book of Darwin's has been reviewed or noticed in almost every newspaper and periodical, while his theories have been the subject of so much criticism and so much dispute, that most educated persons have been able to obtain some general notion of his teachings, often without having read a single chapter of his works,—and very few, indeed, except professed students of science, have read the whole series of them. It has been so easy to learn something of the Darwinian theory at second-hand that few have cared to study it as expounded by its author.

It thus happens that, while Darwin's name and fame are more widely known than in the case of any other modern man of science, the real character and importance of the work he did are as widely misunderstood. The best scientific

[1] This article appeared in the *Century Magazine* of January 1883.

authorities rank him far above the greatest names in natural science—above Linnæus and Cuvier, the great teachers of a past generation—above De Candolle and Agassiz, Owen and Huxley, in our own times. Many must feel inclinded to ask,—What is the secret of this lofty pre-eminence so freely accorded to a contemporary by his fellow-workers? What has Darwin done, that even those who most strongly oppose his theories rarely suggest that he is overrated? Why is it universally felt that the only name with which his can be compared in the whole domain of science is that of the illustrious Newton?

It will be my endeavour in the present chapter to answer these questions, however imperfectly, by giving a connected sketch of the work which Darwin did, the discoveries which he made, the new fields of research which he opened up, the new conceptions of nature which he has given us. Such a sketch may help to clear away some of the obscurity which undoubtedly prevails as to the cause and foundation of Darwin's pre-eminence.

In order to understand the vast and fundamental change effected by the publication of Darwin's most important volume—*The Origin of Species*—we must take a hasty glance at the progress of the science of natural history during the preceding century.

The Century before Darwin

Almost exactly a hundred years before Darwin we find Linnæus and his numerous disciples hard at work describing and naming all animals and plants then discovered, and classifying them according to the artificial method of the great master, which is still known as the Linnæan System; and from that time to the present day a large proportion of naturalists are fully occupied with this labour of describing new species and new genera, and in classifying them according to the improved and more natural systems which have been gradually introduced.

But another body of students have always been

dissatisfied with this superficial mode of studying externals only, and have devoted themselves to a minute examination of the internal structure of animals and plants; and early in this century the great Cuvier showed how this knowledge of anatomy could be applied to the classification of animals according to their whole organisation in a far more natural manner than by the easier method of Linnæus. Later on, when improved microscopes and refined optical and chemical tests became available, the study of anatomy was carried beyond the knowledge of the parts and organs of the body—such as bones, muscles, blood-vessels, and nerves—to the investigation of the tissues, fibres, and cells of which these are composed; while the physiologists devoted themselves to an inquiry into the mode of action of this complex machinery, so as to discover the use of every part, the nature of its functions in health and disease, and, as far as possible, the nature of the forces which kept them all in action.

Down to the middle of the present century the study of nature advanced with giant strides along these separate lines of research, while the vastness and complexity of the subject led to a constantly increasing specialisation and division of labour among naturalists, the result being that each group of inquirers came to look upon *his* own department as more or less independent of all the others, each seemed to think that any addition to *his* body of facts was an end in itself, and that any bearing these facts might have on other branches of the study or on the various speculations as to the "system of nature" or the "true method of classification" that had at various times been put forth was an altogether subordinate and unimportant matter. And, in fact, they could hardly think otherwise. For, while there was much talk of the "unity of nature," a dogma pervaded the whole scientific world which rendered hopeless any attempt to discover this supposed unity amid the endless diversity of organic forms and structures, while so much of it as might be detected would necessarily be speculative and unfruitful. This dogma was that of the original diversity and permanent stability of species, a dogma which the rising generation of naturalists must find it hard to believe was actually held, almost universally, by the great men they look up to as masters in their several departments, and held for the most part with an unreasoning tenacity and scornful arrogance more suited to politicians or theologians than to men of science. Although the doctrine of the special and independent creation of every species that now exists or ever has existed on the globe was known to involve difficulties and contradictions of the most serious nature, although it was seen that many of the facts revealed by comparative anatomy, by embryology, by geographical distribution, and by geological succession were utterly unmeaning and even misleading, in view of it; yet, down to the period we have named, it may be fairly stated that nine-tenths of the students of nature unhesitatingly accepted it as literally true, while the other tenth, though hesitating as to the actual independent creation, were none the less decided in rejecting utterly and scornfully the views elaborated by Lamarck, by Geoffroy St. Hilaire, and at a much later date by the anonymous author of the *Vestiges of Creation*—that every living thing had been produced by some modification of ordinary generation from parents more or less closely resembling it. Holding such views of the absolute independence of each species, it almost necessarily followed that the only aspect of nature of which we could hope to acquire complete and satisfactory knowledge was that which regarded the species itself. This we could describe in the minutest detail; we could determine its range in space and in time; we could investigate its embryology from the rudimental germ, or even from the primitive cell, up to the perfect animal or plant; we could learn every point in its internal structure, and we might hope, by patient research and experiment, to comprehend the use, function, and mode of action of every tissue and fibre, and ultimately of each cell and organic unit. All this was real knowledge, was solid fact. But, so soon as we attempted to find out the relations of *distinct species* to each other, we embarked on a sea of speculation. We could, indeed, state *how* one species differed from another species in every particular of which we had knowledge; but we could draw no sound inferences as to the reason or cause of such differences or resemblances, except by claiming to know the very object and meaning of the creator in producing such diversity. And, in point of fact, the chief inference that was drawn is now proved to be erroneous. It was generally assumed, as almost self-evident, that the ultimate cause of the differences in the forms, structures, and habits of the organic productions of different countries, was that each species inhabiting a country was specially adapted to the physical conditions that prevailed there, to which it was exactly fitted. Even if this theory had been true, it was an unproductive ultimate fact, for it was never pretended that we could discover any reason for the limitation of humming-birds and cactuses to America, of hippopotami to Africa, or of kangaroos and gum-trees to Australia; and we were obliged to believe either that these countries possessed hidden peculiarities of climate or other conditions, or that this was only one out of many unknown and unknowable causes determining the special action of the creative power. All this was felt to be so unsatisfactory that the majority of naturalists openly declared that their sole business was to accumulate facts, and that any attempt to co-ordinate these facts and see what inferences could be drawn from them was altogether premature. In this frame of mind, year after year passed away, adding its quota to the vast mass of undigested facts which were accumulating in every branch of the science. The remotest parts of the globe were ransacked to add to the treasures of our museums, and the number of known species became so enormous that students began to confine themselves not merely to single classes, as birds or insects, but to single orders, as beetles or land-shells, or even to smaller groups, as weevils or butterflies. All, too, were so impressed with the belief in the reality and permanence of species, that endless labour was bestowed on the attempt to distinguish them—a task whose hopelessness may be inferred from the fact that, even in the well-known British flora, one authority describes sixty-two species of brambles and roses, another of equal eminence only ten species of the same groups; and it is by no means uncommon for two, five, or even ten species of one author to be classed as a single species by another. All this time geologists had been so assiduously at work in the discovery of organic remains that the extinct species often equalled, and, in some groups—as the Mollusca—very far exceeded, those now living on the earth, and these were all found to belong to the very same classes and orders as the living forms, and to form part of one great system. Much attention was now paid to the geological succession of the different groups of animals, which were found to exhibit a progressive advancement from ancient to recent times, while the breaks in the series between each great geological formation were held to show that the older forms of life had been destroyed, and were replaced by a new creation of a more advanced organisation suited to the altered conditions of the world.

And thus, perhaps, we might have gone on to this day, ever accumulating fresh masses of fact, while each set of workers became ever more and more occupied in their own departments of study, and, for want of any intelligible theory to connect and harmonise the whole, less and less able to appreciate the labour of their colleagues, had not Charles Darwin made his memorable voyage round the world, and thenceforth devoted himself, as so many had done before him, to a life of patient research in the domain of organic nature. But how different was the object attained! Others have added greatly to our knowledge of details, or created a reputation by some important work; he has given us new conceptions of the world of life, and a theory which is itself a powerful instrument of research; has shown us how to combine into one consistent whole the facts accumulated by all the separate classes of workers, and has thereby revolutionised the whole study of nature. Let us endeavour to see by what means he arrived at this vast result.

The Voyage of the BEAGLE

Passing by the ancestry and early life of Darwin, which have been made known to the whole reading public by many biographical notices and recently by the publication of his

Life and Letters, we may begin with the first event to which we can distinctly trace his future greatness—his appointment as naturalist to the *Beagle,* on the recommendation of his friend and natural-history teacher, Professor Henslow, of Cambridge University. It was in 1831, when Darwin, then twenty-two years of age, had just taken his B.A. degree, that he left England on his five years' voyage in the Southern Hemisphere. It is probably to this circumstance that the world owes the great revolution in our conception of the organic world so well known as the Darwinian theory. The opportunity of studying nature in new and strange lands; of comparing the production of one country with those of another; of investigating the physical and biological relations of islands and continents; of watching the struggle for existence in regions where civilisation has not disturbed the free action and reaction of the various groups of animals and plants on each other; and, what is perhaps more important still, the ample leisure to ponder again and again on every phase of the phenomena which presented themselves, free from the attractions of society and the disturbing excitement of daily association with contemporary men of science,—these are the conditions most favourable to the formation of habits of original thought, and the months and years which at first sight appear intellectually wasted in the companionship of uncivilised man, or in the solitary contemplation of nature, are those in which the seed was sown which was destined to produce in after years the mature fruit of great philosophical conceptions. Let us then first glance over the *Journal of Researches,* in which are recorded the main facts and observations which struck the young traveller, and see how far we can detect here the germs of those ideas and problems to the working out of which he devoted a long and laborious life.

The Journal of Researches

The question of the causes which have produced the distribution and dispersal of organisms seems to have been a constant subject of observation and meditation. At an early period of the voyage he collected infusorial dust which fell on the ship when at sea, and he notes the suggestive fact that in similar dust collected on a vessel 300 miles from land he found particles of stone above the thousandth of an inch square, and remarks: "After this fact, one need not be surprised at the diffusion of the far lighter and smaller sporules of cryptogamic plants." He records many cases of insects occurring far out at sea, on one occasion when the nearest land was 370 miles distant. He paid special attention to the insect and plants inhabiting the Keeling or Cocos, and other recently formed coralline or volcanic islands; the contrast of these with the peculiar productions of the

Galapagos evidently impressed him profoundly; while the remarkable facts presented by this latter group of islands brought out so clearly and strongly in insuperable difficulties of the then accepted theory of the independent origin of species, as to keep this great problem ever present to his mind, and at a later period, led him to devote himself to the patient and laborious inquiries which were the foundation of his immortal work. He again and again remarks on the singular facts presented by these islands. Why, he asks, were the aboriginal inhabitants of the Galapagos created on American types of organisation, though the two countries differ totally in geological character and physical conditions? Why are so many of the species peculiar to the separate islands? He "is astonished at the amount of creative force, if such an expression may be used, displayed in these small, barren, and rocky islands; and still more so at its diverse, yet analogous action on points so near each other."

The variations which occur in species, as well as the modifications of the same organ in allied species, subjects which had been much neglected by ordinary naturalists, were constantly noted and commented on. He remarks on the occasional blindness of the burrowing tucutucu of the Pampas as supporting the view of Lamarck on the gradually acquired blindness of the aspalax; on the hard point of the tail of trigonocephalus, which constantly vibrates and produces a rattling noise by striking against grass and brushwood, as a character varying towards the complete rattle of the rattlesnake; on the small size of the wild horses in the Falkland island, as progressing towards a small breed like the Shetland ponies of the North ; and on the strange fact of the cattle having increased in size, and having partly separated into two differently coloured breeds. While collecting the remains of the great extinct mammals of the Pampas, he was much impressed by the fact that, however huge in size or strange in form, they were all allied to living South American animals, as are those of the cave-deposits of Australia to the marsupials of that country; and he thereon remarks: "This wonderful relationship in the same continent between the dead and the living will, I do not doubt, hereafter throw more light on the appearance of organic beings on our earth, and their disappearance from it, than any other class of facts." He also saw, at this early period, the importance fact that there is some great and constant check to the increase of wild animals, though most of them breed very rapidly, and, of course, would increase in a geometrical ratio were some such check not in constant action. He traces the comparative rarity of a species to less favourable conditions of existence, and extinction to the normal action of still more unfavourable conditions, and compares the destruction of a species by man and its extinction by its natural enemies as being phenomena of the same essential nature. The various classes of facts here referred to seemed to him "to throw some light on the origin of species—that mystery of mysteries, as it has been called by one of our greatest philosophers;" and he tells us that, soon after his return home in 1837, it occurred to him "that something might perhaps be made out on this question by patiently accumulating and reflecting on all sorts of facts which could possibly have any bearing upon it." We know from his own statement that he had already perceived that no explanation but some form of the derivation or development hypothesis, as it was then termed, would adequately explain the remarkable facts of distribution and geological succession which he had observed during his voyage; yet he tells us that he worked on for five years before he allowed himself to speculate on the subject; and then, having formulated his provisional hypothesis in a definite shape during the next two years, he devoted another fifteen years to continuous observation, experiment, and literary research, before he gave to the astounded scientific world an abstract of his theory in all its wide-embracing scope and vast array of evidence, in his epoch-making volume, *The Origin of Species.*

If we add to the periods enumerated above the five years' observation and study during the voyage, we find that this work was the outcome of *twenty-seven* years of continuous thought and labour, by one of the most patient, most truth-loving, and most acute intellects of our age. During all this long period only a very few of his most intimate friends were aware that he had departed from the then beaten tack of biological study, while the great body of naturalists only knew him as a good geologist, as the writer of an interesting book of travels, and the author of an admirable monograph of the cirripeda or barnacles, as well as of a most ingenious explanation of the origin and structure of coral-reefs—a series of volumes which were the direct outcome of his voyage, and which gave him an established reputation. Even when the great work at last appeared, few could appreciate the enormous basis of fact and experiment on which it rested, until, during the succeeding twenty years, there appeared that remarkable succession of works which exhibited a sample (and only a sample) of the exhaustless store of materials and the profound maturity of thought on which his early volume was founded. From these various works, aided by some personal intercourse and a correspondence extending over twenty years, the present writer will endeavour to indicate the nature and extend of Darwin's researches.

Studies of Domestic Animals

Although, as we have said, Darwin had early arrived at the conclusion that allied species had descended from common ancestors by gradual modification, it long remained to him an inexplicable problem how the necessary degree of modification could have been effected, and he adds: "It would thus have remained for ever, had I not studied domestic productions, and thus acquired a just idea of the power of selection." These researches, very briefly sketched in the first and parts of the fifth and ninth chapters or the *Origin of Species,* were published at length (after a delay of nine years, owing to ill health) in two large volumes, with the title *Animals and Plants Under Domestication;* and no one who has not read these can form an adequate idea of the wide range and thorough character of the investigation on which every statement or suggestion in the former work was founded.

The copious references to authorities show us that he must have searched through almost the entire literature of agriculture and horticulture, of horse and cattle breeding, of sporting, of dog, cat, pigeon, and fowl fancying, including endless series of reviews, magazines, journals of societies, and newspapers, besides every scientific treatise bearing in any way on the subject, whether published in the country, on the Continent, or in America. The facts thus laboriously gathered were supplemented by personal inquiries among zoologists and botanists, farmers, gardeners, sporting-men, pigeon-fanciers, travellers, and any one who could possibly afford direct personal information on any of the matters he was investigating. Then came his own observation and experiment, to fill up gaps, to settle doubtful points, or to determine questions the importance of inquiring into which no one had ever suspected; and lastly, there was the power of arrangement and comparison, the originality and depth of thought, which drew out from this vast mass of heterogeneous materials conclusions of the highest value as bearing on the question of the possible change of species, and the means by which it had been brought about.

In order to determine the nature and amount of the variability of domestic productions, he prepared skeletons of all the more important breeds of rabbits, pigeons, fowls, and ducks, as well as of the wild races from which they are known to have been produced, and showed, both by measurements and by accurate drawings, that not only superficial characters, but almost every part of the bony structure varied to such an amount as usually characterises very distinct species or even distinct genera of wild animals. Another set of experiments was made by crossing the different breeds of pigeons and fowls which were most completely unlike the wild race, with the result that in many cases the offspring were more like the wild ancestor than either of the parents. These experiments, supported by a mass of facts observed by other persons, served to establish the principle of the tendency of crosses to revert to the ancestral form; and this principle enabled him to explain the interesting fact of the frequent appearance of stripes on mules, and occasionally on dun-coloured horses, on the hypothesis, supported by a mass of collateral evidence, that the common ancestor of the horse, ass, and zebra tribe was a partially striped and dun-coloured animal.

A number of very important conclusions were deduced from the facts presented by domesticated animals and plants, a few of which may be here referred to. For example, it was proved that the parts most selected or which had already most varied—as the tail in fan-tailed pigeons, which has more tail-feathers than any one of the 8000 different kinds of living birds—were most subject to further variation; and this showed that, when once any part had begun to change, variations became more abundant, thus furnishing materials to render still further change in the same direction comparatively easy. This is the secret of the rapid improvement of breeds or races, and is equally applicable to the formation of species by natural selection. Again, it was found that in many cases, when much variation occurred, there was a tendency to a difference in the sexes which had not before existed. This has been observed in sheep, in fowls, and in pigeons, and it is very interesting as indicating the origin of that wonderful diversity between the two sexes which occurs in several groups of animals. Another curious fact is the correlation of parts which occurs in many animals, such as the tusks and bristles of swine, and the hair and teeth in some dogs, both increasing or becoming lost together; the beak and feet of pigeons, both increasing or diminishing together; the colour and size of the leaves and seeds changing simultaneously in some plants; and numerous other instances which serve to explain some of the peculiar characters of natural objects for which we can discover or imagine no direct use.

The effect of disuse in causing the diminution of an organ was exhibited by careful comparison and measurements of tame and wild birds. The sternum, scapulæ, and furcula to which the muscles used in flight are attached, are found to be diminished in domestic pigeons, as were the wing-bones in domestic fowls, the capacity of the skull in tame rabbits, and the size and strength of the wings in silkworm moths. The evidence afforded by the breeds of pigeons (which have been domesticated for so many centuries and in so many parts of the world) of the process of selection, whether unconscious or methodical, is very clearly set forth, and serves as a typical example with which to compare the various phenomena presented by allied species in a state of nature; and in concluding its discussion, he thus replies to some objections:—

"I have heard it objected that the formation of the several domestic races of the pigeon throws no light on the origin of the wild species of the columbidæ, because their differences are not of the same nature. The domestic races, for instance, do not differ, or hardly at all, in the relative lengths and shapes of the primary wing-feathers, in the relative length of the hind toe, or in habits of life, as in roosting and building on trees. But the above objection shows how completely the principle of selection has been misunderstood. It is not likely that characters selected by the caprice of man should resemble differences preserved under natural conditions, either from being of direct service to each species, or from standing in correlation with other modified and serviceable structures. Until man selects birds differing in the relative length of the wing-feathers or toes, etc., no sensible change in these parts should be expected. ... With respect to the domestic races not roosting or building in trees, it is obvious that fanciers would never attend to or select such changes in habits."

Studies of Cultivated and Wild Plants

Still more remarkable, perhaps, is the collection of facts afforded by plants, which can be so much more easily cultivated and experimented upon than animals, while the general phenomena they present are strikingly accordant in the two kingdoms. As an example of the great mass of facts afforded by horticulture, he records that three hundred distinct varieties were produced, in the course of fifty years, from a single wild rose (Rosa spinosissima). We find in these volumes enormous collections of facts on bud-variation, or the occurrence of changes in the flower or leaf-buds of full-grown plants, from which new varieties can be and often are produced; and, after a most full and interesting discussion of the cases, it is shown that some are probably due to reversion to an ancestral form, other to reversion to one parent when the plant has been derived from a cross, and others, again, to that spontaneous variability which seems to be the universal characteristic of all living organisms.

Three very interesting chapters are then devoted to the subject of inheritance, and a host of strange and heretofore inexplicable facts are brought together, compared, and classified, and shown to be in accordance with a few general principles. Then follow five chapters on crossing and hybridism, perhaps the most important in the whole work, since they afford the clue to so much of the varied structure and complex relations of animals and plants. Notwithstanding the enormous mass of facts and observations here given, the portion relating to plants is often but an abstract of the results of his own elaborate experiments, carried on for a long series of years, and given at length in three separate volumes on *The Fertilisation of Orchids,* on *Cross and Self-Fertilisation of Plants,* and on *The Forms of Flowers.* These works may be said to have revolutionised the science of botany, since, for the first time, they gave a clear and intelligible reason for the existence of that wonderful diversity in the form, colours, and structure of flowers, on the details of which the systematic botanist had founded his generic and specific distinctions, but as to whose meaning or use he was, for the most part, profoundly ignorant. The investigation of the whole subject of crossing and hybridity had shown that, although hybrids between distinct species usually produced sterile offspring, yet crosses between slightly different varieties led to increased fertility; and, during some experiments on this subject, Darwin found that the produce of these crosses were also remarkable for vigour of growth. This led to a long series of experimental researches, the general result of which was to establish the important proposition that cross-fertilisation is of the greatest importance to the health, vigour, and fertility of plants. The fact that the majority of flowers are hermaphrodite, and appear to be adapted for self-fertilisation, seemed to be opposed to this view, till it was found that, in almost every case, there were special arrangements for ensuring, either constantly or occasionally, the transference of pollen from the flowers of one plant to those of another of the same species. In the case of orchids, it was shown that those strange and beautiful flowers owed their singular and often fantastic forms and exceptional structure to special adaptations for cross-fertilisation by insects, without the agency of which most of them would be absolutely sterile. Many of the species are so minutely adapted to particular species or groups of insects, that they can be fertilised by no others; and careful experiment and much thought was often required to find out the exact mode in which this was effected. In some instance the structure of the flowers seemed adapted to prevent fertilisation altogether, till it was at length discovered that a particular insect entering the flower in one particular way caused the pollen to stick to some part of its body, which was always the exact part which the insect, on visiting another flower, would being in contact with the stigma, and thus fertilise it. These investigations explained a host of curious facts which had hitherto been facts only without meaning, such as the twisting of the ovary in most of our wild orchids, which was found to be often necessary to bring the flower into a proper position for fertilisation,—the existence of sacs, cups, or spurs, the latter often of enormous length, but shown to be each adapted to the structure of some particular insect, and often serving to prevent other insect from reaching the nectar which they might rob without fertilising the flower,—the form, size, position, rugosities, or colour of the lip, serving as a landing-place for insects and a guide to the nectar-secreting organs,—the varied odours, sometimes emitted by day, and sometimes by night only, according as the fertilising insect was diurnal or nocturnal, and other characters too numerous to refer to here, so that it became evident that every peculiarity of these wonderful plants, in form or structure, in colour or marking, in the smoothness, rugosity, or hairiness of parts of the flowers, in their times of opening, their movements, or their odours, had every one of them a purpose, and were, in some way or other, adapted to secure the fertilisation of the flower and the preseration of the species.

Researches on the Cowslip, Primrose, and Loosestrife

The next set of observations, on some of our commonest English flowers of apparently simple structure, were not less original and instructive. The cowslip (Primula veris) has two kinds of flowers in nearly equal proportions: in the one the stamens are long and the style short, and in the other the reverse, so that in the one the stamens are visible at the mouth of the tube of the flower, in the other the stigma occupies the same place, while the stamens are half-way down the tube. This fact had been known to botanists for seventy years, but had been classed as a case of mere variability, and therefore considered to be of no importance. In 1860 Darwin set to work to find out what it meant, since, according to his views, a definite variation like this *must* have a purpose.

After a considerable amount of observation and experiment, he found that been and moths visited the flowers, and that their probosces became covered with pollen while sucking up the nectar, and further, that the pollen of a long-*stamened* plant would be most surely deposited on the stigma of the long-*syled* plants, and *vice versâ.* Now followed a long series of experiments, in which cowslips were fertilised either with pollen from the same kind or from a different kind of flower, and the invariable result was that the crosses between the two different kinds of flowers produced more good capsules, and more seeds in each capsule; and as these crosses would be most frequently effected by insects, it was clear that this curious arrangement directly served to increase the fertility of this common plant.

The same thing was found to occur in the primrose, and in many other species of primulaceæ, as well as in flax (Linum perenne), lungworts (Pulmonaria), and a host of other plants, including the American partridge-berry (Mitchella repens). These are called dimorphic heterostyled plants.

Still more extraordinary is the case of the common loose-strife (Lythrum salicaria), which has both stamens and styles of three distinct lengths, each flower having two sets of stamens and one style, all of different lengths, and arranged in three different ways: (1) a short style, with six medium and six long stamens; (2) a medium style, with six short and six long stamens; (3) a long style, with six medium and six short stamens. These flowers can be fertilised in eighteen distinct ways, necessitating a vast number or experiments, the result being, as in the case of the cowslip, that flowers fertilised by the pollen from stamens of the same length as the styles, gave on the average a larger number of capsules and a very much larger number of seeds than in any other case. The exact correspondence in the length of the style of each form with that of the stamens in the two other forms ensures that the pollen attached to any part of the body of an insect shall be applied to a style of the same length on another plant, and there is thus a triple chance of the maximum of fertility. Some other species of lythrum, of oxalis, and pontederia, were also found to have three-formed stamens and styles; and in the case of the oxalis, experiments were made showing that crosses between flowers with stamens and styles of unequal length were always nearly barren. During these experiments 20,000 seeds of Lythrum salicaria were counted under the microscope. For several years a further supplementary series of experiments were carried out, showing that the seeds produced by the illegitimates crosses (as he terms them) were not only very few, but, when sown, always produced comparatively weak, small, or unhealthy plants, not likely to exist in competition with the stronger offspring of legitimate crosses. There is thus the clearest proof that these complex arrangements have the important end of securing both a more abundant and more vigorous offspring.

Perhaps no researches in the whole course of the study of nature have been so fertile in results as these. No sooner were they made known that observers set to work in every part of the world to examine familiar plants under this new aspect. With very few exceptions it is now found that every flower presents arrangements for securing cross-fertilisation, either constantly or occasionally, sometimes by the agency of the wind, but more frequently through the mediation of the insects or birds. Almost all the irregularity and want of symmetry in the forms of flowers, which add so much to their variety and beauty, are found to be due to this cause; the production of nectar and the various nectar-secreting organs is directly due to it, as are the various odours and the various colours and makings of flowers. In many case flowers which seem so simply constructed that the pollen *must* fall on the stigma and thus produce self-fertilisation, are yet surely cross-fertilised, owing to the circumstance of the stigma and the anthers arriving at maturity at slightly different periods, so that, though the pollen may fall on the stigma of its own flower, fertilisation does not result; but when insects carry the pollen to another plant the flowers of which are a little more advanced, cross-fertilisation is effected. There is literally no end to the subjects of inquiry thus opened up, since very single species, and even many varieties of flowering plants, present slight peculiarities which modify to some extend their mode of fertilisation. This is well shown by the remarkable observations of the German botanist Kerner, who points out that a vast number of details in the structure of plants, hitherto inexplicable, are due to the necessity of keeping away "unbidden guests," such as snails, slugs, ants, and many other kinds of animals, which would destroy the flowers or the pollen before the seeds were produced. When this simple principle is once grasped, it is seen that almost all the peculiarities in the form, size, and clothing of plants are to be thus explained, as the spines or hairs of the stem and branches, or the glutinous secretion which effectually prevent ants from ascending the stem, the drooping of the flowers to keep out rain or to prevent certain insects from entering them, and a thousand other details which are described in Kerner's most instructive volume. This branch of the inquiry was hardly touched upon by Darwin, but it is none the less a direct outcome of his method and his teaching.

The Struggle for Existence

But we must pass on from these seductive subject to give some indication of the numerous branches of inquiry of which we have the results given us in the *Origin of Species,* but which have not yet been published in detail. The observations and experiments on the relations of species in a state of nature, on checks to increase and on the struggle for existence, were probably as numerous and exhaustive as those on domesticated animals and plants. As examples of this we find indications of careful experiments on seedling plants and weeds, to determine what proportion of them were destroyed by enemies before they came to maturity; while another set observations determined the influence of the more robust in killing out the weaker plants with which they come into competition. This last fact, so simple in itself, yet so much overlooked, affords an explanation of many of the eccentricities of plants distribution, cultivation, and naturalisation. Every one who has tried it knows the difficulty or impossibility of getting foreign plants, however hardy, to take care of themselves in a garden as in a state of nature. Wherever we go among the woods, mountains, and meadows of the temperate zone, we find a variety of charming flowers growing luxuriantly amid a dense vegetation of other plants, none of which seem to interfere with each other. By far the larger number of these plants will grow with equal luxuriance in our gardens, showing that peculiarities of soil and climate are not of vital importance; but not one in a thousand of these plants ever runs wild with us, or can be naturalised by the most assiduous trials; and if we attempt to grow them under natural conditions in our gardens, they very soon succumb under the competition of the plants by which they are surrounded. It is only be constant attention, not so much to them as to their neighbours—by pruning and weeding close around them so as to allow them to get a due proportion of light, air, and moisture, that they can be got to live. Let any one bring home a square foot of turf from a common or hill-top, containing some choice plant growing and flowering luxuriantly, and place it in his garden, untouched, in the most favourable conditions of light and moisture, and in a year or two it almost certainly disappear, killed out by the more vigorous growth of other plants. The constancy of this result, even with plants removed only a mile or two, is a most striking illustration of the preponderating influence of organism on organism, that is, of the struggle for existence. The rare and delicate flower which we find in one field or hedgerow, while for miles around there is no trace of it, maintains itself there, not on account of any specialty of soil or aspect, or there physical conditions being directly favourable to itself, but because in that spot only there exists the exact combination of other plants and animals which alone is not incompatible with its wellbeing, that combination perhaps being determined by local conditions or changes which many years ago allowed a particular set of plants and animals to monopolise the soil and thus keep out intruders. Such consideration teach us that the varying combinations of plants characteristic of almost every separate field or bank, or hillside, or wood throughout our land, is the result of a most complex and delicate balance of organic forces—the final outcome for the time being of the constant struggle of plants and animals to maintain their existence.

Geographical Distribution and Dispersal of Organisms

Another valuable set of experiments and observation are those bearing on the geographical distribution of animals and plants—a branch of natural history which, under the old idea of special creations, had no scientific existence. It is to Darwin that we owe the establishment of the distinction of oceanic from continental islands, while he first showed us the various modes by which the former class of islands have been stocked with life. By a laborious research in all the accounts of old voyages, he ascertained that none of the islands of the great oceans very remote from land possessed either land mammalia or amphibia when first visited; and on examination it is found that all these islands are either of volcanic origin or consist of coral reefs, and are therefore presumably of comparatively recent independent origin, not portions of submerged continents, as they were formerly supposed to be. Yet these same islands are fairly stocked with plants, insects, land-shells, birds, and often with reptiles, more particularly lizards, usually of peculiar species, and it thus becomes important to ascertain how these organisms originally reached the islands, and the comparative powers different groups of plants and animals posses of traversing a wide extend of ocean.

With this view he made numerous observations and some ingenious experiments. He endeavoured to ascertain how long different kinds of seeds will resist the action of salt water without losing their vitality, and the result showed that a large number of seeds will float a month without injury, while some few survived an immersion of one hundred and thirty-seven days. Now, as ocean currents flow on the average thirty-three miles a day, seeds might easily be carried 1000 miles, and in very exceptional cases even 3000 miles, and still grow. Again, it is known that drift-timber is often carried enormous distances, and some of the inhabitants of the remote coral-islands of the Pacific obtain wood by this means, as well as stones fastened among the roots. Now, Darwin examined torn-up trees, and found that stones are often inclosed by the roots growing round them so as to leave closed cavities containing earth behind; and from a small portion of earth thus completely inclosed, he raised three dicotyledonous plants. Again, the seeds that have passed through the bodies of birds germinate freely, and thus birds may carry plants from island to island. Earth often adheres to the feet of aquatic and wading birds, and these migrate to enormous distances and visit the remotest island, and from earth thus attached to birds' feet several plants were raised. As showing the importance of this mode of transport, an experiment was made with six and three-fourths ounces of mud taken from the edge of a little pond, and it was found to contain the enormous number of five hundred and thirty-seven seeds of several distinct species! This was proved by keeping the mud under glass and pulling up each plant as it appeared, and at the end of six months the result was as given above. It was also found that small portions of aquatic plants were often entangled in the feet of birds, and to these as well as to the feet themselves, molluscs or their eggs were found the be attached, furnishing a mode of distribution for such organisms. Experiments were also made on the power of land-shells to resist the action of sea-water; and we have already referred to the observations on volcanic dust carried far out to sea, illustrating the facilities for the wide extension by aerial currents of such plants as have very minute or very light seeds.[1] The facts are of so anomalous and apparently contradictory a character that, on the old hypothesis of the special independent creation of each species, no rational explanation of them could be found; and we may fairly claim that the clear and often detailed explanation which can be given by means of the theories and investigations of Darwin, lend a powerful support to this views, and go far to complete the demonstration of their correctness.

Our space will not permit us to do more than advert to the numerous ingenious explanations and suggestions with which the *Origin of Species* abounds, such as, for example, the strange fact of so many of the beetles of Madeira being wingless, while the same species, or their near allies on the continent of Europe, have full power of flight; and that this is not due to any direct action of climate or physical conditions is proved by the equally curious fact that such species of insects as have wings in Maderia, have them rather larger than usual. Equally new and important is the Darwinian explanation of the form of the bees' cell, which is shown to

[1] This series of observations and experiments, supplemented by those of other observers, have been applied by the writer of this article to explain in some detail the remarkable phenomena presented by the distribution of animals and plants over the chief islands of the globe. See *Island Life*. Macmillan and Co. be due to a few simple instincts which necessarily lead to the exact hexagonal cell with the base formed of three triangular plates inclined at definite angles, on which so much mathematical learning

and misplaced admiration have been expended; and this explanation is no theory, but is the direct outcome of experiments on the bees at work, as original as they were ingenious and convincing.

The Descent of Man and Later Works

We must, however, pass on to the great and important work, *The Descent of Man and on Selection in Relation to Sex,* which abounds in strange facts and suggestive explanations; and for the reader who wishes to understand the character and bearing of Darwin's teachings, this book is the fitting supplement to the *Origin of Species* and the *Domesticated Animals and Plants.* To give any adequate account of this most remarkable book and the controversies to which it has given rise, would require an article to itself. We refer to it here in order to point out what is not generally known, that its publication was entirely out of its due course, and was not anticipated by its author three years before. In the introduction to *Domesticated Animals* (published in 1868), after explaining the scope of that work, he tells us that in a second work he shall treat of "Variation Under Nature," giving copious facts on variation, local and general, on races, subspecies and species, on geometrical increase, on the struggle for existence, with the results of experiments showing that diversity of forms enables more life to be supported on a given area, while the extermination of less improved forms, the formation of genera and families, and the process of natural selection, will be fully discussed. This work would have given all the facts on which chapters ii, to v. of the *Origin of Species* were founded. In a third work he proposed to show, in detail, how many classes of facts natural selection explains, such as geological succession, geographical distribution, embryology, affinities, classification, rudimentary organs, etc., etc., thus giving the facts and reasonings in full on which the latter part of the *Origin of Species* was founded. Unfortunately, neither of these works has appeared, and thus the symmetry and completeness of the body of facets which Darwin has collected have never been made known. The cause is well known to have been the continued pressure of ill-health. The work on *Domesticated Animals* was thus delayed many years, after which came the labour of bringing out a much enlarged edition of the *Origin of Spacies.* The *Descent of Man* was, apparently, at first intended to be a comparatively small book, but a difficulty connected with the origin of the distinctive peculiarities of the two sexes led to an investigation of this subject throughout the animal kingdom. This was found to be of such extreme interest, and to have such important applications, that it development with the completeness characteristic of all the writer's work led to the production of two bulky volumes, followed by another volume on the *Expression of the Emotions in Man and Animals,* not less instructive. None of Darwin's works has excited greater interest or more bitter controversy than that on man; and the correction of the numerous reprints, and of a final enlarged edition in 1874, was found to be so laborious a task as to convince him that any such extensive literary works as those projected and announced six years previously must be finally abandoned. This, however, by no means implied cessation from work. Observation and experiment were the delight and relaxation of Darwin's life,[1] and he now continued and supplemented those numerous researches on plants we have already referred to. A new edition of an earlier work on the *Movements of Climbing Plants* appeared in 1875; a thick volume on *Insectivorous Plants* in the same year; *Cross and Self-Fertilisation* in 1876; the *Forms of Flowers* in 1877; the *Movements of Plants,* embodying much original research, in 1880; and his remarkable little book on *Earthworms* in 1881. This last work is highly characteristic of the author. In 1837 he had contributed to the Geological Society a short paper on the formation of vegetable mould by the agency of worms. For more than forty years this subject of his early studies was kept in view; experiments were made, in one case involving the keeping a field untouched for thirty years,—and every opportunity was taken of collecting facts and making fresh observations, the final results being to elevate one of the humblest and most despised of the animal creation to the position of an important agent in the preparation of the earth for the use and enjoyment of the higher animals and of man.

[1] About this time he said to the present writer: "When I am obliged to give up observation and experiment, I shall die." And he actually did continue his experiments to within a few days of his death.

The sketch now given of Darwin's work, though it may have seemed tedious to the reader by its length, is yet in many respect imperfect, since it has given to account of those earlier importance labours which would alone have made the reputation of a lesser man. None but the greatest geologists have produced more instructive works than the two volumes of *Geological Observations,* and the profound and original essay "On the Structure and Distribution of Coral Reefs"; the most distinguished zoologists and anatomists might be proud of the elaborate "Monograph of the Cirripedia," of which a competent judge says: "The prodigious number and minute accuracy of his dissections, the exhaustive detail with which he worked out every branch of his subject—sparing no pains in procuring every species that it was possible to procure, in collecting all the known fact relating to the geographical and geological distribution of the group, in tracing all the complicated history of the metamorphoses presented by the individuals of the sundry species, in disentangling the problem of the homologies of these perplexing animals, etc.—all combine to show that, had Mr. Darwin chosen to devote himself to a life of morphological work, his name would probably have been second to none in that department of biology,"[1] while the numerous researches on the fertilisation and structure of flowers and the movements of plants, would alone place him in the rank of a profound and original investigator in botanical sicence.

[1] *Nature,* vol. xxvi. p. 99.

Estimate of Darwin's Life-Work

Yet these works, great as is each of them separately, and, taken altogether, amazing as the production of one man, sink into insignificance as compared with the vast body of research and of thought of which the *Origin of Species* is the brief epitome, and with which alone the name of Darwin is associated by the mass of educated men. I have here endeavoured, however imperfectly, to enable non-specialists to judge of the character and extent of this work, and of the vast revolution it has effected in our conception of nature,—a revolution altogether independent of the question whether the theory of "natural selection" is or is not as important a factor in bringing about changes of animal and vegetable forms as its author maintained. Let us consider for a moment the state of mind induced by the new theory and that which preceded it. So long as men believed that every species was the immediate handiwork of the Creator, and was therefore absolutely perfect, they remained altogether blind to the meaning of the countless variations and adaptations of the part and organs of plants and animals. They who were always repeating, parrot-like, that every organism was exactly adapted to its conditions and surroundings by an all-wise being, were apparently dulled or incapacitated by this belief from any inquiry into the inner meaning of what they saw around them, and were content to pass over whole classes of facts as inexplicable, and to ignore countless details of structure under vague notions of a "general plan," or of variety and beauty being "ends in themselves"; while he whose teachings were at first stigmatised as degrading or even atheistical, by devoting to the varied phenomena of living things the loving, patient, and reverent study of one who really had *faith* in the beauty and harmony and perfection of creation, was enabled to bring to light innumerable hidden adaptations, and to prove that the most insignificant parts of the meanest living things had a use and a purpose, were worthy of our earnest study, and fitted to excite our highest and most intelligent admiration.

That he has done this is the sufficient answer to his critics and to his few detractors. However much our knowledge of nature may advance in the future, it will certainly be by following in the pathways he has made clear for us; and for long years to come the name of Darwin will stand for the typical example of what the student of nature ought to be. And if we glance back over the whole domain of science, we shall find none to stand beside him as equals; for in him we find a patient observation and collection of facts, as in Tycho Brahe; the power of using those facts the determination of laws, as in Kepler, combined with the inspirational genius of a Newton, through which he was enabled to grasp fundamental principles, and so apply them as to bring order out of chaos, and illuminate the world of life as Newton illuminated the material universe. Paraphrasing the eulogistic words of the poet, we may say, with perhaps a greater approximation to truth—

"Nature and Nature's laws lay hid in night;
God said, 'Let Darwin be,' and all was light."

www.ingramcontent.com/pod-product-compliance
Lightning Source LLC
Chambersburg PA
CBHW032012120726
47902CB00013B/492

* 9 7 8 1 4 7 3 3 2 9 6 8 3 *